T0260200

THE STEPHEN BECHTEL FUND

IMPRINT IN ECOLOGY AND THE ENVIRONMENT

The Stephen Bechtel Fund has

established this imprint to promote

understanding and conservation of

our natural environment.

The publisher gratefully acknowledges the generous contribution to this book provided by the Stephen Bechtel Fund.

SHOREBIRD ECOLOGY, CONSERVATION, *and* MANAGEMENT

SHOREBIRD ECOLOGY, CONSERVATION, *and* MANAGEMENT

Mark A. Colwell

UNIVERSITY OF CALIFORNIA PRESS
Berkeley   Los Angeles   London

University of California Press, one of the most distinguished
university presses in the United States, enriches lives around the
world by advancing scholarship in the humanities, social sciences,
and natural sciences. Its activities are supported by the UC Press
Foundation and by philanthropic contributions from individuals and
institutions. For more information, visit www.ucpress.edu.

For digital version, see the press website.

University of California Press
Berkeley and Los Angeles, California

University of California Press, Ltd.
London, England

© 2010 by the Regents of the University of California

Library of Congress Cataloging-in-Publication Data

Colwell, Mark A.
    Shorebird ecology, conservation, and management / Mark A. Colwell.
        p.   cm.
    Includes bibliographical references and index.
    ISBN 978-0-520-26640-7 (cloth : alk. paper)
    1. Shore birds.   2. Shore birds—Conservation.   I. Title.

QL696.C4C655   2010
598.3'317—dc22   2010027335

18   17   16   15   14   13   12   11   10
10   9   8   7   6   5   4   3   2   1

The paper used in this publication meets the minimum requirements
of ANSI/NISO Z39.48-1992 (R 1997)(*Permanence of Paper*).

Cover photo: Female American Avocet, Baylands Nature Preserve,
Palo Alto, California, by Peter LaTourrette, www.birdphotography.com.

FOR TAMMIE

CONTENTS

PREFACE AND ACKNOWLEDGMENTS

The natural world is a beautiful place, replete with wondrous events and remarkable living things. I count shorebirds among the most beautiful, wondrous, and remarkable things in nature. They are beautiful in their subtle plumages and diverse mating displays; wondrous in migration as they wend their way between Arctic breeding grounds and southern latitudes; and remarkable in their synchronized aerial acrobatics as they respond collectively in a dense, wheeling flock to the threat of predation. These descriptors merely scratch the surface of the world of shorebirds; with a little study and attention to detail, they become all the more alluring to students of avian ecology. Unfortunately, opportunities to marvel at and study shorebirds are diminishing with their declining populations. Collectively, the beauty of shorebirds and their conservation status call for an increased focus on applied ecology.

I wrote this book for two main reasons. First, I wished to compile what is known about shorebird ecology from the primary literature, so students and professionals alike may have a useful resource to guide their search for information as they seek answers and ask new questions in shorebird ecology. Second, and perhaps more importantly, I hope that the applied emphasis of this book prompts renewed focus by a new generation of biologists, who are faced with the pressing issues of conserving shorebird populations. This book melds the wonder of shorebird biology with the application of our knowledge toward reversing their population decline. As such, shorebirds serve as a metaphor for conservation in general. As we come to know and love something more deeply, we are moved to act to protect and preserve it.

I came to the study of shorebirds nearly 30 years ago when I started graduate school at the University of North Dakota, working with Lewis Oring. I had been a birder for years and wanted to take the next step in trying to create a profession out of a hobby. Lew offered me a summer research opportunity working on Spotted Sandpipers. That first summer, Lew and Dov Lank, his post-doc on the project, trained me in the ins and outs of avian field ecology. I was steeped in behavioral ecology and enthralled by the social workings of a small population of sandpipers that I came to know by name (that is, by the color band combinations they wore). As a graduate student, I continued this emphasis in studying Wilson's Phalaropes and other shorebirds breeding amid the prairie wetlands of Saskatchewan. When I went looking for my first academic job, I was fortunate to land a position at Humboldt State

University, perched on the shore of a large estuary on the Pacific coast. I shifted gears to studying various facets of the ecology of nonbreeding shorebirds with an emphasis on conservation and management, but I maintained an interest in breeding biology, working on a local population of Snowy Plovers. These experiences have strongly shaped the class I teach at HSU that is focused on the ecology, conservation, and management of shorebirds. This book represents the structural outline for that course. It uses the annual cycle to illustrate the conservation challenges that shorebirds face at virtually every point in the annual cycle.

Over the years, I have benefitted greatly from working relationships with diverse people who share a love of shorebirds. I am indebted to Lew Oring, who gave me my first opportunities in graduate school and added greatly to my understanding of shorebirds while I was a post-doc; Lew continues to impress me with his deep knowledge of shorebird behavior. Dov Lank and Connie Smith shepherded me through my first field season as a budding field biologist and as a beginning graduate student;

to them I owe sincere thanks. I have gained immensely from working with several other professionals over the years, especially Sue Haig, Steve Dinsmore, and Nils Warnock. While at HSU, I have had the good fortune of working with more than 30 graduate students on projects that have greatly enhanced my understanding of shorebird ecology; to them I owe special thanks. Many undergraduate students have contributed to this book by offering helpful criticisms of chapter organization and being a captive audience in the course I have taught each autumn. Several colleagues reviewed and discussed selected chapters with me and occasionally suggested topics that I should include where I would have neglected material. Specifically, I thank Jesse Conklin, Suzanne Fellows, Jim Lyons, Lew Oring, and Nils Warnock for careful treatments of individual chapters; whatever errors or misinterpretations remain are my own. Lastly, I am most indebted to my wife, Tammie, who has steadfastly supported me with her enduring love and companionship for as long as I can remember!

PART ONE

Evolutionary Relationships, Anatomy and Morphology, and Breeding Biology

1

Introduction

WHY STUDY SHOREBIRDS? I've occasionally asked myself this question over the 30 years that I've been an avian ecologist. At first blush, the answer may not be that scientific: because they're fascinating! However, the fascination and wonder of shorebirds (or waders as they're known elsewhere in the English-speaking world) stems from a diversity that seems unrivaled by other bird groups. This diversity is evident across scientific disciplines as varied as biogeography, bioenergetics, behavioral ecology, and evolutionary biology. An additional advantage is that, for the most part, shorebirds provide abundant viewing opportunities in a variety of ecological settings. This makes for relatively easy study by scientists and birders alike. Candidly, I suppose that many of the following observations that characterize shorebirds can be applied, with relatively minor changes, to other avian taxa. This portrayal of shorebirds as ideal and wondrous subjects of study is made more relevant by their population status and need for effective management and conservation. Some of the rarest avian species are shorebirds, and even common ones are experiencing population declines. Accordingly, the relevance of applied ecology is immediate and pressing. Still, their attributes make shorebirds especially alluring subjects for study.

DIVERSITY AND DISTRIBUTION

Ornithologists recognize approximately 215 species of shorebird, unevenly distributed among 14 families in the order Charadriiformes (Table 1.1). To some extent, this diversity may be an artificial human construct, because recent molecular studies suggest the group is polyphyletic (van Tuinen et al. 2004). In other words, the various shorebird families come from two distinct evolutionary lineages. These separate

TABLE 1.1
Taxonomic overview of the shorebirds, their diversity, and breeding distributions

FAMILY	SUBFAMILY	TRIBE	GENERA / SPECIES	BREEDING DISTRIBUTION[a]
Ibidorhynchidae			1 / 1	Himalayas and Tibetan plateau
Pluvianellidae			1 / 1	Tierra del Fuego of S. America
Pedionomidae			1 / 1	Interior of Australia
Dromadidae			1 / 1	Coasts of Red Sea; Indian Ocean
Rostratulidae			1 / 3	Tropical S. America; Austral-Asia
Chionididae			1 / 2	Antarctica; sub-Antarctic islands
Thinocoridae			2 / 4	Highlands of southern S. America
Jacanidae			6 / 8	Neotropic, Africa, Austral-Asia
Burhinidae			1 / 9	Palearctic, Africa, Austral-Asia, Neotropic
Glareolidae	Glareolinae		5 / 17	Palearctic, Africa, Austral-Asia
	Cursorinae		1 / 8	Africa, Asia
	Pluvianinae		1 / 1	Africa
Recurvirostridae			4 / 11	Cosmopolitan
Haematopodidae			1 / 12[b]	Cosmopolitan
Charadriidae	Charadriinae		7 / 41	Cosmopolitan
	Vanellinae		2 / 25	Cosmopolitan
Scolopacidae	Scolopacinae	Tringini	3 / 16	Holarctic
		Prosobonii	1 / 2[b]	Pacific Ocean
		Limosini	1 / 4	Holarctic
		Arenariini	1 / 2	Holarctic
		Numeniini	2 / 9[b]	Holarctic
		Limnodromini	1 / 3	Holarctic
		Gallinagonini	3 / 20	Holarctic
		Scolopacini	1 / 7	Holarctic
		Calidridini	5 / 25	Holarctic, Pacific
	Phalaropodinae		1 / 3	Holarctic

Sources: Cramp and Simmons (1983), del Hoyo (1992), Clements (2007), Delany et al. (2009).
[a] Zoogeographic realm or region corresponding to group's principal breeding areas.
[b] Includes one extinct species.

origins may explain the contrasting and diverse life histories and distributions of shorebirds. Regardless, the 14 shorebird families consist of four monotypic families (that is, consisting of a single species) that have restricted breeding distributions and are either nonmigratory (the Plains-wanderer of Australia and the Magellanic Plover of Tierra del Fuego) or undertake relatively short-distance movements between breeding and wintering grounds (Ibisbill and Crab Plover). The most diverse groups, sandpipers and plovers, have broad distributions, spanning hemispheres. Most sandpipers breed in northern regions, and many migrate to the extremes of southern continents. Most plovers, however, are temperate and tropical species, and they are less prone to move long distances between breeding and wintering areas. It is no surprise that families with fewer species tend to be more restricted in their distributions. However, the oystercatchers, stilts, and avocets are nearly cosmopolitan, being absent from only Antarctica and surrounding islands. By contrast, the sheathbills are permanent residents of Antarctica and sub-Antarctic islands; the seedsnipes reside in higher elevations of the Andes of South America.

Thus, shorebirds are a diverse group. At least one recent discovery of a previously undescribed woodcock (Bukidnon Woodcock: Kennedy et al. 2001) and the rediscovery of a plover from Southeast Asia (White-faced Plover: Kennerley et al. 2008) have increased the diversity of the group. Shorebirds occupy open habitat at the extremes of latitude, and they range from sea level to high elevations. By comparison, waterfowl (375 species) and raptors (313 species) are slightly more speciose than shorebirds, but they are arguably more uniform in their foraging ecologies, social organization, and behaviors.

VARIED ECOMORPHOLOGY

A consequence of cosmopolitan distributions is that closely related shorebirds have evolved in diverse habitats, with consequences for morphological adaptations. Nowhere is this more apparent than in the variation exhibited in body size and feeding apparatus. Among the sandpipers, for instance, mass varies from tiny calidridines (<20 g) to large curlews (>500 g); several other species are much larger (such as stone curlews at >800 g). Most shorebirds meet their daily energy requirements consuming a diet of soft-bodied macroinvertebrates; others consume bivalves, which are ground to a pulp by powerful gizzards. Some larger species, such as thick-knees and curlews, occasionally eat small lizards, fishes, and the eggs of other birds. Pratincoles are insectivorous, feeding on the wing like swallows. Plant material is generally uncommon in the diet of most species, although some Arctic species feed extensively on plant material and fruits late in summer. Recent evidence has shown that some small calidridine sandpipers feeding in intertidal habitats lap "biofilm" with brushlike tongues.

To acquire food, shorebirds have evolved diverse bill morphologies and feeding behaviors. Phalaropes use their needlelike bills to facilitate the movement of water droplets that contain prey through the physics of surface water tension (Rubega and Obst 1993). Other bill shapes include spatulate (Spoon-billed Sandpiper), recurved (avocets), and decurved (curlews); only one bird has a laterally asymmetrical bill (Wrybill). Considerable variation and adaptation in bill morphology even exists within species. For instance, individual Eurasian Oystercatchers have one of three bill shapes, each specialized for feeding on different prey. One type is used to chisel open bivalves, a second slits the adductor muscle to open shells, and a third is better suited for probing into soft substrates for invertebrates (Sutherland et al. 1996).

These diverse bill shapes have been the subject of considerable ecological research, fueled by the supposition that species with similar bill morphologies must experience competition for food. Early on, researchers examined habitat use and aggressive interactions as a means of habitat segregation in dynamic tidal habitats of coastal estuaries (e.g., Recher 1966). More recent analyses of the patterns derived from competition theory have examined mixed species flocks of migrant sandpipers and the minimum size ratio of bills of closely related species (Eldridge and Johnson 1988). Finally, some of the finest examinations of the functional and numerical responses of predators to variation in the density of prey have come from studies of shorebirds, especially large-bodied species that afford easy quantification of intake rate in association with prey density. Notable among shorebirds is the Eurasian Oystercatcher (Goss-Custard 1996), which is an ideal subject for understanding the interrelationships between food availability, foraging behavior, and population size owing to its ease of observation and relatively simple diet of bivalves, mussels, and marine worms.

DIVERSE SOCIAL SYSTEMS

The social systems of shorebirds, their mating relationships, and their patterns of parental care during the breeding season as well as their flocking tendencies during the nonbreeding season are intriguingly diverse. This diversity has been the subject of several reviews (e.g., Oring 1982, 1986; Myers 1984; Goss-Custard 1985). Mating systems run the gamut from extreme polygyny in the lek-breeding Buff-breasted Sandpiper and Ruff to classic polyandry in the phalaropes, jacanas, painted snipes, and various sandpipers. However, the mating systems of most species are characterized by monogamy and varying degrees of shared parental care of eggs and chicks. In the sandpipers, monogamy is the rule, but biparental care of eggs and chicks is highly variable. Females typically depart from breeding areas at variable times after their young have hatched and leave parental care to their mates. Coupled with this diversity of social systems is a pattern of reversed size dimorphism, with females substantially larger than male (Jönsson and Alerstam 1990). This pattern is shared with raptors, which has spawned considerable discourse on the evolution of reversed size dimorphism (Jehl and Murray 1986).

Once they depart from their breeding grounds, most shorebirds typically become much more gregarious. They migrate and winter together in flocks of varying size and density; some species, however, remain solitary throughout the nonbreeding season. The varied flocking tendencies exist both within and among species. As a result, shorebirds have been popular research subjects in weighing the costs and benefits of group living. The attributes that make them ideal subjects include that (1) they are readily observed in open habitats where their foraging behavior is easily quantified, (2) they are common prey of raptors, and (3) they frequently and increasingly interfere with one another's foraging as the flock's size increases. Interestingly, when they are not feeding, shorebirds form dense, mixed-species flocks at roosts. This observation alone suggests that predation has played a strong selective role in shaping this facet of their behavior.

GLOBE-TROTTING MIGRANTS

Most shorebirds—and nearly all sandpipers— breed in northern latitudes where environmental conditions are favorable for breeding. There is a seasonal pulse of food in tremendous abundance, which fuels reproduction by adults and the rapid growth of their young. Northern latitudes also tend to have lower predation pressure on the ground nests of shorebirds. But these environs quickly become inhospitable, and shorebirds must depart to temperate latitudes for the winter.

Most shorebirds undertake short- to long-distance migrations. The exceptions are temperate and tropical species in which some or all individuals in a population migrate short distances. During migration, individuals may spend days, weeks, or months in purposeful, directed movement, with staging at freshwater wetlands and coastal estuaries where they refuel for the next step in their journey. The distance of individual legs of the journey is probably influenced most by the geographical barriers that confront them. For instance, oceanic crossings and passages across inhospitable desert regions (such as the Sahara) are accomplished in a single non-stop flight. When geography and landscape features permit, birds move comparatively short distances among estuaries scattered along continental shorelines. Occasionally, the adaptations for long-distance migrants are nothing short of phenomenal. In the case of the Bar-tailed Godwit, virtually all individuals that breed in western Alaska will stage on the Alaskan peninsula and await favorable weather conditions to carry them nonstop over 11,000 km to wintering areas in New Zealand (Gill et al. 2005). As an adaptation for weight conservation, the godwits' guts atrophy in the weeks just before departure; they regain their mass and function after arrival in their winter quarters (Piersma and Gill 1998).

The nature of shorebird migrations varies within species as well. For instance, the various subspecies of Dunlin migrate along distinct flyways. Such information argues strongly for the conservation of separate populations (subspecies) of Dunlin rather than for a single world population. Even within populations of some species, individuals vary in their migratory nature, yielding classic examples of partial migrants. For example, the population of the Snowy Plover breeding along the Pacific coast of North America consists of both migrants and year-round residents (Stenzel et al. 2008).

WETLAND DEPENDENCE

For much of the year, most shorebirds are intimately tied to open habitats, especially wetlands. In the Arctic, they breed amid tundra wetlands; in temperate regions they are intimately associated with freshwater and hypersaline wetlands. During migration, they concentrate at coastal estuaries and interior wetlands where they rely on food resources to fuel subsequent movements. Worldwide, wetlands are some of the most threatened habitats. The reliance of shorebirds on wetlands has focused conservation (Senner and Howe 1984; Myers et al. 1987) and management (e.g., Eldridge 1992; Helmers 1992) strategies on these valuable habitats. Accordingly, various international treaties

(Ramsar Convention of 1971), federal laws (Migratory Bird Treaty Act, North American Wetland Conservation Act), and programs spearheaded by nongovernmental organizations (Western Hemisphere Shorebird Reserve Network) have been enacted or developed to enhance wetland conservation and management. The recent literature also has produced an abundance of papers addressing the management of wetlands for shorebirds and the integration of their needs with those of other wildlife. Humans have degraded wetland habitats through development of port facilities, pollution, overharvesting of bait and shellfish, and other activities that disrupt the normal activity patterns of shorebirds. Because shorebirds occupy these threatened habitats, management and conservation of wetlands are imperative for the maintenance of viable shorebird populations. In some cases, agricultural lands, pasturelands, and commercial salt production ponds provide important habitats that may be functionally equivalent to seminatural wetlands for large numbers of shorebirds.

CONSERVATION AND MANAGEMENT

The remarkable features that characterize shorebirds are rivaled by the dire conservation status of many species. Several species, such as the Black Stilt of New Zealand and the Spoon-billed Sandpiper of northeastern Russia, are among the rarest birds in the world. Conservationists, ornithologists, and birders worldwide still hold out hope that the Eskimo Curlew persists somewhere in the Canadian Arctic. In the Palearctic, similar hopes prevail for the Slender-billed Curlew, although there have been very few recent sightings. One subspecies of Red Knot was recently proposed for "emergency listing" as endangered under the U.S. Endangered Species Act owing to the most precipitous population decline witnessed in the history of avian conservation (Baker et al. 2004; Niles et al. 2008). Although other populations are quite abundant, their numbers are declining worldwide. In fact, nearly 50% of the world's shorebird populations with known trends are in decline (Zöckler et al. 2003; Thomas et al. 2006). Collectively, these observations have created a growing interest in applied ecology directed at ameliorating the limiting factors that are responsible for the small and declining populations of shorebirds.

RATIONALE FOR AND ORGANIZATION OF THIS BOOK

Over the years, the biology of individual shorebird species has been detailed in countless scientific papers and numerous books (e.g., D. Nethersole-Thompson 1973; D. Nethersole-Thompson and M. Nethersole-Thompson 1979; Goss-Custard 1996; Byrkjedal and Thompson 1998). Additional details on species exist in various regional (Cramp and Simmons 1983; *Birds of North America* accounts) or global compendia (e.g., Johns-gard 1981; del Hoyo et al. 1992), which are invaluable data sources. In the recent past, several general texts have been published, often with multiauthored chapters covering specialized facets of shorebird biology (e.g., Hale 1980; Burger and Olla 1984a, 1984b; Evans et al. 1984). Several ageing and sexing (Prater et al. 1977; Pyle 2008) or field identification (e.g., Hayman et al. 1986; Paulson 2005; O'Brien et al. 2006) guides exist. One recent compilation (van de Kam et al. 2004) has received considerable praise for melding biology with beautiful images that detail the annual cycle of shorebirds. Surprisingly, how-ever, no text or reference has been compiled for shorebirds. By contrast, both waterfowl and raptors have been the subject of books addressing their ecology, conservation, and management (e.g., Bellrose 1976; Ferguson-Lees and Christie 2005; Balldassarre and Bolin 2006). Hence, there seems a clear need for this book.

I wrote this book based on what I perceived as the lack of a general source of information on the ecology, conservation, and management of shorebirds. I organized the book around a semester-long course that I have taught at Humboldt State University for much of the past 20 years. I begin with a general treatment of the evolutionary relationships of shorebirds, their fossil history, and their contemporary distributions. I then define shorebirds by detailing their anatomy, morphology, and physiology. The discussion of breeding includes chapters on facets of breeding biology that have fascinated biologists for decades, including mating systems, courtship behavior, egg laying, incubation, and nesting ecology. Next is a discussion of migration, with a treatment of flyways and staging areas, and the evolution of migration strategies. The chapters on winter ecology cover foraging behavior, roosting ecology, social organization, and population ecology. The final chapters cover applied ecology with topics on wetland management, managing predation during the breeding season, and disturbance by humans.

LITERATURE CITED

Baker, A. J., P. M. González, T. Piersma, L. J. Niles, I. de L. S. do Nascimento, P. W. Atkinson, N. A. Clark, C. D. T. Minton, M. K. Peck, and G. Aarts 2004. Rapid population decline in Red Knots: Fitness consequences of decreased refuelling rates and late arrival in Delaware Bay, *Proceedings of the Royal Society of London, Series B* 271: 875–882.

Baldassarre, G. A., and E. G. Bolen. 2006. *Waterfowl ecology and management*. 2nd ed. New York: Wiley.

Belrose, F. C. 1976. *Ducks, geese and swans of North America*. Harrisburg, PA: Stackpole Books.

Burger, J., and B. L. Olla, eds. 1984a. *Shorebirds: Breeding behavior and populations*. New York: Plenum Press.

———. 1984b. *Shorebirds: Migration and foraging behavior*. New York: Plenum Press.

Byrkjedal, I., and D. Thompson. 1998. *Tundra plovers: The Eurasian, Pacific and American Golden Plovers and Grey Plover*. London: T & AD Poyser.

Clements, J. F. 2007. *The Clements checklist of birds of the world*. 6th ed. Ithaca, NY: Cornell University Press.

Cramp, S., and K. E. L. Simmons, eds. 1983. *Birds of the western Palearctic*. Oxford: Oxford University Press.

Delany, S., D. Scott, T. Dodman, and D. Stroud, eds. 2009. *An atlas of wader populations in Africa and western Asia*. Wageningen, the Netherlands: Wetlands International.

del Hoyo, J., A. Elliot, J. Sargatal, and N. J. Collar, eds. 1992. *Handbook of birds of the world*. Vol. 3. Barcelona: Lynx Edicion.

Eldridge, J. 1992. *Management of habitat for breeding and migrating shorebirds in the Midwest*. Leaflet 13.2.14. Washington, DC: Fish and Wildlife Service.

Eldridge, J. L., and D. H. Johnson. 1988. Size differences in migrant sandpiper flocks: Ghosts in ephemeral guilds. *Oecologia* 77: 433–444.

Evans, P. R., J. D. Goss-Custard, and W. G. Hale. 1984. *Coastal waders and wildfowl in winter*. Cambridge: Cambridge University Press.

Ferguson-Lees, J., and D. A. Christie. 2005. *Raptors of the world*. Princeton, NJ: Princeton University Press.

Gill, R. E., Jr., T. Piersma, G. Hufford, R. Servranckx, and A. Riegen. 2005. Crossing the ultimate ecological barrier: Evidence for a 11000-km-long nonstop flight from Alaska to New Zealand and eastern Australia by Bar-tailed Godwits. *The Condor* 107: 1–20.

Goss-Custard, J. D. 1985. Foraging behaviour of wading birds and the carrying capacity of estuaries. In *Behavioural ecology*, ed. R. M. Sibly and R. H. Smith, 169–188. Oxford: Blackwell.

———, ed. 1996. *The oystercatcher*. Oxford: Oxford University Press.

Hale, W. G. 1980. *Waders*. London: Collins.

Hayman, P., J. Marchant, and T. Prater. 1986. *Shorebirds: An identification guide to the waders of the world*. Boston: Houghton Mifflin.

Helmers, D. L. 1992. *Shorebird management manual*. Manomet, MA: Western Hemisphere Shorebird Reserve Network.

Jehl, J. R., Jr., and B. G. Murray, Jr. 1986. The evolution of normal and reverse sexual size dimorphism in shorebirds and other birds. *Current Ornithology* 3: 1–86.

Johnsgard, P. A. 1981. *The plovers, sandpipers, and snipes of the world*. Lincoln: University of Nebraska Press.

Jönsson, P. E., and T. Alerstam. 1990. The adaptive significance of parental care role division and sexual size dimorphism in breeding shorebirds. *Biological Journal of the Linnean Society* 41: 301–314.

Kennedy, R. S., T. H. Fisher, S. C. B. Harrap, A. C. Diesmos, and A. S. Manamtam. 2001. A new species of woodcock (Aves: Scolopacidae) from the Philippines and a re-evaluation of other Asian/Papuasian woodcock. *The Forktail* 17: 1–12.

Kennerley, P. R., D. N. Bakewell, and P. D. Round. 2008. Rediscovery of a long-lost Charadrius plover from South-East Asia. *The Forktail* 24: 63–79.

Myers, J. P. 1984. Spacing behavior of nonbreeding shorebirds. In *Shorebirds: Migration and foraging behavior*, ed. J. Burger and B. L. Olla, 271–321. New York: Plenum Press.

Myers, J. P., R. I. G. Morrison, P. Z. Anatas, B. A. Harrington, T. E. Lovejoy, M. Sallaberry, S. E. Senner, and A. Tarak. 1987. Conservation strategy for migratory species. *American Scientist* 75: 19–26.

Nethersole-Thompson, D. 1973. *The Dotterel*. London: Collins.

Nethersole-Thompson, D., and M. Nethersole-Thompson. 1979. *Greenshank*. Vermillion, SD: Buteo Books.

Niles, L. J., H. P. Sitters, A. D. Dey, P. W. Atkinson, A. J. Baker, K. A. Bennett, R. Carmona, et al. 2008. *Status of the Red Knot (Calidris canutus rufa) in the Western Hemisphere*, ed. C. D. Marti. Studies in Avian Biology No. 36. Camarillo, CA: Cooper Ornithological Society.

O'Brien, M., R. Crossley, and K. Karlson. 2006. *The shorebird guide*. New York: Houghton Mifflin.

Oring, L. W. 1982. Avian mating systems. In *Avian Biology*. Vol. 6, ed. J. Farner and J. King, 1–92. New York: Academic Press.

———. 1986. Avian polyandry. In *Current ornithology*, ed. R. Johnston, 309–351. New York: Academic Press.

Paulson, D. 2005. *Shorebirds of North America: The photographic guide*. Princeton, NJ: Princeton University Press.

Piersma, T., and R. E. Gill, Jr. 1998. Guts don't fly: Small digestive organs in obese Bar-tailed Godwits. *The Auk* 115:196–203.

Prater, A. J., J. H. Marchant, and J. Vuorinen. 1977. *A guide to the identification and ageing of Holarctic waders*. Tring, United Kingdom: British Trust for Ornithology.

Pyle, P. 2008. *Identification guide to North American birds, Part II*. Point Reyes Station, CA: Slate Creek Press.

Recher, H. F. 1966. Some aspects of the ecology of migrant shorebirds. *Ecology* 47: 393–407.

Rubega, M. A., and B. S. Obst. 1993. Surface-tension feeding in phalaropes: Discovery of a novel feeding mechanism. *The Auk* 110: 169–178.

Senner, S. E., and M. A. Howe. 1984. Conservation of Nearctic shorebirds. In *Shorebirds: Breeding behavior and populations*, J. Burger and B. L. Olla, 379–421. New York: Plenum Press.

Stenzel, L. E., G. W. Page, J. C. Warriner, J. S. Warriner, D. E. George, C. R. Eyster, B. A. Ramer, and K. K. Neuman. 2008. Survival and natal dispersal of juvenile Snowy Plovers (*Charadrius alexandrinus nivosus*) in central coastal California. *The Auk* 124: 1023–1036.

Sutherland, W. J., B. J. Ens, J. D. Goss-Custard, and J. B. Hulscher. 1996. Specialization. In *The oystercatcher,* ed. J. D. Goss-Custard, 56–76. Oxford: Oxford University Press.

Thomas, G. H., R. B. Lanctot, and T. Székely. 2006. Can intrinsic factors explain population declines in North American breeding shorebirds? A comparative analysis. *Animal Conservation* 9: 252–258.

van de Kam, J., B. Ens., T. Piersma, and L. Zwarts. 2004. *Shorebirds: An illustrated behavioural ecology.* Utrecht, the Netherlands: KNNV.

van Tuinen, M., D. Waterhouse, and G. J. Dyke. 2004. Avian molecular systematics on the rebound: A fresh look at modern shorebird phylogenetic relationships. *Journal of Avian Biology* 35: 191–194.

Zöckler, C., S. Delany, and W. Hagemeijer. 2003. Wader populations are declining—how will we elucidate the reasons? *Wader Study Group Bulletin* 100: 202–211.

2

Systematics, Phylogeny, and Phylogeography

CONTENTS

T HE EVOLUTIONARY RELATIONSHIPS within and among taxa are of immense value in understanding many facets of species' ecologies. Phylogenies are an essential tool for understanding the evolution of life history traits. Phylogenetic analyses are important because they estimate patterns of ancestry and descent, thus providing a historical framework upon which to test hypotheses regarding the evolution of character traits (Chu 1994). Accordingly, recent analyses of shorebird phylogeny have fostered productive comparative analyses in diverse areas that include the evolution of delayed

plumage maturation (Chu 1994), migration (Joseph et al. 1999), and mating systems (Székely and Reynolds 1995; Reynolds and Székely 1997). Evolutionary relationships also serve as the foundation for management and conservation actions directed at evolutionary significant units (such as populations, subspecies, or species). Several recent decisions regarding the distinctness and conservation status of shorebird populations were informed by genetic information, and at the foundation of the flyway concept is the notion that distinct populations should be managed separately. In short, there is good reason to understand the evolutionary origins of taxa, whether it be for theoretical or applied purposes.

Scientists have long debated the evolutionary relationships among shorebirds. From early studies emphasizing morphological and anatomical characters to contemporary research incorporating molecular techniques, we are closer to understanding the affinities within and among the taxa that comprise the shorebirds. Contemporary shorebirds often have distinct populations that occupy disjunctive ranges, and a treatment of recent geological history is necessary to understand the origins of these subspecies.

In this chapter, I summarize what is known about the fossil history of shorebirds and their evolutionary relationships, and how Pleistocene glaciation isolated populations and contributed to the phylogeographic variation observed today among Holarctic-breeding shorebirds.

FOSSIL HISTORY

A long and heated debate has raged over the origins of birds from reptiles, including the timing and nature of the diversification of modern birds (Neoaves: Feduccia 2003; Chiappe and Dyke 2006; Ericson et al. 2006; James and Pourtless 2009). To some paleontologists, notably Feduccia (1999, 2003), modern birds radiated rapidly into contemporary taxa after the Cretaceous/Tertiary (K-T) boundary, approximately 65 million years ago and coincident with a mass extinction event. According to Feduccia, shorebirds hold a unique place in this explosive diversification, referred to as the "big bang" theory. He suggested that the earliest fossils in the lineage of modern birds were "transitional shorebirds" from which all modern birds evolved. Moreover, this ancestral shorebird resembles the present-day thick-knees. To support this, Feduccia drew attention to similarities in the postcranial features of skulls of extant thick-knees that are shared with these ancient shorebirds. Subsequent to this, shorebirds radiated during the early Eocene to form the diverse group they are today, which includes the gulls, terns, and alcids. At least one recent analysis combining molecular and fossil data (Ericson et al. 2006) lends some support to the timing of rapid bird diversification in the early Tertiary; it does not, however, address whether shorebirds were the root of the tree of modern birds. Others, however, contest these views on the timing of diversification. Notably, van Tuinen et al. (2003) criticized Feduccia (2003) for vague descriptions of these ancestral shorebirds. They go on to state that "molecular, morphological and fossil data all indicate that the early history of modern birds began in the Cretaceous and did not involve 'transitional shore-birds'." This view is supported by at least two recent molecular analyses (Paton et al. 2001; Hackett et al. 2008).

A BRIEF HISTORY OF SHOREBIRD SYSTEMATICS

Over the past two and a half centuries, beginning with Linnaeus (1758), taxonomists have analyzed new character sets to reexamine the phylogenetic hypotheses of the affinities of the various groups comprising the shorebirds (Table 2.1). Several authors have provided excellent historical perspectives and reviews of the phylogenetic affinities of shorebirds, notably Sibley and Ahlquist (1990) and Dove (2000). The following is a summary based on these works.

MORPHOLOGICAL AND ANATOMICAL CHARACTERS

Not surprisingly, early systematists mostly used external morphology and anatomy to draw conclusions regarding the affinities of shorebirds. These efforts commonly grouped today's shorebirds with other waterbirds. Early taxonomists (Linnaeus 1758) used bill and foot structure, among other external morphological characters, to conclude that shorebirds were closely related to flamingos, storks, herons, rails, and coots. Others used skeletal features to distinguish shorebirds and other taxa. Huxley (1867), for example, identified two families (essentially plovers and sandpipers) based on the schizognathous structure of the palate. A short time later, Seebohm (1888) used osteological characters to argue that modern-day shorebirds are closely allied with larids and alcids, a conclusion that has been affirmed by recent analyses using molecular data. Even later, Gadow (1892) established a classification based on "40 characters from various organic systems" (including integument, skeleton, muscle, and digestive organs) that is essentially unchanged from today's phylogenies based on molecular data. This classification grouped today's shorebirds in a suborder (Limicolae) separate from a suborder that included the gulls and alcids.

TABLE 2.1

Summary of recent studies assessing shorebird phylogenetic affinities

AUTHORS (YEAR)	METHOD/CHARACTERS	MAJOR FINDINGS AND COMMENTS
Fain and Houde (2007)	nuDNA and mtDNA	Confirmed three clades: lari, scolopaci, and charadrii; buttonquails related, embedded in lari; the Egyptian Plover a basal plover, not related to coursers and pratincoles.
Baker et al. (2007)	Bayesian analysis of mtDNA and nuDNA	Confirmed many relationships of previous analyses; established older origins for scolopaci and charadrii.
van Tuinen et al. (2004)	Consensus phylogeny based on published molecular data	Crab Plover, pratincoles, and coursers grouped with gulls, terns, and alcids; Plains-wanderer kin to seedsnipes; plovers, avocets/stilts, and oystercatchers grouped and close relatives of thick-knees, sheathbills, and Magellanic Plover.
Dove (2000)	Downy feather morphology and pigmentation; inclusion of Strauch's skeletal data set	Sandpipers, plovers, seedsnipes, and pratincoles are a well-defined clade; Crab Plover, Egyptian Plover, and sheathbills part of gull clade; Magellanic Plover with turnstones. Some genera scattered among different groups.
Engelmoer and Roselaar (1998)	Morphometrics of feathers, wings, hind limbs, and culmen; body mass; patterns of molt	Confirmation and elucidation of many subspecies of Holarctic shorebirds.
Sibley and Ahlquist (1990)	DNA-DNA hybridization	Sandgrouse were shorebirds; sandpipers distinct from plovers; painted snipes kin to jacanas; Crab Plover, coursers, and pratincoles grouped with gulls and terns.
Chu (1995)	Reanalysis of Strauch's skeletal characters	Sandpipers split into six groups; plovers, avocets/stilts, thick-knees, oystercatchers aligned with gulls and terns.
Christian et al. (1992)	Protein electrophoresis	Plovers, sandpipers, and gulls as distinct lineages; jacanas uncertain.

TABLE 2.1 *(continued)*

AUTHORS (YEAR)	METHOD/CHARACTERS	MAJOR FINDINGS AND COMMENTS
Strauch (1978)	63 skeletal and 7 musculature features	Confirmed three suborders: scolopaci, charadrii, and alcae.
Ahlquist (1974)	Protein electrophoresis	Sandpipers distinct from gull-auk-plover lineage.
Jehl (1968)	Natal down patterns	Painted snipes related to jacanas; plovers and thick-knees related to avocets/stilts and oystercatchers; sandpipers distinct.

Over the past 50 years, attempts to resolve the evolutionary relationships of shorebirds have used (and reused) new character sets, especially molecular data, to evaluate early phylogenetic hypotheses (see Table 2.1). Additionally, the advent of formal hypothesis testing in the form of phylogenetic analyses provided greater rigor to evaluations of hypothesized evolutionary relationships. Strauch (1978) analyzed 63 osteological and seven myological characters and recognized three suborders (Scolopaci, Charadrii, and Alcae). Others (Mickevich and Parenti 1980; Chu 1995) have reanalyzed Strauch's (1978) data because they disputed his conclusions based on philosophical and methodological grounds. Chu (1995) concluded that there were two main clades: the sandpipers and plovers. Within the sandpipers, five lineages emerged (snipes, tringine sandpipers, calidridine sandpipers, phalaropes, and curlews). Chu (1995) concluded that the Crab Plover (Dromadidae) was closely related to gulls, a conclusion that has been bolstered by recent molecular analyses.

FEATHERS

The patterns of natal down and structure of downy feathers have provided the characters for two studies. Jehl (1968) examined patterns of natal down, which he argued were a conservative character. Using no formal method to group taxa other than a checklist published by Peters (1934),

Jehl concluded that there were five shorebird groups (superfamilies): (1) the Crab Plover, (2) jacanas and painted snipes, (3) sheathbills and seedsnipes, (4) sandpipers, and (5) plovers, avocets/stilts, and thick-knees, coursers/pratincoles, and oystercatchers. Of interest, Jehl (1968) drew attention to the unpatterned egg and chick of the burrow-nesting Crab Plover and concluded that it was an outlier among shorebirds. This appears to be supported by the recent molecular evidence suggesting that the Crab Plover is more closely related to gulls, terns, and alcids than to other clades within the Charadriiformes.

More recently, Dove (2000) examined 38 characters based on microscopic examination of morphology and pigmentation of downy feathers, combined with Strauch's (1978) osteological data. Her initial findings (based on feather structure alone) confirmed some earlier phylogenetic hypotheses regarding the placement of major groups. For example, oystercatchers and avocets formed a clade with gulls and terns; embedded in this group were the Crab Plover and Egyptian Plover. There were, however, several instances in which members of a genus (such as *Vanellus*) were scattered among different clades. These problems disappeared when the feather characters were analyzed with Strauch's skeletal data, which does not constitute an independent test of previous hypotheses based on these osteological data alone.

MOLECULAR CHARACTERS

In 1990, Sibley and Ahlquist (1990) published their monumental work on the phylogeny of birds using DNA-DNA hybridization techniques. Their results corroborated many earlier hypotheses. The sandgrouse (Pteroclididae) were close relatives of shorebirds. The sandpipers and their allies (seedsnipes, the Plains-wanderer, jacanas, and painted snipes) were distinct from the plovers, thick-knees, oystercatchers, avocets, and stilts. The pratincoles, coursers, and the Crab Plover were more closely related to gulls and terns. These distinctions have been supported by recent molecular analyses.

The advent of molecular techniques using nuclear and mitochondrial DNA has fostered a wealth of studies that appear close to resolving the phylogenetic affinities among and within the shorebirds (Fig. 2.1). In 2004, van Tuinen and coworkers published a dendrogram based on molecular data from three other studies (Ericson et al. 2003; Paton et al. 2003; Thomas et al. 2004), which offered "surprising consensus." In short, this phylogeny confirmed that the shorebirds consist of three clades: (1) sandpipers, including the jacanas, painted snipes, seedsnipes, and the Plains-wanderer; (2) pratincoles and the Crab Plover, which are grouped with gulls, terns, and alcids; and (3) plovers, avocets/stilts, and oystercatchers, which are close relatives of the

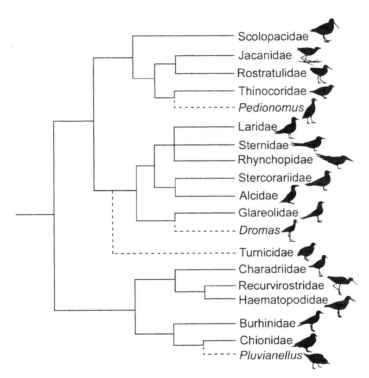

FIGURE 2.1. Consensus phylogeny of modern shorebird evolutionary relationships by van Tuinen et al. (2004) based on molecular data from Ericson et al. (2003), Paton et al. (2003), and Thomas et al. (2004). The phylogenetic position of oddball taxa are indicated by dashed lines. See Sibley and Ahlquist (1990) for placement of the Crab Plover (*Dromas*). This phylogeny is similar to that produced by Chu's (1995) reanalysis of Strauch's (1978) phylogeny based on osteological characters, with the exception of the position of the Alcidae, which Chu considered a sister group to all others. Sibley and Ahlquist (1990) arrived at a very similar phylogeny based on DNA-DNA hybridization. The Ibisbill (Ibidorhynchidae) is absent from this phylogeny; it is considered closely related to the avocets/stilts and oystercatchers by Sibley and Ahlquist (1990). After van Tuinen et al. (2004).

TRINGINE TAXONOMY

Now that much of the higher taxonomic problems of shorebirds have been worked out, some researchers are examining the finer details of species' relationships within lineages. Pereira and Baker (2005) examined the affinities of the tringine sandpipers using three independent data sets: mitochondrial DNA, nuclear DNA, and Strauch's (1978) set of 70 osteological and myological characters. Their analysis indicated that the genera *Xenus* and *Actitis* were two basal lineages, distinct from the other 12 members of the shank (*Tringa*) lineage. The remaining species were an amalgam, with the tattlers (*Heteroscelus*) and Willet (*Catoptrophorus*) embedded in the shank lineage. Based on this finding, the American Ornithologists' Union (Banks et al. 2006) changed these genera to *Tringa*. Pereira and Baker (2005) also showed that the Nearctic yellowlegs (*T. melanoleuca* and *T. flavipes*) and Palearctic redshanks (*T. totanus* and *T. erythropus*) were not each other's closest relative, as was widely believed based on similarities in external morphology (including size, bill shape, and leg color). The Greater Yellowlegs is most closely related to the Palearctic Common Greenshank (*T. nebularia*), and the Lesser Yellowlegs is sister to the Willet. Similarly, the Spotted Redshank's closest relatives are the Greater Yellowlegs and Common Greenshank, whereas the Common Redshank is most closely related to a suite of Palearctic species. Lastly, the diver-

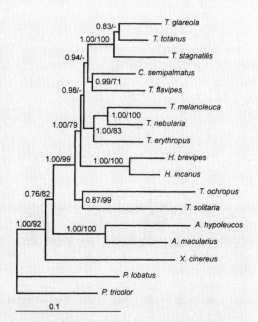

A consensus Bayesian tree obtained by the partitioned likelihood of the combined mitochondrial and nuclear data sets. The numbers at the nodes are the Bayesian posterior probability/maximum parsimony bootstrap value. The bar represents the expected number of substitutions under the mixed model of DNA substitution. This topology is identical to one of three equally parsimonious trees inferred from the same data set. From Pereira and Baker (2005).

gence times of the shank lineage were estimated at Middle and Late Miocene, roughly 14 to 18 million years ago, during a period of marked climate variation across the globe.

thick-knees, sheathbills, and the Magellanic Plover.

The debate over the evolutionary affinities of shorebird taxa continues. Baker et al. (2007) used Bayesian methods to analyze 90 (of 96) genera, including gulls, terns, and buttonquails. They reached similar conclusions regarding the phylogenetic relationships of major taxa. The shorebirds could be split into two suborders (clades): (1) Scolopaci, including sandpipers, curlews, godwits, dowitchers, snipes, jacanas, painted snipes, and the Plains-wanderer; and

(2) Charadrii, including stilts and avocets, plovers, thick-knees, sheathbills, and the Magellanic Plover; the Egyptian Plover (*Pluvianus*) is distantly related to this second group. The most interesting results suggest the plovers are polyphyletic, with large plovers (*Pluvialis*) more akin to stilts and avocets than smaller plovers and dotterels. Baker et al. (2007) also used molecular clock dating to estimate that the 14 major shorebird lineages diverged during the late Cenozoic, approximately 93 million years ago. In other words, these lineages survived the K-T

boundary (approximately 65 million years ago, in the late Cretaceous), which suggests that many more lineages had originated before the last major extinction event on Earth. The current shorebird lineages likely radiated during the Eocene, when global warming yielded productive ecosystems and a flourishing of Earth's biodiversity (Baker et al. 2007).

Hackett et al. (2008) reported on a phylogenomic analysis of all birds based on a 32 kb sequence of DNA representing 19 nuclear loci on 15 different chromosomes in the chicken genome. This analysis found that that shorebirds (and other Charadriiforms) are a sister group to landbirds. However, the relationships among various shorebird families appear quite similar to the results presented by van Tuinen et al. (2004). Collectively, these findings have prompted biologists to refer to shorebirds as a polyphyletic group: shorebirds have a mixed evolutionary origin, and the group is not a cohesive taxon. It also highlights the arbitrary decision to focus this book on "traditional" shorebirds while excluding gulls, terns, and alcids!

It seems clear now, after nearly 250 years of evaluation of morphological, anatomical, behavioral, and molecular data sets, that the shorebirds are a polyphyletic group embedded in a clade that includes the gulls, terns, and alcids. Several other taxa, including the sandgrouse (Pteroclididae) and buttonquails (Turnicidae), may be close relatives of shorebirds. In summary, phylogenetic analyses of multiple independent data sets confirm that the order Charadriiformes is monophyletic, consisting of three clades that are commonly recognized as suborders: Lari (gulls, terns, and alcids; pratincoles and coursers, Crab Plover, and buttonquails), Scolopaci (sandpipers, jacanas, painted snipes, seedsnipes, and the Plains-wanderer), and Charardii (plovers, oystercatchers, thick-knees, avocets and stilts, and sheathbills).

PHYLOGEOGRAPHY

Plate tectonics and changing climates have strongly influenced the Earth's contemporary patterns of biodiversity, especially in northerly realms (Avise and Walker 1997; Hewitt 2004). Over the last 2 million years of the Pleistocene epoch (1.8 million years ago to 10,000 years ago), shifting climates altered the breeding habitats of shorebirds, especially across northern latitudes. During glacial maxima, ice sheets covered much of the continental landmasses of the north, but remnant tundra persisted in refugia in several locations. During intervening interglacials, warmer climates melted ice sheets, allowing forests to advance north and causing sea levels to rise. Each of these changes effectively reduced the area of tundra habitat in coastal regions of the Arctic. The outcome of these geological events on shorebirds is evident in the genetic differences among populations that were geographically isolated and hence experienced reduced gene flow. These genetic differences often correlate with morphological variation. Today, we see the product of this isolation in the phylogeography (that is, the geographical distributions of genealogical lineages; Avise et al. 1987) of many subspecies of Arctic-breeding shorebird.

A brief geologic history is essential to understanding shorebird phylogeography. Paleoclimatological data indicate that, until recently, the earth had been cooling for the past 60 million years, with increasingly severe ice ages occurring at 100,000-year intervals (Hewitt 2004). The Pleistocene epoch began 1.8 million years ago and ended 10,000 years ago. During the Pleistocene, earth experienced alternating periods of cooling and warming. Pronounced cooling produced immense glaciers and ice sheets that covered northern landmasses. Some estimates suggest that approximately 30% of the Earth's surface was covered in ice. These vast inhospitable regions altered the distribution of many boreal and Arctic-breeding species by pushing them south. Even in temperate and tropical regions, however, habitat changes associated with ice ages had an effect on the distribution of biotas (Hewitt 2004).

Curiously, amid this Pleistocene landscape of ice were several northern areas that remained ice-free. These biotic refugia were sufficiently disjunct to facilitate allopatric speciation, especially among shorebirds. During the last major ice age (Wisconsinan/Weichselian), two large Arctic refugia existed. From 10,000 to 110,000 years before present, Beringia spanned an immense area of northeast Siberia, northwest Alaska, and the Bering Strait. A second, smaller refugium existed in the western Palearctic. During the last glacial maximum (approximately 10,000 to 30,000 years before present), ice covered much of Beringia but several smaller refugia existed. Nested within these 100,000-year ice ages were shorter oscillations, which have been especially cold. Changes of 7° to 15°C occasionally occurred over decades. At least two periods of warming, called interglacials, occurred during the past 250,000 years of the Pleistocene. During the Eemian interglacial (approximately 125,000 years before present) and again in the early Holocene (8,000 years before present), summer temperatures in the Arctic were 4° to 8°C warmer than today. This warming caused a northward shift in boreal forest, nearly to the edge of the Arctic Ocean, at the expense of tundra. Additionally, an increased sea level associated with melting ice, and the isostatic downpressing of the land associated with former effects of glaciation caused substantial flooding in coastal areas. The effect of this alternating warming and cooling on shorebirds probably differed among the species with different habitat preferences. For instance, those favoring wet coastal tundra may have increased in abundance, whereas others breeding in drier habitats may have declined in population size. Collectively, the changing climate, varying coverage by ice masses, and refugia created ideal conditions for allopatric speciation among shorebirds (Kraaijeveld and Nieboer 2000).

The effects of Pleistocene ice sheets on the differentiation of shorebirds probably varied with species' breeding latitude and habitat (Kraaijeveld and Niebor 2000). Many species of shorebird breed in the low Arctic zone, and the isolating effects of refugia during glacial periods have influenced their differentiation. By con-

trast, high Arctic breeding species may have been more strongly affected by climate and habitat during interglacials, when tundra was pushed north and was restricted in its extent.

The Dunlin is arguably the best example of a Holarctic shorebird whose phylogeography has been resolved within the context of variation in Pleistocene climates and changing tundra habitats (Wenink et al. 1993, 1994, 1996; Kraaijeveld and Nieboer 2000; Wennerberg 2001; Buehler and Baker 2005) (Fig. 2.2). Yet remarkably there is still considerable disagreement regarding the number of subspecies. Molecular ecologists have identified five major Holarctic clades (based on mitochondrial DNA haplotypes) that are mostly distributed in low Arctic tundra. Within these groups, ornithologists have further recognized as many as 9 to 13 subspecies based on morphometrics and breeding location (Prater et al. 1977; Cramp and Simmons 1983; Hayman et al. 1986; del Hoyo 1992; Clements 2007). Some of these subspecies are remarkably different in mass. For instance, the largest Nearctic subspecies (*pacifica*) is approximately twice the mass of the smallest Palearctic subspecies (*schinzii*) (Cramp and Simmons 1983). Within the European lineage, the three subspecies (*alpina, schinzii*, and *arctica*) are readily distinguished based on morphology (Engelmoer and Roselaar 1998; Kraaijeveld and Nieboer 2000).

Recent molecular analyses indicate that the various Dunlin subspecies diverged at different times during the relatively recent past (Kraaijeveld and Nieboer 2000). Divergence times span approximately 58,000 to 194,000 years before present, which precedes the last major Pleistocene glaciation. The oldest lineage, *hudsonia* from Canada, dates from approximately 225,000 years before present, coincident with an interglacial (warming). In subsequent glacials *hudsonia* was isolated in eastern North America. The next split in the Dunlin lineage occurred approximately 120,000 years before present when the European clade (*alpina*) split from those occupying Siberia and western Alaska. This split also coincided with an interglacial. The remain-

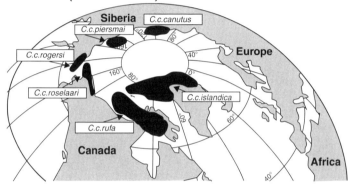

A. Dunlin (*Calidris alpina*)

Siberian lineage
Beringian lineage
Alaskan lineage
Canadian lineage

Siberia
Europe
European lineage
Canada
Africa

FIGURE 2.2. Holarctic distributions of (A) five lineages (based on mitochondrial DNA haplotypes) of Dunlin and (B) six subspecies of Red Knot. For both species, the subspecies migrate to wintering grounds along distinct flyways, lending support to the interpretation that they are evolutionarily significant units worthy of conservation. After Buehler and Baker (2005).

B. Red Knot (*Calidris canutus*)

Siberia
C.c.canutus
C.c.piersmai
C.c.rogersi
C.c.roselaari
Europe
C.c.islandica
C.c.rufa
Canada
Africa

ing three groups (*centralis, sakhalina,* and *pacifica*) diverged still more recently, approximately 70,000 to 80,000 years before present. This split coincided with a warming between two cold spells of the early Wisconsinan/Weichselian (Kraaijeveld and Nieboer 2000).

Buehler and Baker (2005) suggest that this divergence is indicative of low contemporary gene flow among populations. In the Palearctic, the migration patterns of these lineages have been characterized as "parallel," with individuals from the various lineages (such as *schinzii, alpina,* and *sakhalina*) migrating along flyways to distinct wintering areas (Wennerberg 2001). A similar pattern exists for the principal Nearctic subspecies: *arcticola* breeds across coastal tundra from northwest Alaska to northwest Canada and winters along the western Pacific; *pacifica* breeds in western Alaska and migrates south to winter along the Pacific Coast of North Amer-

ica; whereas *hudsonia* breeds in the central Canadian Arctic and moves south to winter along the Atlantic Seaboard. Clearly, Dunlin subspecies represent distinct evolutionary units that differ in morphology, behavior, and genetics. As such, the populations should be managed independently rather than as a global population of one species.

The Red Knot is another example of an Arctic-breeding sandpiper with subspecies, but its phylogeography differs somewhat from the Dunlin (Kraaijeveld and Nieboer 2000). Surprisingly, the six subspecies of Red Knot (Fig. 2.2) are not as well differentiated, and their origins appear to be more recent than the Dunlin lineages. Red Knots appear to have diverged approximately 20,000 years before present, or since the Wisconsinan glaciation. Like the Dunlin, Red Knot subspecies migrate along separate flyways to distinct wintering areas (Buehler and Baker 2005). As high Arctic breeders, the various subspecies of Red Knot likely were isolated and hence diverged during interglacials, when warming temperatures in the Arctic resulted in restricted tundra habitats (Kraaijeveld and Nieboer 2000).

The Rock Sandpiper provides another fine example of the effects of Pleistocene climate on shorebird phylogeographical differentiation, albeit on an even finer spatial scale. Pruett and Winker (2005) sequenced genomic DNA to assess the phylogeography of this Beringian endemic. Ornithologists have recognized four subspecies of Rock Sandpiper, principally distinguished by morphology and breeding range, although the distributional limits of some are unclear. Genetic data confirm that the Rock Sandpiper's sister taxon is the Purple Sandpiper of the Palearctic, from which it diverged approximately 1.8 million years ago, near the start of the Pleistocene. The four subspecies of Rock Sandpiper are, however, of recent origin. The various haplotypes are scattered within the species' range, suggesting a series of range expansions and contractions during the Wisconsinan glaciation and interglacials (Pruett and Winker 2005). It is likely that other Holarctic waders have similar phylogeographic patterns. A simple summary of geographic variation within shorebird taxa (Table 2.2) indicates substantial variation among groups in geographic variation, as evidenced by the number of polytypic species (that is, those with one or more subspecies).

Not all shorebirds, however, are characterized by phylogeographic structure. The American Woodcock, which breeds at temperate latitudes of North America, exhibits little divergence among individuals migrating along central and eastern flyways. Based on this finding, Rhymer et al. (2005) suggested that considerable gene flow has existed throughout the species' breeding population since the woodcock expanded its range after the last glacial maximum. Similar hypotheses have been proposed for the genetic similarity among various disjunct populations of Mountain Plover in the western Great Plains of North America. Oyler-McCance et al. (2005)

suggested that the Mountain Plover had probably undergone a recent (postglacial) range expansion. Lastly, the Ruddy Turnstone is recognized as having Nearctic and Palearctic subspecies (Cramp and Simmons 1983), based on plumage. But Wenink et al. (1994) reported low sequence divergence in mitochondrial DNA genotypes, suggesting that turnstones too have recently expanded their range from a bottlenecked refugial population.

HYBRIDIZATION IN SHOREBIRDS

The degree to which bird species hybridize varies widely among major taxa. Hybridization is well-known among hummingbirds and waterfowl, for instance (McCarthy 2006). Occasional instances of hybridization are evident in shorebirds, often between congeners (e.g., Jehl 1985; Jonsson 1996) or members of the same family (e.g., Principe 1977). In some cases, these hybridizations have led to the erroneous identification of a new species. S. F. Baird is credited with describing the first specimen of Cooper's Sandpiper (*Calidris cooperi*) based on a specimen collected in 1833 on Long Island, New York. A similar specimen was collected in New South Wales, Australia, in 1981. Evidence suggests that these specimens were hybrids between the Curlew Sandpiper and the Sharp-tailed Sandpiper (Cox 1990a, 1990b). More recently, Cox's Sandpiper (*Calidris paramelanotos*) was first described in 1982 based on two specimens collected by J. B. Cox in northwestern Australia (Hayman et al. 1986). Subsequent comparison of mitochondrial DNA (which is maternally inherited) with samples from other calidridine sandpipers (Pectoral Sandpiper, Curlew Sandpiper, White-rumped Sandpiper, and Ruff) determined that Cox's Sandpiper was, in fact, a hybrid resulting from a cross between female Curlew Sandpipers and male Pectoral Sandpipers (Christidis et al. 1996).

BIOGEOGRAPHY AND COMMUNITIES

The divergence and diversification of shorebirds in response to geographic isolation spawned by

TABLE 2.2
Summary of geographic variation within and among the world's shorebirds

TAXON	NUMBER OF SPECIES	MEAN (±SD) NUMBER OF SUBSPECIES	MONO-TYPIC SPECIES	NUMBER OF SUBSPECIES						
				2	3	4	5	6	7	8
Jacanidae	8	1.5±2.3	5		2			1		
Rostratulidae	2	1.0±1.4	1	2						
Dromadidae	1	0	1							
Pedionomidae	1	0	1							
Chionididae	2	3.0±2.8	1							
Haematopodidae	11	1.3±1.7	6	3	1		1			
Ibidorhynchidae	1	0	1							
Recurvirostridae	7	0.7±1.9	6				1			
Burhinidae	9	2.3±2.1	3	2	1	2		1		
Glareolidae	17	1.5±2.2	9	4	2	1				1
Charadriidae										
Vanellus	23	0.9±1.4	15	5	1	2				
Pluvialis	4	0	4							
Charadrius/Thinornis	36	1.1±1.7	23	6	4		2	1		
Pluvianellidae	1	0	1							
Thinocoridae	4	3.3±1.7	1		1	1	1			
Scolopacidae										
Limosini	4	1.8±1.3	1							
Numeniini	9	0.9±1.5	6	2		1				
Tringini	18	0.7±1.5	14	3				1		
Arenariini	2	1.0±1.0	1	1						
Phalaropodini	3	0	3							
Scolopacini	6	0.7±1.0	3	3						
Gallinagonini	18	0.7±1.1	13	3	2					
Limnodromini	3	1.0±1.7	2		1					
Calidridini	24	0.7±1.6	20	1		2		1		

Sources: Cramp and Simmons (1983); Hayman et al. (1986); del Hoyo et al. (1992); O'Brien et al. (2006); Clements (2007).

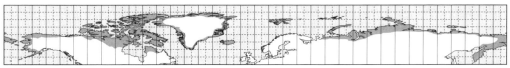

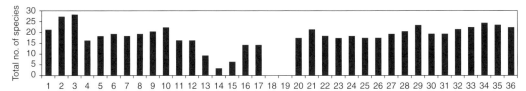

FIGURE 2.3. Patterns of shorebird diversity (total number of species) in 10° longitudinal bands projected across the Arctic. After Henningsson and Alerstam (2005).

long-term variation in the Earth's climate has produced two striking biogeographical patterns based on breeding distributions: (1) diversity is positively correlated with latitude (Järvinen and Väisänen 1978; van Tuinen et al. 2004), which is counter to the pattern for virtually all other avian taxa (except alcids and penguins) and for biodiversity in general; and (2) within Arctic regions, diversity varies appreciably across longitudinal bands (Henningsson and Alerstam 2005).

On a global scale, shorebirds are a notable exception to the prevailing latitudinal pattern of higher diversity in tropical compared with temperate and Arctic regions. This pattern is largely driven by the most diverse family, the sandpipers, which are a northern taxon. The second most speciose group, the plovers, is distributed mostly in temperate and tropical realms. On a finer scale, Järvinen and Väisänen (1978) showed that the number of breeding species of shorebird increased with latitude across a limited geographical area in Fennoscandia. They suggested that this pattern was associated with increased habitat heterogeneity in northerly areas.

Within the Arctic, the longitudinal pattern of shorebird distribution (Fig. 2.3) is almost entirely driven by sandpipers (Henningsson and Alerstam 2005). Diversity is particularly high (28 species) in Beringia, whereas the Arctic of eastern Canada and Greenland are regions depauperate of species. This geographical varia-

tion was attributable to a number of contemporary ecological factors. Diversity was highest in areas of high net primary productivity, where the length of the snow-free period was long, and where diverse migratory flyways connected breeding and wintering areas. Henningsson and Alerstam (2005) also showed that contemporary diversity (in Beringia, for instance) was associated with the extent of tundra area or ice-free refugia during the last glacial maximum. Clearly, past geological events have had a profound effect on the phylogeography of individual species (such as the Dunlin, Red Knot, and Rock Sandpiper), which are evident in the diversity of the Arctic shorebird communities we see today. It would be interesting to apply similar approaches to understand the distributions of temperate-breeding shorebirds worldwide, because the climate changes that affected Arctic regions also altered habitats elsewhere on Earth (Hewitt 2004). In North America, for instance, the breeding ranges of up to nine species of shorebird overlap in the prairie regions of the midcontinent, which was affected by ice sheets during the last Pleistocene glaciation.

CONSERVATION IMPLICATIONS

The 215 species that comprise the shorebirds are actually a more diverse group because the geologic events of the Pleistocene epoch gave rise

THE WESTERN SNOWY PLOVER

Until recently, the Kentish and Snowy Plovers were considered conspecific and among the most widely distributed and phylogeographically diverse shorebirds (Cramp and Simmons 1983). The nominate subspecies breeds across Eurasia and northern Africa; *dealbatus* occurs in the eastern Palearctic and *seebohmi* breeds in Ceylon. Three subspecies occur in the New World: *nivosus* breeds across North America, *tenuirostris* in the Caribbean, and *occidentalis* in western South America. The delineation of the Nearctic subspecies was recently resolved by Funk et al. (2007). Küpper et al. (2009) analyzed the genetic and phenotypic characters to conclusively show that these two species deserve recognition as separate entities, the Kentish (*C. alexandrinus*) and the Snowy Plover (*C. nivosus*).

In North America, The U.S. Fish and Wildlife Service (USFWS) listed the Pacific Coast population of the Western Snowy Plover as threatened under the federal Endangered Species Act in 1993 (U.S. Department of the Interior [DOI] 1993). The USFWS based the decision on a decline in the number of occupied breeding sites and a decrease in population size, based on work by Gary Page and Lynne Stenzel of Point Reyes Bird Observatory (see U.S. Fish and Wildlife Service 2007). The listing decision also recognized the coastal population (within 50 miles of the Pacific Ocean) as distinct, based on the observation that there was very limited evidence of marked birds dispersing between coastal and interior populations (DOI 1993). The USFWS designated critical habitat in 2005 after a lengthy review and economic analysis (DOI 2005).

In 2003, a private group (Pacific Legal Foundation) challenged the notion that the coastal population was distinct, based on preliminary genetic evidence published in a master's thesis (Gorman 2000). The challenge initiated a review of the listing of the coastal population segment. A detailed analysis (Funk et al. 2007) subsequently found that the coastal and interior populations of the western United States were genetically similar. Moreover, this analysis indicated that Snowy Plovers breeding across North America should be considered one species (*nivosus*), and distinct from a small population of plovers breeding in the Caribbean (*tenuirostris*). Despite this, the USFWS denied the delisting proposal on the basis of ecological, behavioral, and other evidence that the populations were distinct (DOI 2006b). More recently, the USFWS proposed an amendment to the 4(d) rule of the Endangered Species Act (DOI 2006a), which was intended to encourage local management agencies to continue to work toward recovery of local populations. This proposal would relax restrictions on incidental take of the plover in counties that had achieved breeding populations of sufficient size for 5 years.

to many subspecies. These subspecies are morphologically and genetically recognizable. In fact, most shorebirds studied in detail have two or more subspecies, with geographically disjunct breeding and wintering ranges, and separate migratory flyways. As such, they should be managed as evolutionarily significant units (Myers et al. 1987). Additionally, the phylogeographic structure resulting from changing Pleistocene climates undoubtedly will be repeated, albeit on an abbreviated schedule, by human-induced climate change. The Arctic is likely to experience more pronounced warming than other regions, with consequences for habitats and the shorebirds that rely on them for the brief Arctic breeding season.

LITERATURE CITED

Ahlquist, J. E. 1974. The relationships of the shorebirds. PhD diss., Yale University.

Avise, J. C., and D. Walker. 1997. Pleistocene phylogeographic effects on avian populations and the speciation process. *Proceedings of the Royal Society of London B* 265: 457–463.

Avise, J. C., J. Arnold, R. M. Ball, E. Bermingham, T. Lamb, J. E. Neigel, C. A. Reeb, and N. C. Saunders. 1987. Intraspecific phylogeography: The mitochondrial DNA bridge between population genetics and systematics. *Annual Review Ecology Systematics* 18: 489–522.

Baker, A. J., S. L. Pereira, and T. A. Paton. 2007. Phylogenetic relationships and divergence times of Charadriiformes genera: Multigene evidence for the Cretaceous origin of at least 14 clades of shorebirds. *Biology Letters* 3: 205–209.

Banks, R. C., C. Cicero, J. L. Dunn, A. W. Kratter, P. C. Rasmussen, J. V. Remsen, Jr., J. D. Rising, and D. F. Stotz. 2006. Forty-seventh supplement to the American Ornithologists' Union checklist of North American birds. *The Auk* 123: 926–936.

Buehler, D. M., and A. J. Baker. 2005. Population divergence times and historical demography of Red Knots and Dunlins. *The Condor* 107: 497–513.

Chiappe, L. M., and G. J. Dyke. 2006. The early evolutionary history of birds. *Journal of the Paleontological Society of Korea* 22: 133–151.

Christian, P. D., L. Christidis, and R. Schodde. 1992. Biochemical systematics of the Charadriiformes (shorebirds): Relationships between Charadrii, Scolopaci and Lari. *Australian Journal of Zoology* 40: 291–302.

Christidis, L., K. Davies, M. Westerman, P. D. Christian, and R. Schodde. 1996. Molecular assessment of the taxonomic status of Cox's Sandpiper. *The Condor* 98: 459–463.

Chu, P. C. 1994. Historical examination of delayed plumage maturation in the shorebirds (Aves: Charadriiformes). *Evolution* 48: 327–350.

———. 1995. Phylogenetic reanalysis of Strauch's osteological data set for the Charadriiformes. *The Condor* 97: 174–196.

Clements, J. F. 2007. *The Clements checklist of birds of the world.* 6th ed. Ithaca, NY: Cornell University Press.

Cox, J. B. 1990a. The enigmatic Cooper's and Cox's Sandpiper. *Dutch Birding* 12: 53–64.

———. 1990b. The measurement of Cooper's Sandpiper and the occurrence of a similar bird in Australia. *South Australian Ornithologist* 30: 169–181.

Cramp, S., and K. E. L. Simmons, eds. 1983. *Birds of the western Palearctic.* Oxford: Oxford University Press.

del Hoyo, J., A. Elliot, J. Sargatal, and N. J. Collar, eds. 1992. *Handbook of birds of the world.* Vol. 3. Barcelona: Lynx Edicion.

Dove, C. J. 2000. A descriptive and phylogenetic analysis of plumulaceous feather characters in Charadriiformes. *Ornithological Monographs* 51: 1–163.

Engelmoer, M., and C. S. Roselaar. 1998. *Geographical variation in waders.* Dordrecht, the Netherlands: Kluwer Academic.

Ericson, P. G. P., I. Envall, M. Irestedt, and J. A. Norman. 2003. Inter-familial relationships of the shorebirds (Aves: Charadriiformes) based on nuclear DNA sequence data. *BMC Evolutionary Biology* 3: 1–14.

Ericson, P. G. P., C. L. Anderson, T. Britton, A. Elzanowski, U. S. Johansson, M. Källersjö, J. I. Ohlson, T. J. Parsons, D. Zuccon, and G. Mayr. 2006. Diversification of Neoaves: Intergration of molecular data and fossils. *Biology Letters* 2: 543–547.

Fain, M. G., and P. Houde. 2007. Multilocus perspectives on the monophyly and phylogeny of the order Charadriiformes (Aves). *BMC Evolutionary Biology* 7: 35.

Feduccia, A. 1999. *The origin and evolution of birds.* New Haven, CT: Yale University Press.

———. 2003. "Big bang" for tertiary birds. *Trends in Ecology and Evolution* 18: 172–176.

Funk, W. C., T. D. Mullins, and S. M. Haig. 2007. Conservation genetics of Snowy Plovers (*Charadrius alexandrinus*) in the Western Hemisphere: Population structure and delineation of subspecies. *Conservation Genetics* 8: 1287–1309.

Gadow, H. 1892. On the classification of birds. *Proceedings of the Zoological Society of London* 1892: 229–256.

Hackett, S. J., R. T. Kimball, S. Reddy, R. C. K. Bowie, E. L. Braun, M. J. Braun, J. L. Chojnowski, et al. 2008. A phylogenomic study of birds reveals their evolutionary history. *Science* 320: 1763–1768.

Hayman, P., J. Marchant, and T. Prater. 1986. *Shorebirds: An identification guide to the waders of the world.* Boston: Houghton Mifflin.

Henningsson, S. S., and T. Alerstam. 2005. Patterns and determinants of shorebird species richness in the circumpolar Arctic. *Journal of Biogeography* 32: 383–396.

Hewitt, G. M. 2004. The structure of biodiversity—insights from molecular phylogeography. *Frontiers in Zoology* 1: 1–16.

Huxley, T. H. 1867. On the classification of birds; and on the taxonomic value of the modifications of certain of the cranial bones observable in that class. *Proceedings of the Zoological Society of London* 1867: 415–472.

James, F. C., and J. A. Pourtless, Jr. 2009. Cladistics and the origin of birds: A review and two new analyses. *Ornithological Monographs* 66: 1–78.

Järvinen, O., and R. A. Väisänen. 1978. Ecological zoogeography of north European waders, or why do so many waders breed in the north? *Oikos* 30: 495–507.

Jehl, J. R., Jr. 1968. Relationships in the Charadrii (shorebirds): A taxonomic study based on color patterns of the downy young. *San Diego Society of Natural History Memoir* 3: 1–54.

———. 1985. Hybridization and evolution of oystercatchers on the Pacific Coast of Baja California. *Ornithological Monographs* 36: 484–504.

Jonsson, L. 1996. Mystery stint at Groote Keeten: First known hybrid between little and Temminck's stint? *Dutch Birding* 18: 24–28.

Joseph, L., E. P. Lessa, and L. Christidis. 1999. Phylogeny and biogeography in the evolution of migration: Shorebirds of the *Charadrius* complex. *Journal of Biogeography* 26: 329–342.

Kraaijeveld, K., and E. N. Nieboer. 2000. Late quaternary paleogeography and evolution of Arctic breeding waders. *Ardea* 88: 193–205.

Küpper, C., J. Augustin, A. Kosztolányi, T. Burke, J. Figuerola, and T. Székely. 2009. Kentish versus Snowy Plover: Phenotypic and genetic analyses of *Charadrius alexandrinus* reveal divergence of Eurasian and American subspecies. *The Auk* 126: 839–852.

Linnaeus, C. 1758. *Systema naturae per regna tria naturae.* 10th ed. Rev. 2 vols. L. Salmii, Homiiae.

McCarthy, E. M. 2006. *Handbook of avian hybrids of the world.* Oxford: Oxford University Press.

Mickevich, M. F., and L. R. Parenti. 1980. Review of "The phylogeny of the Charadriiformes (Aves): A new estimate using the method of character compatibility analysis." *Systematic Zoology* 29: 108–113.

Myers, J. P., R. I. G. Morrison, P. Z. Antas, B. A. Harrington, T. E. Lovejoy, M. Sallaberry, S. E. Senner, and A. Tarak. 1987. Conservation strategy for migratory species. *American Scientist* 75: 19–26.

O'Brien, M., R. Crossley, and K. Karlson. 2006. *The shorebird guide.* New York: Houghton Mifflin.

Oyler-McCance, S. J., J. St. John, F. L. Knopf, and T. W. Quinn. 2005. Population genetic analysis of mountain plover using mitochondrial DNA sequence data. *The Condor* 107: 353–362.

Paton, T., O. Haddrath, and A. J. Baker. 2001. Complete mitochondrial DNA genome sequences show that modern birds are not descended from transitional shorebirds. *Proceedings of the Royal Society of London, Series B* 269: 839–846.

Paton, T. A., A. J. Baker, J. G. Groth, and G. F. Barrowclough. 2003. RAG-1 sequences resolve phylogenetic relationships with Charadriiform birds. *Molecular Phylogeny Evolution* 29: 268–278.

Pereira, S. L., and A. J. Baker. 2005. Multiple gene evidence for parallel evolution and retention of ancestral morphological states in the shanks (Charadriiformes: Scolopacidae). *The Condor* 107: 514–526.

Peters, J. L. 1934. *Check-list of Birds of the World.* Vol. 2. Cambridge, MA: Harvard University Press.

Prater, A. J., J. H. Marchant, and J. Vourinen. 1977. *A guide to the identification and ageing of Holarctic waders.* Tring, United Kingdom: British Trust for Ornithology.

Principe, W. L., Jr. 1977. A hybrid American Avocet×Black-necked Stilt. *The Condor* 79: 128–129.

Pruett, C. L., and K. Winker. 2005. Biological impacts of climate change on a Beringian endemic: Cryptic refugia in the establishment and differentiation of the Rock Sandpiper (*Calidris ptilocnemis*). *Climate Change* 68: 219–240.

Reynolds, J. D., and T. Székely. 1997. The evolution of parental care in shorebirds: Life histories, ecology, and sexual selection. *Behavioral Ecology* 8: 126–134.

Rhymer, J. M., D. G. McAuley, and H. L. Ziel. 2005. Phylogeography of the American Woodcock (*Scolopax minor*): Are management units based on band recovery data reflected in genetically based management units? *The Auk* 122: 1149–1160.

Seebohm, H. 1888. An attempt to diagnose the suborders of the great Gallino-Gralline group of birds, by the aid of osteological characters alone. *The Ibis* 30: 415–435.

Sibley, C. G., and J. E. Ahlquist. 1990. *Phylogeny and classification of birds.* New Haven, CT: Yale University Press.

Strauch, J. G., Jr. 1978. The phylogeny of the Charadriiformes (Aves): A new estimate using the method of character compatibility analysis. *Transactions of the Zoological Society of London* 34: 263–345.

Székely, T., and J. D. Reynolds. 1995. Evolutionary transitions in parental care in shorebirds. *Proceedings of the Royal Society of London, Series B* 262: 57–64.

Thomas, G. H., M. A. Wills, and T. Székely. 2004. A supertree approach to shorebird phylogeny. *BMC Evolutionary Biology* 4: 1–18.

U.S. Department of the Interior. 1993. Threatened status for the Pacific coast population of the Western Snowy Plover. *Federal Register* 58: 12864–12874.

———. 2005. Designation of critical habitat for the Pacific Coast population of the Western Snowy Plover. *Federal Register* 70: 56970–57119.

———. 2006a. Proposed special rule pursuant to Section 4(d) of the Endangered Species Act for the Pacific Coast distinct population segment of the Western Snowy Plover. *Federal Register* 77: 20625–20626.

———. 2006b. Twelve-month finding on a position to delist the Pacific Coast population of the Western Snowy Plover. *Federal Register* 71: 20607–20608.

U.S. Fish and Wildlife Service. 2007. *Recovery plan for the Pacific Coast population of the Western Snowy Plover (Charadrius alexandrinus nivosus)*. 2 vols. Sacramento, CA: California/Nevada Operations Office, U.S. Fish and Wildlife Service.

van Tuinen, M., T. Paton, O. Haddrath, and A. Baker. 2003. "Big Bang" for tertiary birds: A reply. *Trends in Ecology and Evolution* 18: 442–443.

van Tuinen, M., D. Waterhouse, and G. J. Dyke. 2004. Avian molecular systematics on the rebound: A fresh look at modern shorebird phylogenetic relationships. *Journal of Avian Biology* 35: 191–194.

Wenink, P. W., A. J. Baker, H-U. Rosner, and M. G. J. Tilanus. 1996. Global mitochondrial DNA phylogeography of Holarctic breeding Dunlins (*Calidris alpina*). *Evolution* 50: 318–330.

Wenink, P. W., A. J. Baker, and M. G. J. Tilanus. 1993. Hypervariable-control-region sequences reveal global population structuring in a long-distance migrant shorebird, the Dunlin (*Calidris alpina*). *Proceedings of the National Academy of Sciences of the United States of America* 90: 94–98.

———. 1994. Mitochondrial control-region-sequences in two shorebird species, the Turnstone and the Dunlin, and their utility in population genetic studies. *Molecular Biology and Evolution* 11: 22–31.

Wennerberg, L. 2001. Breeding origin and migration patterns of Dunlin (*Calidris alpina*) revealed by mitochondrial DNA analysis. *Molecular Ecology* 10: 1111–1120.

3

Morphology, Anatomy, and Physiology

CONTENTS

BIRDS ARE REMARKABLY UNIFORM in many features of their external morphology and internal anatomy. They all have integuments covered with feathers and scales; all are oviparous; and all have circulatory and respiratory systems that are almost universally adapted for flight by rapidly delivering oxygen and energy to the tissues of the body. Still, there is considerable interspecific variation in other systems, principally the morphology of the bill and the digestive system, which have evolved to acquire, handle, and process different types of food.

Shorebirds as a group share many features associated with their migratory nature and their use of open, wetland habitats. Their digestive systems are adapted for feeding on soft-bodied invertebrate prey; their wing morphologies are shaped by the distances they migrate; and their precocial young hatch from energy-rich eggs. Still, these commonalities are balanced by substantial interspecific (and occasionally intraspecific) variation in bill morphology associated with the varied diets among different habitats, migration strategies involving short-term atrophy and hypertrophy of the gut as an adaptation for conserving weight during long-distance flights, and sexual dimorphism associated with varying sex roles in diverse mating systems. In this chapter, I review the essentials of shorebird anatomy, morphology, and physiology, emphasizing the underlying ecological conditions that have shaped particular adaptations.

SKELETAL AND MUSCLE SYSTEM

Shorebirds are united with most other extant avian taxa by the shared structure of the neognathae palate; only the ratites and tinamous have a paleognathae palate (Sibley and Ahlquist 1990; Gill 2007). As evidenced in Chapter 2, shorebirds are polyphyletic. Thus, even the most basic features of the skeletal system vary somewhat among taxa such as families. This variation is greatest in the features associated with feeding.

Understandably, variation in skeletal features and musculature is most pronounced in those morphological features associated with feeding. Burton (1974) offered a detailed treatment of interspecific variation in the musculature and skeletal features of the skull. The bill is used to acquire prey. Given the wide range of foraging maneuvers employed by shorebirds, it is not surprising that there is substantial interspecific variation in osteological features and musculature of the skull. In particular, shorebirds vary greatly in bill kinesis or the articulation of the upper jaw with the braincase. All birds exhibit PROKINESIS, which entails flexion of the upper jaw with the narrow strip of flexible bone at the frontonasal hinge. However, many shorebirds exhibit RHYNCHOKINESIS, which involves flexion of the upper jaw occurring at varying distances down the length of the bill (Burton 1974). The intraspecific variation in kinesis correlates with a continuum of feeding maneuvers ranging from pecking to probing. Rhynchokinesis is least developed in plovers, where flexion occurs very close to the frontonasal hinge. Plovers detect prey visually, typically foraging by picking food from the surface. By contrast, various sandpipers use tactile cues to detect prey in soft sediments and then extract and handle their food using flexible, rhynchokinetic bills. This latter adaptation is especially noticeable in snipes and curlews. Burton (1974) suggested that highly rhynchokinetic upper jaws had evolved independently in at least seven shorebird lineages, most of them sandpipers. These groups include large-bodied species (such as curlews) that probe deeply into substrates and smaller taxa that feed very rapidly from substrate surfaces and occasionally probe for prey.

In addition to kinesis, the underlying skeletal features associated with different bill mor-phologies exhibit a wide range of adaptations, from short, straight bills to recurved and strongly decurved forms. These varying skeletal features are correlated with the principal physical forces acting on the bill as individuals feed. For instance, the straight, blunt bill and hypertrophied jaw musculature of the Eurasian Oystercatcher are adaptations for feeding on bivalves. Burton (1974) provided a detailed comparative treatment of other facets of the musculature and skeletal features of the skull as they relate to feeding.

Hindlimb structure also strongly influences how shorebirds feed in wetlands and other simple habitats where they feed. Barbosa and Moreno (1999) examined the hindlimb osteology and myology of 16 Palearctic species and correlated interspecific variation with stride length and frequency, as recorded using video images and focal observations in the field. Their results suggest two evolutionary strategies of locomotion in shorebirds. First, an increase in the length of the leg segments, especially the femur, correlates with increased stride length. Not surprisingly, long-legged waders (such as stilts and avocets) have comparatively long strides, whereas smaller calidridines have short strides. These results are especially interesting among vertebrates because the study controlled for the effects of body size. So it is not merely that larger species have longer strides, but rather that they have long strides in addition to being large. A second finding was that variation in the structure of certain muscles affects the force or speed of contraction, and that this in turn influences stride frequency. Specifically, interspecific variation in the muscle (M. iliotibialis cranialis) that causes femur protraction is associated with changes in the force developed by the muscle (Barbosa and Moreno 1999). The forward movement of the femur is powered by this muscle, and variation in stride frequency across species is directly related to variation in the structure of this and other muscles originating on the iliac crest.

Other skeletal features are seemingly more uniform across shorebird groups. For example, the arrangement of toes is anisodactyl, with three digits pointing forward and one (the hallux) directed posteriorly. An exception to this skeletal uniformity occurs within taxa that are otherwise considered closely related. For example, in the 23 species of lapwing, 11 possess a hallux (Johnsgard 1981); in the 24 species of calidridine sandpiper, the hallux is present in all but the Sanderling. Most shorebirds feed amid wetland habitats, and many possess hindlimb adaptations that serve them well while wading or swimming. Aquatic taxa frequently have partial or complete webbing between the toes (Johnsgard 1981). For instance, the three species of phalarope possess scalloped webbing to varying extents; it is most evident in Red-necked and Red Phalaropes and is least apparent in Wilson's Phalarope, the most terrestrial of the three species. Avocets have webbed toes, but their close relatives the stilts do not. Several sandpipers and plovers also have partial webbing, as evidenced by their common or scientific names, such as the Semipalmated Sandpiper, the Semipalmated Plover, and *Tringa semipalmata* (the Willet).

INTEGUMENTARY SYSTEM

All shorebirds have 11 primaries (Sibley and Ahlquist 1990), although the eleventh is rudimentary (Prater et al. 1977; Pyle et al. 2008). There is, however, substantial interspecific variation in the number of secondaries (8 to 14) and tertiaries (5 to 8). Most species have 12 rectrices, although some snipes (*Gallinago*) have as many as 28 (Prater et al. 1977; Johnsgard 1981). The contour feathers of the body exist in distinct feather tracts (pterylae), separated by bare skin (apteria). Shorebirds possess a bilobed uropygial gland (Sibley and Ahlquist 1990). In sandpipers, the secretions of the uropygial gland vary seasonally between monoesters and diesters, the latter increasing during the time of incubation (Reneerkens et al. 2002, 2005).

As in other birds, the sequence of MOLTS that produces successive plumages in shorebirds proceeds in a predictable pattern, from newly hatched chicks to adults. There is, however, considerable interspecific and even intraspecific variation in the seasonal timing with which particular molts take place. In this generalized summary, I follow the terminology outlined by Humphrey and Parkes (1959) where the terms BASIC and ALTERNATE replace winter/nonbreeding and breeding/nuptial, respectively. The following summary is derived from Prater et al. (1977) and Pyle (2008).

In northern climes, young shorebirds hatch fully feathered in a natal down that offers some insulative properties as well as enhanced crypsis of chicks hiding from predators (Jehl 1968). Immediately upon hatching and continuing through fledging, the prejuvenal molt is initiated whereby the down is replaced by a juvenal plumage. This interval may be as short as a couple of weeks in small, rapidly growing calidridines to nearly 2 months in larger species. The prejuvenal molt is complete in that all contour and flight feathers are newly grown; these flight feathers may be retained for more than a year. In most sandpipers, the natal down and juvenal plumage appears to be patterned in a manner so as to increase the crypsis of young birds. Similarly, in species such as plovers, which nest in open habitats, plumages of grayish white feathers blend with unvegetated substrates. Exceptions to this rule are the golden-plovers, whose young are covered in a beautiful mix of golden yellow, black, and white feathers. Some feathers, especially the inner median coverts, of the juvenal plumage are retained well into the first winter. The brownish and buffy edges of these retained coverts provide a useful cue to age sandpipers (as juveniles) when they are captured well into their first winter (Prater et al. 1977; Pyle 2008).

The first prebasic molt is initiated during late summer and continues into the subsequent nonbreeding season. In many shorebirds, this molt replaces only the body feathers. The timing of the prebasic molt probably varies among individuals of different species owing to the length of the breeding season and duration of time over which the young reach independence. At one end of the continuum are various Arctic species with a short breeding season and a molt that occurs predictably in a rather limited window of time. For example, the prebasic molt is probably more uniform across populations of northern sandpipers than in temperate-breeding species. In the subspecies of Dunlin (*pacifica*) that breeds in western Alaska, virtually all birds molt at northern staging sites before arriving in wintering areas in midautumn (Warnock and Gill 1996). By contrast, in the Pacific Coast population of the Snowy Plover, which has a protracted breeding season, the early-hatching and late-hatching chicks may differ in age by more than 120 days. Consequently, juveniles may initiate their first basic molt over an equally long period that spans late summer through early winter (Page et al. 1995).

In reality, shorebirds exhibit several patterns of flight-feather molt associated with the first prebasic molt (Gratto and Morrison 1981). In most species, juveniles do not replace their remiges during their first nonbreeding season; in other words, the flight feathers are retained for more than a year. In others species, all remiges are replaced. In others still, juveniles exhibit a partial wing molt, which typically involves only the outer primaries and inner secondaries. In the Semipalmated Sandpiper, for instance, most yearlings retain some of the old inner primaries, secondaries, and underwing coverts grown as juveniles (Gratto and Morrison 1981). On the breeding grounds, these feathers often appear quite worn; hence, they are useful in ageing birds as yearlings compared with older breeders. These older birds have undergone a wing molt more recently and thus have newer feathers. Partial postjuvenile wing molt appears to be most common in calidridines and tringines (Gratto and Morrison 1981). The Bristle-thighed Curlew is unique among shorebirds in experiencing a prebasic molt of the remiges over several months, during which individuals become flightless. Marks et al. (1990) suggest that this molt is an adaptation to wintering on predator-free islands in the Pacific.

Before breeding, shorebirds initiate a prealternate molt. This molt replaces the dull-colored contour feathers of the body with a brighter, more patterned and contrasting plumage. Relatively little has been published on the timing and details of this molt. The quality of feathers and completeness of this molt may vary with age, gender, and across species. For example, many shorebirds oversummer in their southern wintering areas as yearlings and even older. These individuals, often males, molt partially at approximately the correct time of year as the same-age conspecifics that migrate to breeding areas.

Subsequent to breeding, the sequence of molts and plumages tends to follow a predictable pattern. A complete prebasic molt of the body and flight feathers replaces the alternate plumage with a definitive basic plumage. Before migration, a prealternate molt replaces the body plumage during spring.

The following synopsis of a generalized prebasic molt sequence is based on Prater et al. (1977) and Pyle (2008); the details vary greatly among species. Molt begins with the innermost primary and progresses outward. When primary molt is about half complete, the outermost secondary is shed and molt proceeds toward the body. The tertials are replaced soon after, and, in most species, the inner secondaries are then renewed, and molt proceeds in two directions toward the center of the secondaries. The rectrices are replaced during secondary molt, usually by progression from the middle feathers outward. Primary molt usually spans the entire period of the replacement of the remiges and rectrices. Tertials are also frequently replaced during the partial body molt.

There are three basic patterns of remex molt (Prater et al. 1977). In most species, adults have a complete, uninterrupted molt during the autumn or winter, related to the date, period, and distance of migration. In many species, adults start to molt in northern areas, suspend, and then complete it on the wintering grounds. As pointed out by Prater et al. (1977), the term SUSPEND is preferential to *arrest* because virtually all species that interrupt molt in this way

PLUMAGE POLYMORPHISM IN THE RUFF

The unusual lek-based mating system of the Ruff has intrigued biologists for more than a half century. Early descriptions (Hogan-Warburg 1966; van Rhijn 1991) of the behaviors of displaying males identified two plumage types associated with a behavioral dimorphism: resident males were darker plumaged and occasionally shared their courts with lighter plumaged, subordinate males (Lank et al. 1995). The resident (dominant) morph is most common in the population, whereas the satellite strategy represents approximately 16% of males. In an exciting development, Jukema and Piersma (2006) identified a third, female-like ("faeder") mating strategy of even lower frequency (approximately 1%) in the population of males. This latter strategy represents a "sneaky" mating strategy, with males appearing female-like but with testes approximately 2.5 times the size of males of other morphs. The plumage/behavioral polymorphism for the resident and satellite morphs stems from a single-locus autosomal dominant-recessive allele system (Lank et al. 1995). Pedigree analysis supports a genetic model whereby the satellite allele is dominant to the resident allele. These alleles could reflect alternative sets of loci maintained in linkage disequilibrium by a chromosomal inversion polymorphism (Lank et al. 1999).

Two of three alternate plumages of male Ruffs include (A) the dominant, dark-plumaged bird and (B) a subordinant, light-plumaged individual. A third morph is a female-like plumage (not shown). Photo credit: Noah Burrell.

will complete it during the following months. Only the largest species, such as the Eurasian Oystercatcher, sometimes show unmolted outer primaries, which may be retained for up to 2 years (for instance, arrested molt). Finally, a fairly common pattern among first-winter birds that migrate to the southern hemisphere involves the replacement of outer primaries and inner secondaries only. Some first-winter birds undertake a complete normal descendent primary molt.

The progress of molt is a valuable piece of information used in ageing individuals of many species; consequently, it is of great value in

MOLT MIGRATION
IN WILSON'S PHALAROPE

In some birds, notably waterfowl, individuals undertake a molt migration, which is defined as a postbreeding, one-way movement to a restricted area where individuals complete an extensive and intensive molt, especially of the tail and flight feathers (Salomonsen 1968). These molts almost always consist of a complete prebasic molt, during which birds shed most or all of their body feathers and some flight feathers. In many waterfowl, the simultaneous molting of flight feathers renders birds flightless and vulnerable to predators for extended periods. Only one shorebird, the Bristle-thighed Curlew, is known to undertake a comparable molt, which takes place on isolated, predator-free islands in the Pacific Ocean (Marks et al. 1990). In general, however, molt migration is relatively rare in shorebirds, occurring in just a few sandpipers (such as the Wood and Green Sandpipers). Probably the best example of molt migration in shorebirds comes from Wilson's Phalarope (Jehl 1987). The nonterritorial, polyandrous mating system of phalaropes sets the stage for the early departure of females from the breeding grounds. Because males are solely responsible for care of eggs and chicks, their arrival at molting locations is delayed by a few weeks to more than a month.

Most of the world's population of Wilson's Phalarope undertakes an early summer movement to hypersaline lakes in the western United

During early summer, Wilson's Phalaropes stage and molt in large numbers at hypersaline lakes in the western United States. Photo credit: Mark Colwell.

States where they initiate a complete molt of body feathers and most of the remiges. At the same time, individuals nearly double their weight prior to a nonstop migratory flight to high elevation lakes in the Andes of South America and inland marshes of Argentina. The simultaneous occurrence of the energetically demanding processes of molt and premigratory fattening are made possible by the warm climate and abundant food present in hypersaline lakes of the western Great Basin. At Mono Lake, California, for instance, Wilson's Phalaropes arrive as early as late June and remain for 3 to 6 weeks. During this time, the body molt is so intense that they look "bedraggled" (Jehl 1987). At the same time, phalaropes replace all of their rectrices and varying numbers of their primaries, secondaries, and tertiaries.

demographic studies. The pattern of wear along a completely full-grown set of primaries can be informative. In the wing of a juvenile, where all primaries are grown at the same time, wear tends to increase from the protected inner primaries toward the outer ones. An excellent example of the application of molt to understanding age differences in breeding comes from the Semipalmated Sandpiper. Yearling Semipalmated Sandpipers retain outer primaries when returning to breed for the first time at La Perouse

Bay, Manitoba (Gratto and Morrison 1981), which has proved useful in demonstrating that yearlings have lower reproductive success than older birds (Gratto et al. 1983).

Feathers may be compromised by normal wear and tear associated with migratory and daily movements. Accordingly, shorebirds spend considerable time tending to their feathers at all times of the year. In the nonbreeding season, Long-billed Curlews foraging during low-tide intervals will interrupt feeding bouts and spend nearly 5% of their time involved in maintenance behavior (including preening and stretching) (Leeman et al. 2001). Piping Plovers wintering in coastal Louisiana preen at slightly less than 10% of their daily activity budget. The percentage time preening, however, is inversely related to the time spent foraging. The greatest amount of time spent preening occurs during late winter, when birds are molting into alternate plumage (Johnson and Baldassarre 1988). American Avocets spend similar amounts of their activity budget involved in maintenance behavior during the day (3%) and night (5%) (Kostecke and Smith 2003). An assemblage of 11 species of Palearctic wader wintering in coastal Kenya exhibit varying patterns of maintenance behavior, with 2% to 23% of the daily activity pattern devoted to "comfort" behaviors (Fasola and Biddau 1997).

The integumentary system of birds provides rich resources for a host of ectoparasites that feed on feathers, skin, and blood. The evolutionary relationships among ectoparasitic mites of shorebirds have been shown to follow Fahrenholz's rule, which states that parasite phylogeny mirrors its host phylogeny. This is a strong argument for coevolution (Mironov and Dabert 1999). Overall, however, few studies have documented the ectoparasites of shorebirds (Hunter and Colwell 1994). No study has examined the effects on individual survival and reproductive success, despite the likelihood that ectoparasites are readily transmitted between individuals in the dense flocks that occur at roosts.

SENSORY APPARATUS, FORAGING, AND DIGESTION

Shorebirds are remarkably diverse in the external morphological features associated with foraging, especially the bill and hindlimb. The bill is deeply recurved in avocets, especially females; slightly recurved in some tringine sandpipers and godwits; long and decurved in curlews, with a diminutive version occurring in many calidridine sandpipers; short and straight

ISOTOPES AND TRACE ELEMENTS IN MIGRATORY SHOREBIRDS

Stable isotopes and trace elements are incorporated into the tissues of all organisms, and their geographic and habitat signatures vary with isotope and the life history of a species. Increasingly, stable isotopes have been used to understand the origins and movements of birds. Shorebirds are well-known for their migratory nature, and stable isotopes grown in the feathers of young birds on breeding grounds or adults molting in nonbreeding habitats indicate the origins of individuals. Norris et al. (2007) analyzed 42 trace elements in Western Sandpiper feathers grown at five wintering sites along the Pacific Coast of North America. The chemical composition of feathers differed among sites, even those less than 3 km apart, indicating the utility of this tool in understanding migratory movements on a fine spatial scale. Evans Ogden et al. (2006) used stable isotopes to show that Dunlin acquire a significant element of their diet from terrestrial agricultural habitats adjacent to intertidal areas in the Fraser River Delta of British Columbia.

in many plovers; and stout and broad-based in the turnstones. This variation in bill morphology correlates with the foraging maneuvers of the species. Avocets scythe their recurved bill through shallow water or fine sediments to feed tactilely on invertebrates (Hamilton 1975). The decurved bill of curlews is used to extract large burrowing shrimp, shore crabs, small bivalves, large polychaetes, and small fishes (Leeman et al. 2001). By contrast, the shorter decurved bill of many calidridine sandpipers is adapted for shallow probing for small invertebrates. Turnstones are named from their habit of lifting and turning small rocks, under which they acquire prey. Smaller shorebirds, especially phalaropes, may routinely use surface tension of water and its adhesion to the bill to transport

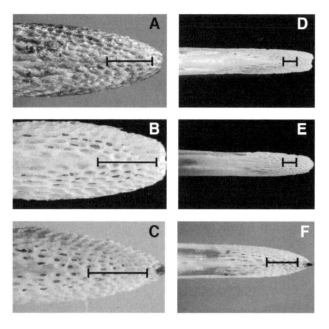

FIGURE 3.1. Sensory pits in the distal bill of several calidridine sandpipers attest to the importance of tactile sensory apparatus in foraging. Scale=1mm. (A) Western Sandpiper female, lower bill, bleached. (B) Western Sandpiper female, upper bill, bleached. (C) Western Sandpiper male, lower bill, bleached. (D) Dunlin female, lower bill, bleached. (E) Dunlin male, upper bill, bleached. (F) Least Sandpiper male, upper bill, bleached. From Nebel et al. (2005).

small prey into the mouth (Rubega and Obst 1993; Rubega 1997; Estrella et al. 2007).

These differences in bill morphology are broadly associated with variation among taxa in tactile and visual foraging. For example, sensory pits exist in the distal portions of some sandpiper bills (Bolze 1969; Fig. 3.1). Godwits and dowitchers, which feed predominantly by probing in substrates, extract prey using tactile cues that are detected by Herbst corpuscles embedded in the tips of the bill. By contrast, plovers search visually for prey at the surface of substrates. There are differences between tactile and visual foragers in the anatomical features of the brain associated with these feeding modes. For instance, scolopacid sandpipers have a pronounced enlargement of the trigeminal area of the brain, which is associated with tactile sensation of the bill tip. Plovers lack this enlargement (Pettigrew and Frost 2008).

Piersma et al. (1998) reported on a novel method of tactile foraging in Red Knots feeding on bivalves in soft sediments. Knots are capable of detecting immobile bivalves buried up to 3 cm deep in soft, wet substrates, and their success rates while foraging often exceed what is expected based on prey density. Using an ex-

perimental design with individual knots feeding in captivity, Piersma et al. (1998) suggested that knots must use pressure differences between the sensory pits at the distal tip of the bill and hard objects (such as prey) in wet substrates when foraging. These findings have implications for how scientists estimate the quality of foraging habitats based on prey density. Specifically, knots feeding in intertidal habitats are better at detecting prey than previously thought. Therefore, simple measures of prey density are inadequate (underestimate) the amount of food available to individuals and that can support an overwintering population.

Interspecific differences in bill morphology are accompanied by additional variation in the structure of the tongue. Recent evidence indicates a novel feeding method in the Western Sandpiper and the Dunlin associated with a brushlike tongue (Elner et al. 2005; Kuwae et al. 2008). Both species have papillae and keratinized spines on lateral portions of the tongue, although these structures are denser in the Western Sandpiper than in the Dunlin. This structure (combined with feeding observations) indicates that both species feed directly on surficial biofilm by lapping with their brushlike tongues. It is unlikely that this biofilm is consumed incidentally to foraging on epifauna (Elner et al. 2005).

There is considerable interspecific variation in the extent to which shorebirds forage nocturnally and diurnally (McNeil et al. 1992). Many species (such as oystercatchers, plovers, and sandpipers) forage both day and night; others (such as painted snipes, the Crab Plover, and thick-knees) feed principally at night; pratincoles, which are aerial insectivores, are mostly crepuscular. The visual sensory apparatus associated with these patterns are related to how species detect and acquire prey. Thomas et al. (2006) examined variation in shorebird eye size (as gauged by eye socket diameter) using comparative analyses that controlled for phylogenetic relationships (the observation that greater variation in eye size exists among than within taxa owing to shared evolutionary histories). Species

PREEN GLANDS AND OLFACTORY CRYPTICISM

The uropygial or preen gland of birds produces oils and waxes that waterproof individuals and inhibit microbial and ectoparasites from degrading integumentary tissues such as the skin and feathers. Recent evidence (Reneerkens et al. 2002, 2005) examining the chemical composition of shorebird preen waxes indicates a remarkable shift in wax composition that coincides with breeding in a variety of sandpipers. The change in preen gland composition is from lower molecular weight monoesters to higher molecular weight diesters, which are less volatile. The original finding of this pattern in Red Knots was broadened to include 19 additional sandpipers from Arctic and boreal habitats. Reneerkens et al. (2002, 2005) have suggested that this compositional change occurs before incubation to reduce the detectability of incubating adults to mammalian predators, which search for prey using olfactory cues. They support this conjecture with experimental evidence showing that olfactory searching dogs have greater difficulty locating diesters than monoesters.

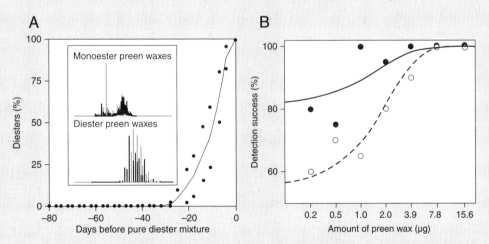

Red Knots shift from monoester to diester preen waxes, the latter being more difficult to detect by a sniffer dog. (A) The shift from monoester to diester waxes in spring takes place within 1 month in individual captive Red Knots (n = 14 individuals: 95% confidence interval around the mean value of the percentage of diester is indicated by dots). (B) The likelihood of successful detection is a function of the type and amount of preen wax. Each data point represents detection success during 20 sessions (monoesters: black circles; diesters: open circles). From Reneerkens et al. (2005).

that forage at night have larger eyes than those that feed during daylight periods. However, species that feed visually have similar-sized eyes to those using tactile cues to locate prey. Moreover, there is no relationship between the tendency to feed visually and diurnal foraging. Eye size may be only one adaptation to varying light levels. Rojas de Azuaje et al. (1993, 1999) examined the retinal structure (for instance, rod:cone densities) of several species that differ in foraging behavior. The retinas of plovers and stilts are characterized by a high density of receptors associated with greater visual acuity and a tendency to use visual foraging maneuvers both day and night. By contrast, tactile-feeding sandpipers have a lower density of retinal receptors.

Most birds are thought to have a poor sense of smell, and shorebirds are not special in this regard. Some evidence, however, suggests that olfaction may be important in the foraging of some shorebirds (Gerritsen et al. 1983). Healy and Guilford (1990) examined olfactory bulb size across a large sample of species, including members of the Charadriidae and Scol-

opacidae. After controlling for body size, they showed that nocturnal or crepuscular plovers have larger olfactory bulbs than their diurnal counterparts. This result was obtained for 11 of 12 other taxonomic comparisons, the exception being sandpipers.

DIGESTIVE SYSTEM

The diets of shorebirds vary greatly (see Chapter 7) both among and within species. This variation correlates with associated structures necessary to digest soft-bodied and hard-bodied prey (for instance, polychaetes versus bivalves, respectively). Posterior to the buccal cavity, most of the anatomical variation in the digestive system occurs in the region of the proventriculus and gizzard, the sections responsible for glandular and mechanical digestion, respectively. In virtually all species studied by Piersma et al. (1993), proventriculi represented a much smaller proportion of the stomach mass than gizzards. Considerable variation in stomach mass (proventriculus + gizzard) is isometrically related to

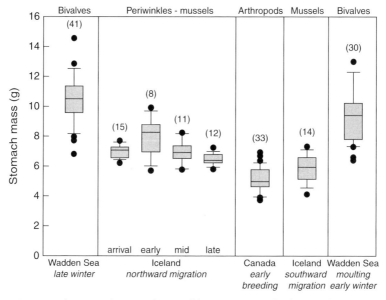

FIGURE 3.2. Changes in the stomach mass of digestive organs of Red Knots of the subspecies *islandica* throughout the annual cycle, associated with general diet and activity. Sample sizes are given above each box. Values are mean, standard deviation, and interquartile range. After Battley and Piersma (2005).

body mass for 17 species (Piersma et al. 1993); in other words, stomach mass increases proportionate to body size, and there is no difference in the relative size of stomachs between groups such as plovers and sandpipers. Of interest, however, is the considerable intraspecific variation in stomach mass associated with seasonal changes in diet (Fig. 3.2). Captive Red Knots fed soft-bodied prey had much smaller stomachs than conspecifics that ate large bivalves, suggesting that diet influences the structure of the gut. Curiously, however, some of the smallest stomachs have been obtained from wild-caught knots staging just before long-distance migrations. The stomachs of these individuals had undergone seasonal atrophy as an adaptive response anticipating long-distance migration (Piersma et al. 1993; Battley and Piersma 1995; van Gils et al. 2005). In other words, their stomachs had shrunk as a weight conservation measure for reducing wing-loading and effecting greater flight efficiency. These findings have been confirmed with other species such as the Bar-tailed Godwit, which is the champion long-distance migrant of all birds (Piersma and Gill 1998).

ENERGETICS AND THERMOREGULATION

Shorebirds appear to be unusual in the way in which they meet energetic challenges of thermoregulation. Compared with other birds, shorebirds use more energy, have greater daily energy expenditures, and maintain higher metabolic rates (Castro 1987; Kersten and Piersma 1987; Castro et al. 1992; Lindström and Klaassen 2003; Piersma et al. 2003). Basal metabolic rates of various shorebird species are roughly 40% higher than that reported for other nonpasserines of comparable size. Kersten and Piersma (1987) suggested that these energetic properties are a consequence of the relatively sparse and hence low insulative properties of their plumages, as well as the tendency for shorebirds to occupy open habitats (such as Arctic tundra and coastal wetlands). In many regions of the world, such

habitats often offer little in the way of environmental features to aid in thermoregulation via habitat selection. For example, wintering shorebirds experience cool, wet, and windy conditions when feeding in tidal reaches of northern estuaries, and they often roost in open habitats that offer little shelter from the elements. Consequently, shorebirds "run their metabolic engines at higher rpms" than other species to effect thermoregulation (Kersten and Piersma 1987).

Interspecific and intraspecific comparisons have shed additional light on shorebird energetics from a variety of habitats. During the breeding season, shorebirds have especially high energetic demands associated with egg production and parental care. Piersma et al. (2003) showed that incubating shorebirds in the Arctic have especially high daily energy expenditure. In fact, the values for seven arctic sandpipers and one plover were up to 50% higher than in temperate-breeding shorebirds of comparable size. The importance of cold environments in shaping shorebird energetics is evident in their morphology. Many shorebirds, especially sandpipers, breed in the north where heat conservation is at a premium even during the summer. Thermal advantages are accrued by individuals that reduce their exposure to cool, windy conditions, and this may be accomplished by having a lower profile. Cartar and Morrison (2005) correlated the hindlimb length of 17 Nearctic shorebirds with weather data from stations across northern Canada. They controlled statistically for differences in body size and wing length of the various sandpipers (14) and plovers (3), and found that species breeding in colder regions tend to have shorter legs than those from warmer areas. This suggests that natural selection has acted on leg length specifically, which in turn suggests that selection is strong, given the short interval of the annual cycle that these species spend in northern regions.

During the nonbreeding season, shorebirds occupy habitats that vary greatly in thermal environment. Purple and Rock Sandpipers, each other's closest relative, winter in coastal areas of the far north (approximately 70° north). By contrast, many shorebirds winter in temperate and tropical regions that are much more thermally benign. These different thermal environments have shaped the energetic patterns of shorebirds. For instance, a comparison of species indicates that basal metabolic rates are lower in shorebirds wintering in tropical Africa compared with temperate regions (Klaassen et al. 1990; Kersten et al. 1998). Intraspecific comparisons offer additional insights into the energy demands of wintering in regions of varying thermal environment. During the nonbreeding season, Sanderlings winter across nearly 100° of latitude, and individuals experience very different thermal environments at the extremes of their range. Castro et al. (1992) studied daily energy expenditure in Sanderlings wintering at four locations (New Jersey, Texas, Panama, and Peru) that differed substantially in thermal environment. Sanderlings wintering in North America have significantly higher daily energy expenditure than those wintering in the tropics, and this variation is largely explained by colder temperatures in northern areas.

Although existence in cold, open habitats is the rule for most shorebirds, some occasionally face significant thermal challenges associated with "wintering" in tropical regions. This challenge, however, is of a very different nature— that of dissipating heat in hot environments when their activity patterns and physiology require significant energy. Battley et al. (2003) studied Great Knots staging at Roebuck Bay, Australia, and summarized the energetic challenges facing birds while fueling for spring migration. Briefly, knots have problems dissipating heat due to a number of factors:

- They breed in Arctic regions and possess many of the previously discussed physiological features of shorebirds.
- They winter in the tropics.
- They cannot seek shade to enhance cooling owing to the heightened predation risk in obstructed habitats.

- They experience increased metabolism as they accumulate the premigratory fat loads necessary to fuel a nonstop migration of approximately 6,000 km to the next staging site.
- They may be compromised in their ability to lose heat when laden with subcutaneous fat.
- They molt into a darker, alternate plumage that absorbs more solar energy.
- They may increase foraging to fuel these activities.

In support of the notion of the increased thermal challenges of dissipating heat, Great Knots increase heat-reduction behavior as they attain darker plumages, especially when solar radiation is high. Specifically, darker plumaged knots more frequently forage with raised feathers on their back, which allows for greater convective loss of heat from the exposed skin. Moreover, these facets of energy balance may affect the habitat choice of birds. In another study of knots at Roebuck Bay, Rogers and coworkers (Rogers, Battley et al. 2006; Rogers, Piersma, Hassell 2006) showed that choice of roosts by Red and Great Knots is affected by heat loads. During the day, roosting knots prefer shallowly flooded salt pans where heat loss is greater than drier habitats where heat loads are higher.

OSMOREGULATION

At some point during the annual cycle, most shorebirds frequent environments where their osmoregulatory abilities are challenged by salty or hypersaline conditions (pelagic ocean, coastal wetlands, or hypersaline lakes). In these circumstances, individuals must rid the body of excess salt. In addition to the kidneys, many shorebirds have well-developed salt glands (also known as nasal or supraorbital glands) that extract excess salt from the blood (Staaland 1967). These glands, which lie in a skeletal furrow above the orbits of the eye, consist of a countercurrent system of arteries and veins interwoven with secretory tubules. This system passively (via a concentration gradient) and actively (via a potassium pump) pulls salts from the blood and concentrates them in the lumen of the gland. Concentrated droplets of saltwater occasionally can be seen dripping from bill tips of shorebirds that feed in marine environs.

Salt glands do not occur in all shorebirds. They are absent from many species, especially those that frequent freshwater habitats. The size and function of salt glands also vary with the season and the associated exposure to saline conditions; this attests to the energetic costs of maintaining this expensive tissue. Growth of kidneys and salt glands in neonate shorebirds was studied by Rubega and Oring (2004), who collected American Avocet chicks from a freshwater wetland at 1, 7, and 14 days after they had hatched. The chicks had hatched with salt glands comparable in size to those of marine birds. However, the relative size of salt glands decreased as the chicks aged; by contrast, kidney mass increased. These findings suggest that American Avocet chicks hatch prepared for the osmotic challenge of a salty environment but that development of extrarenal tissue is arrested under freshwater conditions (Rubega and Oring 2004).

The size of salt glands varies seasonally with the movement of individuals among habitats varying in salinity (Staaland 1967). In the Great Basin of the western United States, several large hypersaline lakes serve as important staging areas for migrant shorebirds. These salt-laden wetlands pose an osmotic challenge to species, which is balanced by a superabundant food resource in the form of brine flies (*Ephydra hians*) and shrimp (*Artemia monica*). Virtually the entire world population of Wilson's Phalarope stages at a few hypersaline lakes, including Great Salt Lake, Utah, Mono Lake, California, and Lake Abert, Oregon (Jehl 1988). Phalaropes and avocets staging at Mono Lake show no signs of excessive salt loading, which suggests that they gain most of their freshwater needs from their prey and tolerate the salty environment well (Mahoney and Jehl 1985).

REPRODUCTIVE SYSTEM

All birds lay hard-shelled eggs external to the body and engage in some form of behavior that incubates the developing embryo. Given this rather invariant nature of the avian reproductive system (Whittow 2001), it is not surprising that the reproductive anatomy of shorebirds is similar to virtually all other birds. Females have a left functional ovary that atrophies during the nonbreeding season. This is commonly viewed as an adaptation for weight conservation associated with flight. In many species, females possess sperm storage tubules in the lower oviduct, specifically folds of the lamina propria at the junction of the uterus and vagina (Rivers and Briskie 2003). The presence of stored sperm in these tubules does not appear until females have arrived on breeding grounds, however. Thus, it appears unlikely that females fertilize eggs before arrival on the breeding grounds, despite observations of courtship and copulations during spring migration in some species (Rivers and Briskie 2003). Once eggs are fertilized, females must lay them. Females ovulate at intervals of approximately 1 day (or greater), which gives rise to minimum egg-laying intervals of either 1 or 2 or more days (Colwell 2006).

Perhaps the most notable variation in the reproductive system of male shorebirds occurs in testes size, which correlates with the mating system. Cartar (1985) examined interspecific variation in testes size of a variety of sandpipers and found that males of polygynous sandpipers have larger testes than those of monogamous congeners. Additionally, the sperm of polygynous shorebirds tends to have morphologies (such as long tails and midpieces that house the mitochondria powering movement) that correlate with selection for increased motility (Johnson and Briskie 1999). Sperm tail length correlates positively with testis size. Together, these observations indicate that sperm competition varies among species of shorebird and may be intense in polygynous taxa.

CONSERVATION IMPLICATIONS

Detailed knowledge of a species' morphology, anatomy, and physiology is a prerequisite for effective conservation and management actions. A few examples illustrate the point. First, understanding patterns of molt is critical to quantifying age-related patterns of survival and reproductive success, which is the foundation of many demographic analyses. Second, the strong sexual size dimorphism of many shorebirds has consequences for habitat use (Figuerola 1999). In some species, for instance, males and females may behave (in use of habitats and foraging) as different species. As an example, male and female Eurasian Curlews effectively segregate between terrestrial and intertidal habitats, respectively, during the winter (Townshend 1981). These differences may necessitate a two-pronged approach to management of tidal flats and pastures to ensure population persistence in the face of anthropogenic habitat change. Similar arguments may apply to other species. Lastly, knowledge of sperm storage capabilities and extra-pair fertilizations refines our understanding of variance in reproductive success, with consequences for estimating effective population size.

LITERATURE CITED

Barbosa, A., and E. Moreno. 1999. Hindlimb morphology and locomotor performance in waders: An evolutionary approach. *Biological Journal of the Linnaean Society* 67: 313–330.

Battley, P. F., and T. Piersma. 1995. Adaptive interplay between feeding ecology and features of the digestive tract in birds. In *Physiological and ecological adaptations to feeding in vertebrates,* ed. J. M. Starck and T. Wang, 201–228. Enfield, NH: Science Publishers.

Battley, P. F., D. I. Rogers, T. Piersma, and A. Koolhaus. 2003. Behavioural evidence for heat-load problems in Great Knots in tropical Australia fuelling for long-distance flight. *The Emu* 103: 97–103.

Bolze, G. 1969. Anordnung und Bau der Herbstchen Körperschen in Limikolen-schnäbeln im Zusammenhang mit der Nahrungsfindung. *Zoologischer Anzeiger* 181:313–355.

Burton, P. J. K. 1974. *Feeding and the feeding apparatus in waders: A study of anatomy and adaptations in the Charadrii.* British Museum (Natural History) Publication No. 719. London: Trustees of the British Museum.

Cartar, R. V. 1985. Testis size in sandpipers: The fertilization frequency hypothesis. *Naturwissenschaften* 72: 157–158.

Cartar, R. V., and R. I. G. Morrison. 2005. Metabolic correlates of leg length in breeding Arctic shorebirds: The cost of getting high. *Journal of Biogeography* 32: 377–382.

Castro, G. 1987. High basal metabolic rates in Sanderlings. *Wilson Bulletin* 92: 267–268.

Castro, G., J. P. Myers, and R. E. Ricklefs. 1992. Ecology and energetics of Sanderlings migrating to four latitudes. *Ecology* 73: 833–844.

Colwell, M. A. 2006. Egg-laying intervals in shorebirds. *Wader Study Group Bulletin* 111: 50–59.

Elner, R. W., P. G. Beninger, D. L. Jackson, and T. M. Potter. 2005. Evidence of a new feeding mode in Western Sandpiper (*Calidris mauri*) and Dunlin (*Calidris alpina*) based on bill and tongue morphology and ultrastructure. *Marine Biology* 146: 1223–1234.

Estrella, S. M., J. A. Masero, and A. Perez-Hurtado. 2007. Small-prey profitability: Field analysis of shorebird use of surface tension of water to transport prey. *The Auk* 124: 1244–1253.

Evans Ogden, L. J., K. A. Hobson, D. B. Lank, and S. Bittman. 2006. Stable isotope analysis reveals that agricultural habitat provides an important dietary component for nonbreeding Dunlin. *Avian Conservation and Ecology* 1: 3.

Fasola, M., and L. Biddau. 1997. An assemblage of wintering waders in coastal Kenya: Activity budget and habitat use. *African Journal of Ecology* 35: 339–350.

Figuerola, J. 1999. A comparative study on the evolution of reversed size dimorphism in monogamous waders. *Biological Journal of the Linnean Society* 67: 1–18.

Gerritsen, A. F. C., Y. M. van Heezik, and C. Swennen. 1983. Chemoreception in two further *Calidris* species (*Calidris maritima* and *C. canutus*) with a comparison of the relative importance of chemoreception during foraging in *Calidris* species. *Netherlands Journal of Zoology* 33: 485–496.

Gill, F. A. 2007. *Ornithology*. 3rd ed. New York: W.H. Freeman.

Gratto, C. L., F. Cooke, R. I. G. Morrison. 1983. Nesting success of yearling and older breeders in the Semipalmated Sandpiper *Calidris pusilla*. *Canadian Journal of Zoology* 61: 1133–1137.

Gratto, C. L., and R. I. G. Morrison. 1981. Partial postjuvenal moult of the Semipalmated Sandpiper (*Calidris pusilla*). *Wader Study Group Bulletin* 33: 33–37.

Hamilton, R. B. 1975. *Comparative behavior of the American Avocet and the Black-necked Stilt*. Ornithological Monographs No. 17. Washington, DC: American Ornithologists' Union.

Healy, S., and T. Guilford. 1990. Olfactory-bulb size and nocturnality in birds. *Evolution* 44: 339–346.

Hogan-Warburg, A. L. 1966. Social behavior of the Ruff, *Philomachus pugnax* (L). *Ardea* 54: 109–229.

Humphrey, P. S., and K. C. Parkes. 1959. An approach to the study of molts and plumages. *The Auk* 76: 1–31.

Hunter, J. E., and M. A. Colwell. 1994. Ectoparasite (Phtheriptera) infestation of six shorebird species. *Wilson Bulletin* 94: 400–403.

Jehl, J. R., Jr. 1968. Relationships in the Charadrii (shorebirds): A taxonomic study based on color patterns of the downy young. *Memoirs of the San Diego Society of Natural History* 3: 1–54.

———. 1987. Moult and moult migration in a transequatorially migrating shorebird: Wilson's Phalarope. *Ornis Scandinavica* 18: 173–178.

———. 1988. Biology of the Eared Grebe and Wilson's Phalarope in the nonbreeding season: A study of adaptations to saline lakes. *Studies in Avian Biology* 12: 1–74.

Johnsgard, P. A. 1981. *The plovers, sandpipers, and snipes of the world*. Lincoln: University of Nebraska Press.

Johnson, C. M., and G. A. Baldassarre. 1988. Aspects of the winter ecology of Piping Plovers in coastal Alabama. *Wilson Bulletin* 100: 214–223.

Johnson, D. D. P., and J. V. Briskie. 1999. Sperm competition and sperm length in shorebirds. *The Condor* 101: 848–854.

Jukema, J., and T. Piersma. 2006. Permanent female mimics in a lekking shorebird. *Biology Letters* 2: 161–164.

Kersten, M., L. W. Bruinzeel, P. Wiersma, and T. Piersma. 1998. Reduced metabolic rate of migratory waders wintering in coastal Africa. *Ardea* 86: 71–80.

Kersten, M., and T. Piersma. 1987. High levels of energy expenditure in shorebirds: Metabolic adaptations to an energetically expensive way of life. *Ardea* 75: 175–187.

Klaassen, M., M. Kersten, and B. J. Ens. 1990. Energetic requirements for maintenance and premigratory body mass gain of waders wintering in Africa. *Ardea* 78: 209–220.

Kostecke, R. M., and L. M. Smith. 2003. Nocturnal behavior of American avocets in playa wetlands of the southern high plains of Texas, USA. *Waterbirds* 26: 192–195.

Kuwae, T., P. G. Beninger, P. Decottignies, K. J. Mathot, D. R. Lund, and R. W. Elner. 2008. Biofilm grazing in a higher vertebrate: The Western Sandpiper, *Calidris mauri*. *Ecology* 89: 599–606.

Lank, D. B., M. Coupe, and K. E. Wynne-Edwards. 1999. Testosterone-induced male traits in female ruffs (*Philomachus pugnax*): Autosomal inheritance and gender differentiation. *Proceedings of the Royal Society of London, Series B* 266: 2323–2330.

Lank, D. B., C. M. Smith, O. Hanotte, T. Burke, and F. Cooke. 1995. Genetic polymorphism for alternative mating behaviour in lekking male ruff (*Philomachus pugnax*). *Nature* 378: 59–62.

Leeman, L. W., M. A. Colwell, T. S. Leeman, and R. L. Mathis. 2001. Diets, energy intake, and kleptoparasitism of non-breeding Long-billed Curlews in a northern California estuary. *Wilson Bulletin* 113: 194–201.

Lindström, A., and M. Klaassen. 2003. High metabolic rates of shorebirds while in the Arctic: A circumpolar view. *The Condor* 105: 420–427.

Mahoney, S. A., and J. R. Jehl, Jr. 1985. Adaptations of migratory shorebirds to highly saline and alkaline lakes: Wilson's Phalarope and American Avocet. *The Condor* 87: 520–527.

Marks, J. S., R. L. Redmond, P. Hendricks, R. B. Clapp, and R. E. Gill, Jr. 1990. Notes on longevity and flightlessness in Bristle-thighed Curlews. *The Auk* 107: 779–781.

McNeil, R., P. Drapeau, and J. D. Goss-Custard. 1992. The occurrence and adaptive significance of nocturnal habits in waterfowl. *Biological Review* 67: 381–419.

Mironov, S. V., and J. Dabert. 1999. Phylogeny and co-speciation of feather mites of the subfamily Avenzoariinae (Analgoidea: Avenzoariidae). *Experimental and Applied Acarology* 23: 525–549.

Nebel, S., D. L. Jackson, and R. W. Elner. 2005. Functional association of bill morphology and foraging behavior in calidrid sandpipers. *Animal Biology* 55: 235–243.

Norris, D. R., D. B. Lank, J. Pither, D. Chipley, R. C. Ydenberg, and T. K. Kyser. 2007. Trace element profiles as unique identifiers of Western Sandpiper (*Calidris mauri*) populations. *Canadian Journal of Zoology* 85: 579–583.

Page, G. W., J. S. Warriner, J. C. Warriner, and P. W. C. Paton. 1995. Snowy Plover (*Charadrius alexandrinus*). In *The birds of North America*, ed. A. Poole and F. Gill, no. 154. Philadelphia/Washington, DC: Academy of Natural Sciences/American Ornithologists' Union.

Pettigrew, J. D., and B. J. Frost. 2008. A tactile fovea in Scolopacidae? *Brain, Behavior and Evolution* 26: 185–195.

Piersma, T., and R. E. Gill, Jr. 1998. Guts don't fly: Small digestive organs in obese Bar-tailed Godwits. *The Auk* 115: 196–203.

Piersma, T., A. Koolhaas, and A. Dekinga. 1993. Interactions between the stomach structure and diet choice in shorebirds. *The Auk* 110: 552–564.

Piersma, T., A. Lindström, R. H. Drent, I. Tulp, J. Jukema, R. I. G. Morrison, J. Reneerkens, H. Schekkerman, and G. H. Visser. 2003. High daily energy expenditure of incubating shorebirds on high Arctic tundra: A circumpolar study. *Functional Ecology* 17: 356–362.

Piersma, T., R. van Aelst, K. Kurk, H. Berkhoudt, and L. R. M. Maas. 1998. A new pressure sensory mechanism for prey detection in birds: The use of principles of seabed dynamics? *Proceedings of the Royal Society of London, Series B* 265: 1377–1383.

Prater, A. J., J. H. Marchant, and J. Vuorinen. 1977. *Guide to the identification and ageing of Holarctic waders*. Tring, United Kingdom: British Trust for Ornithology.

Pyle, P. 2008. *Identification guide to North American birds, Part II*. Point Reyes Station, CA: Slate Creek Press.

Reneerkens, J., T. Piersma, and J. S. Damste. 2002. Sandpipers (Scolopacidae) switch from monoester to diester preen waxes during courtship and incubation, but why? *Proceedings of the Royal Society of London, Series B* 269: 2135–2139.

Reneerkens, J., T. Piersma, and J. S. Sinninghé Damste. 2005. Switch to diester preen waxes may reduce avian nest predation by mammalian predators using olfactory cues. *Journal of Experimental Biology* 208: 4199–4202.

Rivers, J. W., and J. V. Briskie. 2003. Lack of sperm production and sperm storage by Arctic-nesting shorebirds during spring migration. *The Ibis* 145: 61–66.

Rogers, D. I., P. F. Battley, T. Piersma, J. van Gils, and K. G. Rogers. 2006. High-tide habitat choice: Insights from modelling roost selection by shorebirds around a tropical bay. *Animal Behaviour* 72: 563–575.

Rogers, D. I., T. Piersma, and C. J. Hassell. 2006. Roost availability may constrain shorebird distribution: Exploring the energetic costs of roosting and disturbance around a tropical bay. *Biological Conservation* 133: 225–235.

Rojas de Azuaje, L. M., R. McNeil, T. Cabana, P. Lachapelle. 1999. Diurnal and nocturnal visual capabilities in shorebirds as a function of their feeding strategies. *Brain, Behavior and Evolution* 53: 29–43.

Rojas de Azuaje, L. M., S. Tai, and R. McNeil. 1993. Comparison of rod/cone densities in three species of shorebirds having different nocturnal foraging strategies. *The Auk* 110: 141–145.

Rubega, M. A. 1997. Surface tension prey transport in shorebirds: How widespread is it? *The Ibis* 139: 488–493.

Rubega, M. A., and B. S. Obst. 1993. Surface-tension feeding in phalaropes: Discovery of a novel feeding mechanism. *The Auk* 110: 169–178.

Rubega, M. A., and L. W. Oring. 2004. Excretory organ growth and implications for salt tolerance in hatchling American avocets *Recurvirostra americana*. *Journal of Avian Biology* 35: 13–15.

Salomonsen, R. 1968. The moult migration. *Wildfowl* 19: 5–24.

Sibley, C. G., and J. E. Ahlquist. 1990. *Phylogeny and classification of birds*. New Haven, CT: Yale University Press.

Staaland, H. 1967. Anatomical and physiological adaptations of nasal glands in the Charadriiformes birds. *Comparative Biochemistry and Physiology* 23: 933–944.

Thomas, R. J., T. Székely, R. F. Powell, and I. C. Cuthill. 2006. Eye size, foraging methods and the timing of foraging in shorebirds. *Functional Ecology* 20: 157–165.

Townshend, D. J. 1981. The importance of field feeding to the survival of wintering male and female curlews *Numenius arquata* on the Tees Estuary. In *Feeding and survival strategies of estuarine organisms*, ed. N. V. Jones and W. J. Wolff, 261–273. New York: Plenum Press.

van Gils, J. A., P. F. Battley, T. Piersma, and R. Drent. 2005. Reinterpretation of gizzard sizes of Red Knots world-wide emphasises overriding importance of prey quality at migratory stopover sites.

Proceedings of the Royal Society of London, Series B 272: 2609–2618.

van Rhijn, J. G. 1991. *The Ruff: Individuality in a gregarious wading bird.* London: Academic Press.

Warnock, N. D., and R. E. Gill, Jr. 1996. Dunlin (*Calidris alpina*). In *The birds of North America online*, ed. A. Poole. Ithaca, NY: Cornell Lab of Ornithology. Available online at bna.birds .cornell.edu/bna/species/203 (accessed April 27, 2010).

Whittow, G. C., ed. 2001. *Sturkie's avian physiology.* 5th ed. New York: Academic Press.

Mating Systems

CONTENTS

THE STUDY OF MATING SYSTEMS arguably began with Darwin's (1871) treatise on sexual selection. In his classic work,

Darwin detailed differences between males and females in secondary sexual characteristics and mating behaviors. He compared these characters in a diverse group of birds, including many shorebirds such as the Ruff, lapwings, snipes, phalaropes, painted snipes, and pratincoles. As a group, shorebirds run the gamut of mating system diversity, from extreme polygyny in lek-breeding species through monogamous and polyandrous taxa. This variation probably stems from the nidifugous nature of the young, which sometimes emancipates parents of one gender from parental care and allows them to pursue additional mating opportunities. In this chapter, I summarize the variation in mating systems of shorebirds, and I examine various topics related to their evolutionary origins and the ecological factors influencing these patterns.

DEFINING A MATING SYSTEM

A species' mating system has been characterized by the social relationships of males and females and the means by which individuals control access to resources essential for breeding or to mates directly. The expression of a mating system in a species is, however, greatly influenced by its evolutionary history. Specifically, phylogenetic factors may constrain the behaviors of individuals such that the mating system is limited to a subset (chiefly monogamy) of the full range of possibilities (various forms of polygamy). Two principal phylogenetic factors influence avian mating systems: oviparity and mode of development (Oring 1982). Because all birds lay hard-shelled eggs external to the body, females are not constrained to obligate parental care in the manner of viviparous mammals (which are largely polygynous). This emancipation of females from obligate care of eggs and young creates the opportunity for the evolution of diverse mating systems, including extremes of male parental care and polyandry.

A second phylogenetic factor influencing the expression of a species' mating system acts within birds as a group and offers an explanation for why shorebirds exhibit such diverse mating systems as compared with other taxa (Oring 1982). The development of young birds varies along a continuum from altricial to highly precocial (O'Connor 1984). Passerines, for instance, exhibit altricial development in which females invest less energy and calcium in the egg than precocial species, and chicks hatch comparatively undeveloped. Adults brood and feed nidicolous young, which grow rapidly. In altricial species, care of eggs and young often requires both members of a pair; hence, social monogamy is common (Lack 1968). Shorebirds, by contrast, have precocial, nidifugous young. Females invest comparatively more in the development of the embryo. Therefore, eggs take longer to hatch, but neonates hatch fully feathered with a downy plumage and in most cases are capable of feeding themselves. Adult oystercatchers feed their young for varying durations extending up to a year or more (Winkler and Walters 1984); this is also true of the Crab Plover (del Hoyo 1992). Adults provide important parental care by brooding young and defending them from predators. Given that broods are seldom larger than four, a single adult, either

the male or female, is sufficient to brood and tend young.

Traditionally, a species' mating system is described using several terms (Emlen and Oring 1977; Oring 1982; Ligon 1999; Table 4.1). A principal descriptor emphasizes the social relationships between breeding adults, including the number of mates acquired in a breeding season and the permanence of pair bonds. The terms monogamy, polygyny, and polyandry describe the social pairings of males and females with reference to whether individuals have one versus multiple mates. A second descriptor of mating systems concerns the means by which individuals control access to mates. In many monogamous species, males compete for territories that contain resources essential for breeding. In some polygynous species, males defend no resources within a territory but compete directly through dominance interactions for access to females. Finally, in even fewer polyandrous species, females defend males directly. A final characterization of mating systems involves the timing with which individuals acquire multiple mates, which obviously applies only to polygamous systems. In some mating systems, females or males mate with multiple mates in sequence, whereas in other circumstances females or males attract multiple mates to their territory simultaneously. Accordingly, these polygamous mating systems are often discussed with reference to whether female and males acquire mates sequentially (also serially) or simultaneously.

Shorebirds have diverse mating systems that are not always easily categorized by common descriptors (see Table 4.1). A few well-chosen examples of shorebird mating systems illustrate the variation among and within species and highlight the challenges associated with applying a simple label to a species mating system.

The Spotted Sandpiper is a classic example of a species with a resource-defense polyandrous mating system (Oring and Lank 1982, 1984, 1986; Oring et al. 1983; Lank et al. 1985; Colwell and Oring 1988a). Females defend multipurpose

TABLE 4.1

Principal mating systems observed in shorebirds
Terminology from Emlen and Oring (1977)

MATING SYSTEM	DESCRIPTION	EXAMPLES
Monogamy		
Resource defense	Males defend territories with resources for breeding; males and females share incubation and brood-rearing duties to varying extents.	Dunlin, Killdeer, Long-billed Curlew, Marbled Godwit, Black Oystercatcher
Colonial monogamy	Males and females form pairs, with aggregated nesting to varying degrees in colonies; males and females share parental care.	Black-necked Stilt, American Avocet
Polygyny		
Male dominance	Males display on arenas within a lek and compete via dominance hierarchies for females; females visit lek for mating, and provide sole parental care of eggs and chicks off leks.	Ruff, Buff-breasted Sandpiper, White-rumped Sandpiper
Polyandry		
Resource defense	Females defend territories with resources essential for breeding; multiple males nest in sequence or simultaneously; males provide most parental care for eggs and chicks.	Spotted Sandpiper, Northern Jacana
Mate access	Females compete for males directly; males provide all care of eggs and chicks.	Phalaropes
Polygamy		
Rapid multiple clutch	Females lay two clutches in rapid sequence with a single male and split incubation duties between nests; females occasionally pair with two different males in sequence.	Temminck's Stint, Sanderling, Mountain Plover
Resource defense	Males and females pair monogamously, on territories; over long breeding season, individuals obtain multiple mates in sequence; males and females share incubation, but males tend chicks.	Snowy Plover, Kentish Plover

territories that encompass feeding and nesting habitat. Females lay separate clutches for one to several males in sequence over 2 to 3 months; rarely, females acquire two mates simultaneously. Males provide most parental care by incubating eggs and brooding chicks; females occasionally assist their last mate of the season in rearing young. Females occasionally engage in extra-pair copulations, sometimes with males that may become future mates. In a long-term study, variation in polyandry was not related to differences in territory quality as indexed by food, which was abundant on all territories. Rather, the incidence of polyandry varied in association with breeding density and availability of mates. When predation caused high rates of nest failure, sandpipers renested frequently, and females often obtained multiple mates, especially when population density was high. From the perspective of the individual, older (experienced) females were more likely than yearlings to become polyandrous because they arrived earlier and had more time to compete for territories and attract multiple mates. A good deal of the success of older females lies in arriving early, but they also benefit from pairing with mates from an earlier breeding season. These former mates sometimes tolerate females laying a clutch for a first mate and wait to become the second mate for the female. Females occasionally nest build with the first male and copulate with the second mate, often on sites a short distance away. Male Spotted Sandpipers on the other hand, are constrained by greater parental care duties to a maximum of one successful mating a year. Lastly, the complexity of the mating system was illustrated by the finding that males occasionally cared for young that were not their own (Oring et al. 1992). The likelihood of mixed parentage in broods increased with successive nesting attempts. Males that were the second and third mates of a female were more likely to be cuckolded, and, furthermore, the offspring for whom they cared were most likely to be those of the first male. This suggests females are capable of sperm storage.

The Ruff (female is known as a Reeve) lies at the other extreme of polygamy with a male-dominance, polygynous mating system (Hogan-Warburg 1966; van Rhijn 1973; Widemo 1998; Lank and Smith 1987; Lank et al. 1995, 2002). Males are considerably larger than Reeves. The Ruff is a lek-breeding shorebird; males display to females on arenas within a lek, and females visit display grounds solely for the purpose of mating. Females provide sole parental care for eggs and chicks. Males exhibit considerable variation in plumage and behavior, with individuals assuming one of three strategies that are influenced by a relatively simple autosomal, dominant-recessive allele system. The polymorphism includes (1) behaviorally dominant, dark-plumaged males that display on the lek and acquire the most matings; (2) subordinate, light-plumaged satellites that associate with dark-plumaged males; and (3) "faeder" males that have female-like plumage and mate with females away from the lek (Jukema and Piersma 2006). It was widely thought that females visited the lek once for copulation. However, recent evidence indicates that females care for young of mixed parentage, indicating that they copulate with more than one male (Lank et al. 2002).

Many other shorebirds exhibit a monogamous mating system that lies somewhere between the extremes of polyandry and polygyny that have been summarized. Most large sandpipers, such as godwits and curlews, exhibit resource-defense monogamy. These species appear to have long-term pair bonds, with males and females often reuniting each spring on the breeding grounds. Smaller males perform aerial displays over their large territories, which harbor their nesting and brood-rearing habitat. Individuals occasionally forage off their territories. Male and female godwits, for instance, share nearly equally in incubation and brood care, although both males and females occasionally desert broods late in the brood-rearing period (Gratto-Trevor 2000). The same is true of curlews, although females usually depart breeding grounds earlier than the males, which provide most of the care for chicks. There is no evidence for any shorebird that members of a pair associate with one another during the nonbreeding season.

THE ROLE OF ECOLOGICAL FACTORS

Describing a shorebird's mating system seems straightforward—a species is monogamous, polygynous, or polyandrous; males and females defend territories, compete on leks, or defend mates directly; and mates are acquired simultaneously or in sequence. In reality, however, there often exists considerable intraspecific variation in a species' mating system, especially for some polygamous taxa. This arises because individuals may employ alternative reproductive strategies under different ecological conditions. As a consequence, a species may be socially monogamous in some habitats and tend toward polygamy in other settings.

In their seminal paper on the evolution of mating systems in birds and mammals, Emlen and Oring (1977) recognized this variation and drew attention to the strong influence of environment in shaping the expression of a species' mating system. In particular, they emphasized that a mating system was the outcome of individual responses to varying ecological circumstances that favor polygamy. The "environmental potential for polygamy" (or the proportion of individuals acquiring multiple mates) in a population was influenced by the spatial and temporal distribution of resources necessary to acquire a territory or mates directly. In effect, species were constrained to monogamy, and polygamy increased when the distribution or quality of resources became increasingly patchy or when the temporal distribution of mates created the opportunity for some individuals to acquire multiple mates in sequence. For shorebirds, the environmental potential for polygamy is associated with several ecological variables that vary among populations. These variables are length of breeding season, arrival schedules of males and females to breeding areas, population density, and the distribution of resources (or variation in habitat quality).

LENGTH OF BREEDING SEASON

The amount of time that individuals have to breed can strongly affect the potential for po-lygamy by constraining an individual's ability to breed multiple times in a season. Put simply, the longer the breeding season, the greater the opportunity for multiple breeding attempts. Length of breeding season is strongly associated with latitude, with northerly populations having shorter breeding seasons than those breeding in temperate and tropical regions. For example, many Arctic-breeding shorebirds initiate clutches over a few weeks, which limits opportunities to obtain multiple mates in sequence, let alone breed at all. In years of late snow melt in the Canadian Arctic, Red Knots may forego breeding altogether (Evans and Pienkowski 1984).

By contrast, temperate-breeding species breed over longer intervals, during which individuals have sufficient time to obtain multiple mates. In northern Minnesota, female Spotted Sandpipers initiate clutches over approximately 2 months (May through early July). Females defend territories and attract mates in sequence (rarely simultaneously). Initial pairings result in clutches incubated by males, whereas females normally assist their last mate of the season in tending eggs and chicks (Oring et al. 1983). Some females acquire four mates in sequence.

Along the Pacific Coast of North America, Snowy Plovers initiate clutches over approximately 4 months (mid-March to mid-July). Females and males share incubation for approximately 28 days, but males almost always tend chicks for an additional 28 days. Consequently, females may breed successfully with a maximum of three males; by contrast, extended male parental care means males may fledge a maximum of two broods in a breeding season (Warriner et al. 1986). The plover mating system is characterized as serial polygamy, with males and females both acquiring multiple mates in a breeding season.

SPRING ARRIVAL SCHEDULE

A species' arrival schedule refers to the timing and duration over which individual males and females show up at breeding grounds each spring. It can strongly affect local sex ratios and hence the intensity of competition and potential for polygamy. The greater the asynchrony in arrival of the limiting sex Emlen and Oring 1977), the greater the opportunity for the competing sex to monopolize multiple mates in sequence. For example, at the prairie wetlands of North America, the earliest arriving male Wilson's Phalaropes are accompanied by multiple females that compete vigorously for them as mates. Local sex ratios often are strongly female-biased early in the breeding season (Reynolds et al. 1987). However, females that are successful in competing for mates early on when competition is intense (Colwell and Oring 1988c) are more likely to become polyandrous (Colwell 1986b; Colwell and Oring 1988c). The same is true for early arriving female Spotted Sandpipers (Oring and Lank 1982).

BREEDING DENSITY

Population density may also influence mating system expression by facilitating or hindering encounters between potential mates. At high population densities, individuals may be able to defend high-quality territories that attract multiple mates either simultaneously or in sequence. By contrast, at low population densities, individuals may be less capable of monopolizing mates via territory defense. The incidence of polyandry in the Spotted Sandpiper varies among populations. In a high-density population breeding on an isolated island in a northern lake, females obtained an average of 1.86 mates annually; by contrast, in low-density populations most females obtained a single mate each year, and monogamy was the rule (Hays 1972; Oring and Knudson 1972; Oring and Lank 1986).

RESOURCE DISTRIBUTION

Resources are never evenly distributed across habitats. Rather, high-quality patches of breeding habitat may have particularly rich food resources or attractive nesting sites; by contrast, other habitats may be marginally suitable for breeding. The patchy distribution of resources favors polygamy when some individuals can monopolize high-quality habitat to which they attract multiple mates simultaneously or in sequence. Other individuals, however, may acquire a territory and a single mate or none at all. In lekking species, where the display arenas of males offer no resources to females, the arena position or male quality itself may influence mating success and polygamy.

SOCIAL VERSUS GENETIC RELATIONSHIPS

Early ornithologists developed their understanding of mating systems based on detailed study of the social interactions between individually marked birds. The advent and development of molecular techniques for determining parentage has revolutionized views of mating systems. The classification of mating systems adhered to by ornithologists for decades (e.g., Lack 1968; Emlen and Oring 1977) has been replaced by wide recognition that social relationships among individuals are often complex. In fact, it is much more common for mixed parentage to be demonstrated in a species than not (Griffith et al. 2002). There are several ways in which mating behavior may result in offspring being reared by adults other than their genetic parents: nest parasitism, quasi parasitism, and extra-pair fertilizations. In each of these cases, adults care for young that are not their own, presumably with little or no fitness benefit.

MIXED MATERNITY AND NEST PARASITISM

Nest parasitism is essentially mixed maternity in a clutch or brood. It occurs when offspring do not match those of the putative mother. There are two distinct ways in which nest parasitism occurs in shorebirds. In the first case, a female may lay an egg in the nest of another pair and not participate at all in parental care. This may be a reproductive strategy employed by some females, as suggested for the Pied Avocet (Hötker 2000); in this case, the behavior has evolved as an alternative to the typical pattern of pairing monogamously. Evidence is weak that the egg-laying behaviors of females of any shorebird species correspond to this behavioral strategy.

Alternatively, a female may lay an egg in the nest of a conspecific when her clutch fails during the laying process and before she has completed her own clutch. In essence, the female has an egg forming in her oviduct, and rather than laying it in the failed nest or elsewhere, she oviposits in the nest of a conspecific. This requires that a female has knowledge of conspecific nests nearby, which may be facilitated by high nesting densities (Colwell 1986b; Hötker 2000). This scenario is probably best labeled by the term "egg dumping" as it represents an opportunistic response to the loss of a clutch during the laying process. The success of either of these forms of parasitism is lessened by the short window of time in which shorebirds lay their clutches of four or fewer eggs, the pattern of laying eggs at daily (approximately 27-hour) intervals typical of most shorebirds (Colwell 2006), and the requirement that the parasitizing female lay during the host female's laying interval (Andersson 1984). Otherwise, eggs are unlikely to hatch owing to asynchrony in development of embryos or the compromised hatchability of enlarged clutches.

The second general way in which mixed maternity occurs in shorebird clutches may not truly qualify as nest parasitism (Amat 1998). In this case, the occurrence of extra eggs occurs when two females (or pairs) claim a common nest and each female lays her clutch in a normal fashion. As a result, clutches are often approximately double the modal clutch size for a species. This scenario may account for a large percentage of the cases of nest parasitism reported for shorebirds, especially among colonial species such as avocets and stilts (Gibson 1971; Amat 1998). This scenario may occur rarely in socially polyandrous species as well. In Wilson's Phalarope, for instance, only one unambiguous case of nest parasitism occurred out of several hundred nests over 6 years (Colwell 1986a). At this nest, two females laid a total of seven eggs in a nest, as evidenced by multiple eggs being laid in less than 24 hours.

Nest parasitism is often detected based on three types of evidence. First, a nest may contain an "extra" egg; that is, one more than the modal clutch size for the species. Second, when nests are monitored during the laying stage, eggs may appear to be laid at intervals shorter than the minimum (24 hours) for a species. Third, a single, anomalous egg (that is, differing in shape, background color, or markings) may be present in a clutch. These criteria have been used widely by researchers to present instances of nest parasitism in shorebirds, and, in one case, to evaluate the ecological and social determinants of nest parasitism. In a 9-year study involving 3,182 nests of the Pied Avocet breeding along the Wadden Sea, Hötker (2000) estimated that nest parasitism occurred in 3.3% of clutches and 1.3% of eggs. However, the validity of using these field methods in establishing cases of nest parasitism has been questioned. Specifically, Grønstøl et al. (2006) used genetic methods to show that in three nests that would have been classified as cases of nest parasitism in the Northern Lapwing all revealed no loss of parentage. In other words, these nests were misclassified as resulting from nest parasitism using each of the three commonly used methods. Clearly, observational methods may not always be accurate in representing the rates of nest parasitism in shorebirds.

Compared with other birds, intraspecific nest parasitism is not common in shorebirds (Andersson 1984; Yom-Tov 2001), probably because the small clutch size (hence, short egg-laying period) restricts opportunities for females to synchronize the development of their eggs with a potential host's clutch. Nevertheless, nest parasitism has been observed on occasion when breeding (nest) densities are high. For example, supernormal clutch sizes (eight eggs) have been reported for colonial-breeding avocets and stilts (Gibson 1971; Rowher et al. 1979; Amat 1998). In these examples, females of two different pairs are likely laying a full clutch in a common nest. Instances in other (noncolonial) species are even rarer. Colwell (1986a) reported single cases of nest parasitism for Wilson's Phalarope, Willet, and Marbled Godwit breeding at a wetland in the prairies of Saskatchewan. The unusual occurrence of egg-dumping in this case probably was linked to high population densities and to high nest loss to predators during a period of drought.

QUASI PARASITISM

A variation on the theme of nest parasitism is quasi parasitism, when a female lays an egg in another female's nest when that egg was fertilized by the male partner of the parasitized nest. Quasi parasitism is comparatively rare in birds; it has been reported in just 12 species (Griffith et al. 2004), two of which are shorebirds (Blomqvist, Andersson, et al. 2002a). In the Kentish Plover and Common Sandpiper, 1% and 6% of chicks, respectively, gave evidence of quasi parasitism (Griffith et al. 2004). In these same species, there was no genetic evidence that nest parasitism (as previously described) accounted for mixed parentage in broods. Interestingly, in no case was the clutch larger than the modal size (of three and four, respectively), suggesting that parasitizing females had removed eggs when they laid their own in the nest (Blomqvist, Anderssson, et al. 2002). Alternatively, the physiological mechanism leading to ovulation and laying of the last egg by a host female may have been "turned off" by the presence of an extra egg in the clutch, especially if a female parasitizes a nest when it has two eggs.

MIXED PATERNITY AND EXTRA-PAIR FERTILIZATIONS

The most common behavioral mechanism by which the genetics of offspring do not match their putative male parent is via extra-pair fertilizations. In fact, the classification of mating systems has been revolutionized by the observation that, contrary to Lack's (1968) statement that most birds are monogamous, extra-pair fertilizations are widespread among birds with socially monogamous pair-bonds (Griffith et al. 2002). Extra-pair fertilizations may be even more common among polygamous species. In most monogamous species, males often attempt to copulate with females other than their mate, and females regularly solicit copulations and achieve fertilizations with sperm from males other than their social mate. In polygamous taxa, extra-pair fertilizations appear even more common.

The rate of extra-pair paternity is defined as the percentage of fertilizations resulting from copulations outside the social bonds recognized by traditional mating system classification (Griffith et al. 2002). A growing list of shorebirds (Table 4.2) demonstrates considerable interspecific variation in the extent to which offspring result from extra-pair fertilizations. At one extreme are several socially monogamous species, such as the Purple Sandpiper and Common Ringed Plover, where there is little evidence of mixed paternity in broods. One monogamous species, the Common Sandpiper, has higher rates of extra-pair fertilizations. In the Southern Lapwing, a species with a variable social structure that includes monogamy and small groups of cooperative breeders, mixed parentage in broods always involves immediate members of the group. Mixed paternity is slightly more common in polyandrous species; a notable exception is Wilson's Phalarope, with no evidence of extra-pair fertilizations. In the Spotted Sandpiper, where some females pair with multiple males in sequence, the cases of extra-pair paternity indicate that mismatched young in later broods derive from sperm of first mates of females. This indicates that females store sperm and use it to fertilize eggs in later clutches. At the extreme of mixed paternity is the Ruff, a polygynous species in which females visit leks for copulations and provide sole parental care for chicks; fully half of broods sampled indicated mixed paternity.

The results in Table 4.2 are suggestive of trends across species with increasing polygamy and skewed reproductive success associated with multiple mating partners. However, the interpretation of such a pattern is complicated by low accuracy (the percent error) when the sample size (the number of chicks sampled) is less than 200 (Griffith et al. 2002). Most (3 out of 20) of the studies of shorebirds fall well below this number. Therefore, it is probably safest to conclude that variation in the rate of extra-

TABLE 4.2

Interspecific variation in the occurrence of extra-pair fertilizations among selected shorebirds

SPECIES	EXTRA-PAIR FERTILIZATIONS		SOCIAL PAIRBOND	SOURCE
	% BROODS (N)	% CHICKS (N)		
Comb-crested Jacana	10 (10)	3 (36)	Polyandry	Haig et al. 2003
Wattled Jacana	18 (74)	8 (235)	Polyandry	Emlen et al. 1998
Eurasian Oystercatcher	4 (26)	2 (65)	Monogamy	Heg et al. 1993
Southern Lapwing	19 (16)	10 (41)	Monogamy	Saracura et al. 2008
Common Ringed Plover	0 (21)	0 (57)	Monogamy	Wallander et al. 2001
Semipalmated Plover	4 (24)	5 (85)	Monogamy	Zharikov and Nol 2000
Snowy Plover	5 (65)	3 (170)	Polygamy	Blomqvist et al. 2002a
	8 (89)	4 (229)	Polygamy	Küpper et al. 2004
Eurasian Dotterel	9 (22)	5 (44)	Polyandry	Owens et al. 1995
Common Sandpiper	19 (27)	16 (83)	Monogamy	Mee et al. 2004
Kentish Plover	20 (15)	8 (53)	Monogamy	Blomqvist et al. 2002a
Spotted Sandpiper	21 (34)	11 (111)	Polyandry	Oring et al. 1992
Western Sandpiper	8 (48)	5 (98)	Monogamy	Blomqvist et al. 2002b
Purple Sandpiper	4 (27)	1 (82)	Monogamy	Pierce and Lifjeld 1998
Buff-breasted Sandpiper	40 (47)	—	Polygyny	Lanctot et al. 1997
Ruff	50 (34)	—	Polygyny	Lank et al. 2002
	52 (66)	—	Polygyny	Thuman and Griffith 2005
Wilson's Phalarope	0 (17)	0 (51)	Polyandry	Delehanty et al. 1998
Red Phalarope	9 (18)	7 (70)	Polyandry	Dale et al. 1999
Red-necked Phalarope	6 (63)	2 (226)	Polyandry	Schamel et al. 2004

pair paternity across shorebird species is related to the transience of pair-bonds between males and females, which may be facilitated by sperm storage capabilities (Rivers and Briskie 2003) and the tendency for females to seek matings with males other than their social mate.

The ecological conditions favoring high rates of mixed parentage in broods has been the subject of considerable interest (Andersson 1984; Møller and Birkhead 1993; Westneat and Sherman 1997; Griffith et al. 2002). A common speculation is that high breeding density favors high rates of both nest parasitism and extra-pair fertilizations. In the former case, habitat limitations occasionally may force individuals to breed at high density, which increases the probability that females will successfully find conspecific nests to parasitize (Andersson 1984). Anecdotal evidence (Colwell 1986a) and the tendency for parasitism to be comparatively common in colonial shorebirds (Gibson 1971; Rohwer et al. 1979; Amat 1998), especially when nests are close by (Hötker 2000), support the breeding density hypothesis. Similar arguments have been given to explain variation in extra-pair fertilizations: high breeding density favors higher rates of extra-pair fertilization (Møller and Birkhead 1993). Recent reviews of all birds, however, suggest that there is limited evidence for a direct relationship between breeding density and the rate of extra-pair fertilizations across species (Westneat and Sherman 1997; Griffith et al. 2002).

PARENTAL CARE PATTERNS

Shorebirds exhibit diverse patterns of parental care (Fig. 4.1). They vary from male-only care in polyandrous species (such as jacanas, painted snipes, the Plains-wanderer, and phalaropes), through biparental incubation and predominantly male care of chicks in monogamous species (as with many scolopacids), to female-only care in highly polygynous taxa (such as the Buff-breasted Sandpiper and the Ruff) (Jönsson and Alerstam 1990; Székely and Reynolds 1995; Székely et al. 2006). Clearly, the contributions of males and females to care of eggs and chicks are integral elements of any mating system. For this reason, understanding the parental roles of the sexes is critical to discussions of the evolution and diversity of mating systems within a group such as shorebirds. Parental care influences mating system evolution by setting the stage for emancipation of either males or females. Evolutionarily, this may be followed by mating strategies that involve pursuit of polygamous matings. Clearly, understanding ancestral patterns of parental care is a necessary first step in discussions of the evolution of mating systems.

One method to investigate the evolution of parental care involves the use of established phylogenies to understand transitions from ancestral behavioral states to contemporary patterns. These phylogenies allow for comparative analyses that control for the fact that closely related species share common evolutionary histories and hence are not independent of one another. Consequently, comparisons among higher level taxa (such as tribes, subfamilies, or families) are more valid. Székely and Reynolds (1995) used the comparative method to examine evolutionary transitions in parental care in the various shorebird lineages. Their analysis included 14 families, 64% (32 out of 50) of genera, and 47% (96 out of 203) of species. They used several different outgroups (such as Pteroclididae, Alcidae, and the Laroidea clade within the Charadriiformes), and they validated their findings using alternate phylogenies based on different character states.

A phylogenetic tree illustrating evolutionary transitions for parental care in shorebirds is shown in Figure 4.2 (Székely and Reynolds 1995). Biparental care was determined to be the ancestral condition in all shorebirds; this result was robust to the inclusion of different outgroups. This biparental condition appears to have been retained in the plover clade (Charadrii), including the Ibisbill, thick-knees, avocets and stilts, sheathbills, and Magellanic Plover. With a few exceptions, virtually all members of this clade are represented by a pattern of shared male–female care of eggs and chicks. A few groups, notably the dotterels and sand-plover lineages, have evolved male-only parental care. Even within these groups, however, there is considerable variation. For instance, in the Snowy Plover, males and females normally share equally in incubation, with males providing most care of chicks. Females typically desert their mates shortly after chicks hatch; rarely females tend eggs and young alone. Biparental care is the rule for most other plovers. However, uniparental care by females occurs rarely in some species such as the Northern Lapwing (Blomqvist and Johansson 1994). The sandpiper clade, by contrast, appears to have evolved male parental care early in its evolutionary his-

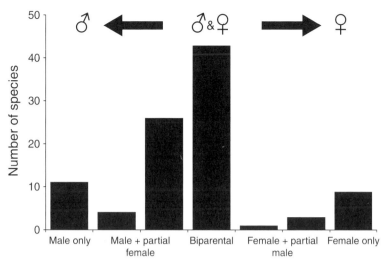

FIGURE 4.1. Distribution of parental care in shorebirds. Male-only care is often associated with polyandrous mating systems, whereas female-only care is associated with polygynous systems. After Székely et al. (2006).

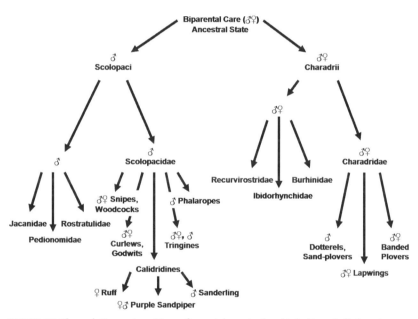

FIGURE 4.2. The evolutionary transitions of parental care in shorebirds. From Székely and Reynolds (1995).

tory. This was followed by multiple independent groups evolving reduced male parental care; it is most extreme in the phalaropes and painted snipes, in which males only develop a brood patch and care for eggs and young. Female parental care is highly variable in the sandpipers, and is most noteworthy in several species of calidridines. In conclusion, male-only parental care is widespread among shorebirds, occurring in numerous lineages, especially the sandpiper lineage.

It is obvious that the ancestral conditions and evolved patterns of parental care in a taxon are intimately related to discussions of the evolution of mating systems. In any lineage, the extent to which males care for young strongly influences the opportunity for females to desert mates and pursue additional mating opportunities. Similarly, when female care has evolved, males are more likely to pursue polygynous matings.

EVOLUTION OF POLYANDRY

Polyandry is an unusual mating system among vertebrates and it occurs chiefly among birds, with shorebirds providing a disproportionate number of examples (Oring 1986). Polyandry in its various forms regularly occurs in 19 species from five shorebird families; incidental observations suggest it occurs occasionally in other species (Oring 1986) (Table 4.3). Accordingly, considerable attention has been paid to the biology and ecology of shorebirds in understanding the evolution of polyandry (Orians 1969; Jenni 1974; Pitelka et al. 1974; Graul et al. 1977; Maynard Smith 1977; Oring 1982, 1986; Erckmann 1983; Lennington 1984; Owens 2002). This discourse has sought common ecological explanations for the evolution of polyandry in diverse taxa. But because shorebirds exhibit varying forms of polyandry, it may be difficult to arrive at a simple scenario describing how this mating system evolved.

Consider the following variations on the theme of polyandry in shorebirds. In some species (jacanas), females pair with multiple mates simultaneously, whereas in others mates are acquired sequentially over a breeding season (such as with the Snowy Plover). In a few calidridine sandpipers (such as Temminck's Stint and the Sanderling), females lay two clutches in rapid sequence with a single mate, with each adult caring for the eggs in separate nests; females occasionally change mates between clutches and are polyandrous. In some shorebirds, females share parental care with males to varying degrees (as with the Spotted Sandpiper); in others, males alone care for eggs and chicks (as with phalaropes and painted snipes). Finally, females may defend mates directly in sometimes intense, scramble competitions (phalaropes) or they may defend territories to which they attract mates (jacanas). These diverse facets make for challenges in understanding the evolution of polyandry, and they suggest that polyandry has evolved independently several times in shorebirds (Erckmann 1983).

Several biological features may predispose shorebirds to paternal care and polyandrous social pairings (Erckmann 1983; Oring 1986). First, as determinate layers of small clutches, females cannot increase their breeding success by simply laying more eggs in a nest. Rather, they are constrained to one, two, three, or four eggs, depending on the species. Therefore, if they are to achieve greater reproductive success, the only recourse is to lay a second clutch and either incubate it themselves or to acquire a second mate. This conclusion, however, was not supported in a comparative analysis of all birds (Owens 2002). There was no tendency for polyandrous taxa to have smaller clutch sizes than other groups.

Second, shorebird young are precocial and nidifugous. Therefore, uniparental care of eggs and chicks by either the male or female may be sufficient to raise young to independence. In shorebirds, males typically assume a larger role than females in parental care, which is associated with reversed size dimorphism in most species (Jönsson and Alterstam 1990; Figuerola 1999). For some polyandrous species (such as phalaropes and Spotted Sandpiper), the pattern of uniparental care can be viewed as a question of what ecological conditions favor increased male parental care and hence the opportunity for females to pursue second mates. However, in other taxa (such as jacanas, painted snipes, and the Plains-wanderer), male parental care appears to be the primitive condition.

Several ecological conditions appear to have favored the evolution of polyandry from a monogamous ancestral condition. Polyandry probably has evolved under some combination of high rate of clutch loss, low food availability influencing the high energetic costs to females of egg formation, and high breeding density. A common scenario envisions that the high rates

TABLE 4.3

*Summary of parental-care patterns of the sexes and methods of mate acquisition in shorebirds
with mating systems that are either classic polyandry or rapid, multiple-clutch polygamy*

FAMILY/SPECIES	PARENTAL CARE[a]		MATE ACQUISITION		
	EGGS	CHICKS	TIMING[b]	METHOD[c]	MATING SYSTEM
Jacanidae					
Lesser Jacana	♂♀	♂♀	Si/Se	Territory	Polyandry
Wattled Jacana	♂	♂	Si	Territory	Polyandry
African Jacana	♂	♂	Si	Territory	Polyandry
Bronze-winged Jacana	♂	♂	Si	Territory	Polyandry
Pheasant-tailed Jacana	♂♀	♂♀	Si	Territory	Polyandry
Comb-crested Jacana	♂	♂	Si	Territory	Polyandry
Rostratulidae					
American Painted-snipe	♂♀	♂♀		Territory	Polyandry
Greater Painted-snipe	♂	♂		Territory	Polyandry
Charadriidae					
Kentish (Snowy) Plover	♂♀	♂	Se	Territory	Polyandry
Eurasian Dotterel	♂	♂	Se	Territory	Polyandry
Mountain Plover	♂♀	♂♀	Se	Territory	Multiple clutch polygamy
Scolopacidae					
Spotted Sandpiper	♂♀	♂♀	Se/Si	Territory	Polyandry
Sanderling	♂♀	♂♀	Si	Territory	Multiple clutch polygamy
Temminck's Stint	♂♀	♂♀	Si	Territory	Multiple clutch polygamy
Little Stint	♂♀	♂♀		Territory	Multiple clutch polygamy
Red Phalarope	♂	♂	Se	Mate defense	Polyandry
Red-necked Phalarope	♂	♂	Se	Mate defense	Polyandry
Wilson's Phalarope	♂	♂	Se	Mate defense	Polyandry
Pedionomidae					
Plains-wanderer	♂♀	♂	Se	Territory	Polyandry

Sources: Erckmann (1983), Oring (1986), Reynolds and Székely (1997).
[a] Parental care: ♂ = male-only incubation or brooding; ♂♀ = male and female share care to varying extents, although male care is often much greater than female.
[b] Timing of mate acquisition: Si = males paired simultaneously with female; Se = sequential pairing.
[c] Method of mate acquisition: Territory = female defends territory to which she attracts mates; mate defense = female defends male directly in scramble competition.

of nest loss to predators or other natural causes place a premium on the ability of females to lay replacement clutches. When nest failure is frequent, and low food availability is coupled with high costs of egg laying, conditions favor a reduction in female incubation duties immediately after completing a clutch. Because female shorebirds lay large eggs relative to their body size and do not rely on endogenous reserves to form eggs (Erckmann 1983), the process of egg formation and oviposition taxes them energetically. Consequently, females must recoup the energy and calcium essential for egg formation to lay replacement clutches. However, females emancipated from their incubation duties are free to pursue additional matings. The probability of polyandrous pairings is enhanced under conditions of high population density.

The next step toward polyandry envisions that these emancipated females pursue second mates while their first mate assumes sole care of young. The interpretation of the relative strengths of food versus predation of clutches acting to influence polyandry varies among researchers. One argument stresses that polyandry was more likely to evolve when food was scarce (limiting) such that poor female body condition during egg laying placed a premium on increased male parental care of eggs and chicks (Graul et al. 1977). In this scenario, females freed from parental care could recoup energy and nutrient reserves immediately after completing a clutch. Under conditions of increased food availability, female emancipation leads to polyandry. Another view combines this energy argument about the importance of female body condition with an emphasis on frequency of clutch loss and replacement clutches (Oring 1986). These arguments suggest that polyandry has evolved under conditions of high predation pressure such that females must readily recoup energy (and calcium) reserves for laying frequent replacement clutches. In either scenario, polyandry evolves from a monogamous system.

Given the diverse forms of polyandry in separate lineages of shorebird, it is certain that it evolved independently multiple times (Székely and Reynolds 1995). Therefore, the ecological circumstances favoring its expression probably vary among groups. Oring (1986) pointed out that a number of conditions were necessary in ancestral populations to favor the evolution of polyandry. First, males provide a significant amount of parental care. Second, food was not limiting, which facilitated uniparental care and production of multiple clutches. Third, uniparental care was effective in coping with threats of predation to eggs and chicks. Fourth, males were as capable as females in providing parental care. Fifth, females were better able to exploit alternative reproductive opportunities. And, sixth, clutch size was fixed such that females could only increase fecundity by laying additional clutches (rather than more eggs in each clutch). Classic polyandry, as is present in Spotted Sandpipers, phalaropes, and jacanas, probably evolved from an ancestral mating system of monogamy (Pitelka et al. 1974; Erckmann 1983; Oring 1986). Others, however, contest this supposition and argue that uniparental care by males was the ancestral condition (Reynolds and Székely 1997) and that classic polyandry has evolved via rapid, multiple-clutch polygamy typical of a few calidridines (Ridley 1978).

VARIANCE IN REPRODUCTIVE SUCCESS

One important consequence of a mating system is that individual males and females vary in reproductive success. This variation is comparatively low in monogamous species, whereas intrasexual variance in reproductive success is greater in polygamous taxa, especially lekking species (Höglund and Alatalo 1995). The relative measures of male and female variance have been used to gauge the intensity of sexual selection (Payne 1984; Trail 1985). In polygynous birds, male variance exceeds that of females; the reverse is true in polyandrous species. In the latter case, variance in female reproductive success probably never approaches the extremes of males in highly polygynous species because the upper limits of female reproductive success are constrained by egg production (Oring, Colwell, and Reed 1991).

However, recent evidence from two lekking shorebirds calls into question the notion that variance in male reproductive success is highly skewed toward a few males. In the Buff-breasted Sandpiper, for instance, observational data showed that a small number of males had disproportionate mating success. But paternity analyses of broods indicated that few males in the population approached the mating success indicated by observed matings. In fact, only one male sired 10 offspring, and most broods were sired by multiple males. Hence, variance in male reproductive success was much lower than expected (Lanctot et al. 1997). Data from the Ruff show that the broods of approximately half of females are sired by multiple males (Lank et al. 2002). The presence of a third male ("faeder" morph) breeding strategy suggests that variance in male mating success may be even lower than previously suspected for this lek-breeding species.

Numerous studies have quantified the annual reproductive success of shorebirds by estimating the average per capita hatching or fledging success of males and females in a population. Given the long-lived nature of even the smallest shorebirds, it is not surprising that few studies have monitored a population long enough to quantify lifetime reproductive success. In a 20-year study of the polyandrous Spotted Sandpiper, Oring and colleagues (Oring, Colwell, and Reed 1991; Oring, Reed, et al. 1991) demonstrated that variance in female reproductive success exceeded that of males, indicating that sexual selection acted more strongly on females. Oring, Colwell, and Reed (1991) examined the influence of five life-history components—longevity, number of mates acquired, eggs laid or tended per mate, proportion of clutches hatching, and proportion of chicks fledging—on variance in lifetime reproductive success. The most important component (that

LIFETIME REPRODUCTIVE SUCCESS IN THE SPOTTED SANDPIPER

The Spotted Sandpiper breeds in early successional habitats across North America. In areas of high population density, females defend territories to which they may attract one to four mates sequentially over a 2-month nesting season. Males are the principal caregivers for the four-egg clutch and chicks. Lewis Oring and colleagues (1983; Oring, Colwell, and Reed 1991; Oring, Reed, et al. 1991) studied a small population on 1.4 hectares of Little Pelican Island, Minnesota, for approximately 20 years. Females and males varied greatly in lifetime reproductive success, and variance in female reproductive success exceeded that of males. For both sexes, the strongest components influencing variance in lifetime reproductive success were hatching and fledging success. In turn, these were associated with the tendency of predators to be present on the island. In other words, when predators (such as the mink, *Mustela vison*) were on the island, they ate virtually all eggs and chicks. In the absence of predators, however, individuals had high reproductive success.

is, explaining the greatest variance in lifetime reproductive success within sexes) was fledging success, which in turn was positively correlated with hatching success. Over the 20 years, individuals that bred during good years (when predators did not consume nearly all eggs and chicks) were the most productive. Longevity was the second most important component affecting lifetime reproductive success. These findings emphasize that annual variation in ecological factors such as predation can strongly influence patterns of reproductive success, even in a population where intrasexual competition can strongly influence mating success and individual fitness.

SIZE DIMORPHISM

Many shorebirds exhibit pronounced sexual size dimorphism (Jehl and Murray 1986; Jönsson and Alerstam 1990; Székely and Reynolds 1995; Reynolds and Székely 1997; Figuerola 1999; Sandercock 2001; Székely et al. 2006), but size difference varies among species. In general, females are larger than males. This is especially true of sandpipers, in which females are slightly larger than males in most (84%) of the 57 species (Jönsson and Alerstam 1990). In the Northern Jacana males are 0.6 times the mass of females (Székely et al. 2006). However, the pattern is reversed in polygynous taxa. In the Ruff, males are 1.7 times the mass of females.

Several evolutionary explanations have been suggested for the pattern of larger female (or smaller male) body size. One plausible explanation for reversed size dimorphism is based on an energy argument associated with male aerial displays during the breeding season. Jehl and Murray (1986) argued that reversed size dimorphism resulted from selection favoring small males in species where males perform aerial displays during the breeding season. It is argued that natural selection has favored small males owing to the energy advantages of lighter wing loading and enhanced aerial display maneuverability. This argument is similar to that applied to raptors in which females are also larger than the males, which also perform acrobatic displays (Andersson and Norberg 1981). Mueller (1989) criticized this interpretation, and Jehl and Murray (1989) countered. Several recent papers using comparative methods that control for phylogenetic relationships offer some support for the acrobatic display hypothesis (Figuerola 1999; Sandercock 2001; Székely et al. 2006). In ground-displaying species, males tended to be larger than females. In species where courting males used acrobatic (aerial) displays, males were significantly smaller than females.

Alternative explanations exist for the preponderance of small males in many shorebirds.

Jönsson and Alerstam (1990) applied an energy argument coupled with the parental care roles of the sexes to explain the pattern of small males in polyandrous and monogamous taxa. They argued that, during the breeding season, selection favored small size when males invested more in parental care. Small males were better able to recoup and maintain energy reserves during incubation and brooding. One outcome of small male size was a shorter bill, which they argued was advantageous for feeding in diverse habitats, including uplands, when rearing chicks. By contrast, during the nonbreeding season, selection favors large female size owing to the advantages of laying down sufficient energy reserves for egg laying. Large body size also carries with it large bill size, which allowed females to feed across a greater range of habitats (such as probing to varying depths of intertidal flats). In polygynous species, females are smaller than males, and the arguments are reversed. Unfortunately, this comparative review was confined to 57 species of sandpiper, which did not control for phylogenetic relationships and makes the results and interpretation open to question.

SEX RATIOS

The number of males and females in a population is intimately related to gender differences in mating opportunities and hence to a species' mating system. Simply put, a biased sex ratio (a preponderance of one sex) establishes conditions whereby the availability of mates limits mating opportunities for individuals of the other gender. When this occurs, individuals will compete among themselves for mates of the limiting sex (the resource in short supply). In most shorebird mating systems, males compete for territories or arenas to which they attract females as mates for varying durations. In a smaller proportion of cases, females compete for territories or for mates directly. These observations suggest that, in most species of shorebird, males outnumber females; conversely, the opposite prediction holds for polyandrous species.

Considerable theoretical (e.g., Fisher 1930) and empirical evidence has been amassed on the subject of population sex ratio. Population sex ratios may be biased for several reasons. In birds, a skewed sex ratio may arise from events occurring during fertilization or egg laying (primary sex ratio), at hatching (secondary sex ratio), or owing to differential mortality of males and females later in life. Very few empirical data have been collected on the population sex ratios of shorebirds, and all of the data concern the Kentish Plover. Székely et al. (2004) reported a slight (0.461 males) but statistically nonsignificant bias in the sex ratio of plover chicks at hatch. Interestingly, the proportion of males varied seasonally. The proportion of males in broods decreased from 0.555 to 0.378 during the roughly 90 days over which chicks hatched. Sex ratios also became increasingly male biased as the chicks aged, which suggests that male chicks survive better than females. Finally, larger plover chicks were significantly more likely to be males, and this relationship held true when controlling for the effects of season. There has been no estimate for the breeding population sex ratio in the large population studied by Székely and his colleagues in Turkey, but the evidence suggests that differences in sex ratio and subsequent survival bias the population toward males. Along the Pacific Coast of North America, Stenzel et al. (2010) used true survival estimates to evaluate sex ratio in a long-term study of a color-marked population of the Snowy Plover. Higher survival of older (2-year-old) males than females produced a population sex ratio that was biased toward males. Thus, it appears that proportionally fewer male plovers hatch but they survive better, which results in a population sex ratio that is increasing skewed toward males. Males typically compete vigorously for territories. Curiously, however, the mating system of the Kentish Plover and Snowy Plover is characterized by sequential polygamy and a tendency for female to abandon their mates to care for broods shortly after the young hatch. As a result, the mating strategies of females are driven by local breeding opportunities.

Local sex ratios often vary from unity owing to chance environmental effects and because of the differential investment of males and females in reproduction. Emlen and Oring (1977) coined the term OPERATIONAL SEX RATIO (OSR) to describe the situation whereby sex ratios may be skewed because males or females are increasingly unavailable as potential mates owing to their parental care duties. For example, OSR may vary seasonally because individuals of one gender (such as male phalaropes or Reeves) are continuously removed from the local population as they become committed to incubation and subsequent brood care. In Wilson's Phalarope, for instance, OSR and the intensity of competition among females for mates varies seasonally (Colwell and Oring 1988b). Early-season sex ratios are often strongly biased toward females, which compete vigorously for the few males. As the bulk of males arrive, the OSR approaches unity, and competition decreases. Rarely, males outnumber females, the OSR becomes male biased, and males may compete for females. However, once these males assume sole care of eggs and chicks, fewer males are available as mates, and the OSR shifts toward more females and an increase in intrasexual competition. The late season variation in OSR, however, may be influenced by the intensity of nest loss to predators in the local population. Males that lose their eggs often recycle into the breeding population. When this is common, OSR and intrasexual competition are often reduced.

In summary, the sex ratio of a population is an integral component of a species' mating system. Environmental circumstances may skew the local OSR, and alter the mating strategies and reproductive success of individuals. It remains unclear, however, whether population sex ratios are skewed toward one gender and whether this is a consequence or cause of the mating system expression.

CONSERVATION IMPLICATIONS

The study of mating systems has been of immense theoretical interest since Darwin hypothesized about sexual selection in 1871. Darwin emphasized individual variance in mating success when he discussed the evolution of secondary sexual characteristics. But variance in reproductive success has immense applied value as well. Interest in the applied value of studying mating systems has increased recently with conservation efforts focused on small and declining populations. From a conservation perspective, there are several good reasons to investigate mating systems and understand variation in reproductive success among individuals of a population.

VARIANCE IN REPRODUCTIVE SUCCESS AND EFFECTIVE POPULATION SIZE

An important reason to study a species' mating system is that it gives insight into patterns of variation in reproductive success among individuals, which may have consequences for the genetic structure of small populations. As a rule, monogamous species exhibit lower variance in reproductive success compared with polygamous species. This arises because a greater percentage of individuals in a monogamous population acquire mates compared with highly polygynous (and to a lesser extent polyandrous) species. In other words, the genetic contributions of males in lekking species are highly skewed toward a small number of individuals who are disproportionately successful at mating; by contrast, mating success is more evenly distributed among individuals in monogamous species.

A detailed understanding of variance in reproductive success allows population geneticists to quantify effective population size (N_e), where N_e is defined as the size of an idealized population that undergoes the same amount of loss of genetic variability as the real population under consideration (Kimura and Crow 1963). For several reasons, N_e is almost always smaller than the censused population size (N_c) (Koenig 1988), including (1) overlapping generations, (2) breeding dispersal, and (3) unequal progeny production. This latter measure uses the ratio of variance to mean reproductive success for a population (of males or females separately) to assess the degree to which mating deviates from a random pattern (where variance equals mean reproductive success). The greater the variance in reproductive success among individuals (for example, the fewer Ruffs who dominate matings on a lek), the greater the reduction in N_e. The effect of unequal progeny production on N_e is magnified when studies track lifetime (as opposed to annual) reproductive success of individuals (Koenig 1988). Conservation geneticists are especially interested in N_e as it applies to rare species because they may suffer a variety of genetic problems associated with small population size, such as drift and inbreeding. Several species of shorebird (for instance, the Black Stilt, Spoon-billed Sandpiper, and Slender-billed Curlew) are rare enough that genetic concerns associated with small N_e are justified.

VITAL RATES, POPULATION GROWTH, AND VIABILITY

Variation in reproductive success (fledging success) is also useful to demographers seeking to model the viability and growth of populations. A population's vital rates consist of survival estimates (for adults and juveniles separately) and per capita reproductive success. Demographers use per capita reproductive success in combination with survival estimates to estimate lambda (λ) for a population. Lambda is used to understand whether a population is increasing ($\lambda > 1.0$), stable ($\lambda = 1.0$), or declining ($\lambda < 1.0$). Formally, λ is estimated as (Adult survival) + (Juvenile survival × Per capita reproductive success).

EVALUATING MANAGEMENT ACTIONS

Monitoring reproductive success offers a straightforward means of assessing the effectiveness of management actions on a population. It is more difficult and time consuming to evaluate the success of management on survival. In particular, researchers often use some measure of reproductive success (such as nest survival estimates or per capita reproductive success) to evaluate the success of management actions, whether they be habitat manipulations, control of invasive species, or predator management. For instance, nest exclosures have been used to lessen the negative impacts of egg predators on the productivity of shorebird populations (see Chapter 12). Evaluating the success of this management action requires estimates of nest survival coupled with per capita fledging success. This is essential because if hatching and fledging success covary positively (e.g., Oring, Colwell, and Reed 1991), then the same predators that prey on eggs may consume nidifugous chicks when they leave nest exclosures. Consequently, exclosures may not be the solution to the negative impacts of predators on low productivity.

LITERATURE CITED

Amat, J. A. 1998. Mixed clutches in shorebird nests: Why are they so uncommon? *Wader Study Group Bulletin* 85: 55–59.

Andersson, M. 1984. Brood parasitism within species. In *Producers and scroungers,* ed. C. J. Barnard, 195–228. New York: Chapman and Hall.

Andersson, M., and R. Å. Norberg. 1981. Evolution of reversed size dimorphism and role partitioning among predatory birds, with a size scaling of flight performance. *Biological Journal of the Linnean Society* 15: 105–130.

Blomqvist, D., and O. C. Johansson. 1994. Double clutches and uniparental care in Lapwing *Vanellus vanellus,* with a comment on the evolution of double-clutching. *Journal of Avian Biology* 25: 77–79.

Blomqvist, D., M. Andersson, C. Küpper, I. C. Cuthill, J. Kis, R. B. Lanctot, B. K. Sandercock, T. Székely, J. Wallander, and B. Kempenaers. 2002a. Genetic similarity between mates and extra-pair parentage in three species of shorebird. *Nature* 419: 613–615.

Blomqvist, D., B. Kempenaers, R. B. Lanctot, and B. K. Sandercock. 2002b. Genetic parentage and mate guarding in the Arctic-breeding Western Sandpiper. *The Auk* 119: 228–233.

Colwell, M. A. 1986a. Intraspecific brood parasitism in three species of prairie-breeding shorebirds. *Wilson Bulletin* 98: 473–475.

———. 1986b. The first documented case of polyandry for Wilson's Phalarope (*Phalaropus tricolor*). *Auk* 103: 611–612.

———. 2006. Egg-laying intervals in shorebirds. *Wader Study Group Bulletin* 111: 50–59.

Colwell, M. A., and L. W. Oring. 1988a. Extra-pair mating in the Spotted Sandpiper: A female mate acquisition tactic. *Animal Behavior* 38: 675–684.

———. 1988b. Sex ratios and intrasexual competition for mates in a sex-role reversed shorebird, Wilson's phalarope (*Phalaropus tricolor*). *Behavioral Ecology and Sociobiology* 22: 165–173.

———. 1988c. Variable female mating tactics in a sex-role reversed shorebird, Wilson's Phalarope (*Phalaropus tricolor*). *National Geographic Research* 4: 426–432.

Dale, J., R. Montgomerie, D. Michaud, and P. Boag. 1999. Frequency and timing of extrapair fertilisations in the polyandrous Red Phalarope (*Phalaropus fulicarius*). *Behavioral Ecology and Sociobiology* 46: 50–56.

Darwin, C. R. 1871. *The descent of man, and selection in relation to sex.* Princeton, NJ: Princeton University Press.

Delehanty, D. J., R. C. Fleischer, M. A. Colwell, and L. W. Oring. 1998. Sex-role reversal and the absence of extra-pair fertilizations in Wilson's Phalarope. *Animal Behavior* 55: 995–1002.

del Hoyo, J., A. Elliot, J. Sargatal, and N. J. Collar, eds. 1992. *Handbook of birds of the world.* Vol. 3. Barcelona: Lynx Edicion.

Emlen, S. T., and L. W. Oring. 1977. Ecology, sexual selection, and the evolution of mating systems. *Science* 197: 215–223.

Emlen, S. T., P. H. Wrege, and M. S. Webster. 1998. Cuckoldry as a cost of polyandry in the sex-role reversed Wattled Jacana, *Jacana jacana. Proceedings of the Royal Society of London, Series B* 265: 2359–2364.

Erckmann, W. J. 1983. The evolution of polyandry in shorebirds: An evaluation of hypotheses. In *Social behavior of female vertebrates,* ed. S. K. Wasser, 113–168. New York: Academic Press.

Evans, P. R., and M. W. Pienkowski. 1984. Population dynamics of shorebirds. In *Shorebirds: Breeding behavior and populations,* ed. J. Burger and B. L. Olla, 83–123. New York: Plenum Press.

Figuerola, J. 1999. A comparative study on the evolution of reversed size dimorphism in monogamous waders. *Biological Journal of the Linnean Society* 67: 1–18.

Fisher, R. 1930. *The genetical theory of natural selection.* Oxford: Oxford University Press.

Gibson, F. 1971. The breeding biology of the American Avocet (*Recurvirostra americana*) in central Oregon. *The Condor* 73: 444–454.

Gratto-Trevor, C. L. 2000. Marbled Godwit (*Limosa fedoa*). In *The birds of North America,* ed. A. Poole and F. Gill, no. 492. Philadelphia: Birds of North America.

Graul, W. D., S. R. Derrickson, and D. W. Mock. 1977. The evolution of avian polyandry. *American Naturalist* 111: 812–816.

Griffith, S. C., B. E. Lyons, and R. Montgomerie. 2004. Quasi-parasitism in birds. *Behavioral Ecology and Sociobiology* 56: 191–200.

Griffith, S. C., I. P. F. Owens, K. A. Thuman. 2002. Extra pair paternity in birds: A review of interspecific variation and adaptive function. *Molecular Ecology* 11: 2195–2212.

Grønstøl, G., D. Blomqvist, and R. H. Wagner. 2006. The importance of genetic evidence for identifying intra-specific brood parasitism. *Journal of Avian Biology* 37: 197–199.

Haig, S. M., T. R. Mace, and T. D. Mullins. 2003. Parentage and relatedness in polyandrous Comb-crested Jacanas using ISSRs. *Journal of Heredity* 94: 302–309.

Hays, H. 1972. Polyandry in the Spotted Sandpiper. *Living Bird* 11: 43–57.

Heg, D., B. J. Ens, T. Burke, L. Jenkins, and J. P. Kruijt. 1993. Why does the typically monogamous oystercatcher (*Haematopus ostralegus*) engage in extra-pair copulations? *Behaviour* 126: 247–289.

Hogan-Warburg, A. J. 1966. Social behaviour of the Ruff, *Philomachus pugnax* (L.). *Ardea* 54: 109–229.

Höglund, J., and R. V. Alatalo. 1995. *Leks.* Princeton, NJ: Princeton University Press.

Hötker, H. 2000. Conspecific nest parasitism in the Pied Avocet *Recurvirostra avosetta. The Ibis* 142: 280–288.

Jehl, J. R., Jr., and B. G. Murray. 1986. The evolution of normal and reversed size dimorphism in shorebirds and other birds. *Current Ornithology* 3: 1–86.

———. 1989. Response: Evolution of sexual size dimorphism. *The Auk* 106: 155–157.

Jenni, D. 1974. Evolution of polyandry in birds. *American Zoologist* 14: 129–144.

Jönsson, P. E., and T. Alerstam. 1990. The adaptive significance of parental care role division and sexual size dimorphism in breeding shorebirds. *Biological Journal of the Linnean Society* 41: 301–314.

Jukema, J., and T. Piersma. 2006. Permanent female mimics in a lekking shorebird. *Biology Letters* 2: 161–164.

Kimura, M., and J. F. Crow. 1963. The measurement of effective population number. *Evolution* 17: 279–288.

Koenig, W. D. 1988. On determination of viable population size in birds and mammals. *Wildlife Society Bulletin* 16: 230–234.

Küpper, C., J. Kis, A. Kosztolányi, T. Székely, I. C. Cuthill, and D. Blomqvist. 2004. Genetic mating system and timing of extra-pair fertilizations in the Kentish Plover. *Behavioral Ecology and Sociobiology* 57: 32–39.

Lack, D. 1968. *Ecological adaptations for breeding in birds.* London: Methuen.

Lanctot, R. B., K. T. Scribner, B. Kempenaers, and P. J. Weatherhead. 1997. Lekking without a paradox in the Buff-breasted Sandpiper. *American Naturalist* 149: 1051–1070.

Lank, D. B., L. W. Oring, and S. J. Maxson. 1985. Mate and nutrient limitation of egg-laying in a polyandrous shorebird. *Ecology* 66: 1513–1524.

Lank, D. B., and C. M. Smith. 1987. Conditional lekking in the ruff. *Behavioral Ecology and Sociobiology* 20: 137–145.

Lank, D. B., C. M. Smith, O. Hannotte, T. Burke, and F. Cooke. 1995. Genetic polymorphism for alternative mating behaviour in lekking male in ruff *Philomachus pugnax. Nature* 378: 59–62.

Lank, D. B., C. M. Smith, O. Hanotte, A. Ohtonen, S. Bailey, and T. Burke. 2002. High frequency of polyandry in a lek mating system. *Behavioral Ecology* 13: 209–215.

Lennington, S. 1984. The evolution of polyandry in shorebirds. In *Shorebirds: Breeding behavior and populations,* ed. J. Burger and B. L. Olla, 149–167. New York: Plenum Press.

Ligon, D. J. 1999. *The evolution of avian breeding systems.* Oxford: Oxford University Press.

Maynard Smith, J. 1977. Parental investment: A prospective analysis. *Animal Behavior* 25: 1–9.

Mee, A., D. P. Whitfield, D. B. A. Thompson, and T. Burke. 2004. Extrapair paternity in the Common Sandpiper, *Actitis hypoleucos*, revealed by DNA fingerprinting. *Animal Behaviour* 67: 333–342.

Møller, A. P., and T. R. Birkhead. 1993. Cuckoldry and sociality: A comparative study of birds. *American Naturalist* 142: 118–140.

Mueller, H. C. 1989. Aerial agility and the evolution of reversed size dimorphism (RSD) in shorebirds. *The Auk* 106: 154–155.

O'Connor, R. J. 1984. *The growth and development of birds.* Chichester, United Kingdom: John Wiley and Sons.

Orians, G. H. 1969. On the evolution of mating systems in birds and mammals. *American Naturalist* 103: 589–603.

Oring, L. W. 1982. Avian mating systems. In *Avian biology,* Vol. 6, ed. J. Farner and J. King, 1–92. New York: Academic Press.

———. 1986. Avian polyandry. In *Current ornithology,* ed. R. Johnston, 309–351. New York: Academic Press.

Oring, L. W., M. A. Colwell, and J. M. Reed. 1991. Lifetime reproductive success in the Spotted Sandpiper (*Actitis macularia*): Sex differences and variance components. *Behavioral Ecology and Sociobiology* 28: 425–432.

Oring, L. W., R. C. Fleischer, J. M. Reed, and K. E. Marsden. 1992. Cuckoldry through stored sperm in the sequentially polyandrous Spotted Sandpiper. *Nature* 35: 631–633.

Oring, L. W., and M. L. Knudson. 1972. Monogamy and polyandry in the Spotted Sandpiper. *Living Bird* 11: 59–73.

Oring, L. W., and D. B. Lank. 1982. Sexual selection, arrival times, philopatry and site fidelity in the polyandrous Spotted Sandpiper. *Behavioral Ecology and Sociobiology* 10: 185–191.

———. 1984. Breeding area fidelity, natal philopatry and the social systems of sandpipers. In *Shorebirds: Breeding behavior and populations,* ed. J. Burger and B. L. Olla, 125–147. New York: Plenum Press.

———. 1986. Polyandry in Spotted Sandpipers: The impact of environment. In *Ecological aspects of social evolution,* ed. D. I. Rubenstein and R. W. Wrangham, 21–42. Princeton, NJ: Princeton University Press.

Oring, L. W., D. B. Lank, and S. J. Maxson. 1983. Population studies of the polyandrous Spotted Sandpiper. *The Auk* 100: 272–285.

Oring, L.W., J. M. Reed, M. A. Colwell, D. B. Lank, and S. J. Maxson. 1991. Factors regulating annual mating success and reproductive success in Spotted Sandpipers (*Actitis macularia*). *Behavioral Ecology and Sociobiology* 28: 433–442.

Owens, I. P. F. 2002. Male-only care and classical polyandry in birds: Phylogeny, ecology and sex differences in remating opportunities. *Philosophical Transactions of the Royal Society B: Biological Sciences* 357: 283–293.

Owens, I. P. F., A. Dixon, T. Burke, and D. B. A. Thompson. 1995. Strategic paternity assurance in the sex-role reversed Eurasian Dotterel (*Charadrius morinellus*): Behavioral and genetic evidence. *Behavioral Ecology* 6: 14–21.

Payne, R. B. 1984. Sexual selection, lek and arena behavior, and sexual size dimorphism in birds. *Ornithological Monographs* 33: 1–53.

Pierce, E. P., and J. T. Lifjeld. 1998. High paternity without paternity-assurance behavior in the Purple Sandpiper, a species with high paternity investment. *The Auk* 115: 602–612.

Pitelka, F. A., R. T. Holmes, and S. F. MacLean, Jr. 1974. Ecology and evolution of social organization in Arctic sandpipers. *American Zoologist* 14: 185–204.

Reynolds, J. D., M. A. Colwell, and F. Cooke. 1987. Sexual selection and spring arrival times of Red-necked and Wilson's phalaropes. *Behavioral Ecology and Sociobiology* 18: 303–310.

Reynolds, J. D., and T. Székely. 1997. The evolution of parental care in shorebirds: Life histories, ecology, and sexual selection. *Behavioral Ecology* 8: 126–134.

Ridley, M. 1978. Paternal care. *Animal Behaviour* 26: 904–932.

Rivers, J. W., and J. V. Briskie. 2003. Lack of sperm production and sperm storage by Arctic-nesting shorebirds during spring migration. *The Ibis* 145: 61–66.

Rohwer, S., D. F. Martin, and G. G. Benson. 1979. Breeding of the Black-necked Stilt in Washington. *The Murrelet* 60: 67–71.

Sandercock, B. K. 2001. What is the relative importance of sexual selection and ecological processes in the evolution of sexual size dimorphism in monogamous shorebirds? *Wader Study Group Bulletin* 96: 64–70.

Saracura, V., R. H. Macedo, and D. Blomqvist. 2008. Genetic parentage and variable social structure in breeding Southern Lapwings. *The Condor* 110: 554–558.

Schamel, D., D. M. Tracy, D. B. Lank, and D. F. Westneat. 2004. Mate guarding, copulation strategies and paternity in the sex-role reversed, socially polyandrous Red-necked Phalarope *Phalaropus lobatus. Behavioral Ecology and Sociobiology* 57: 110–118.

Stenzel, L. E., G. W. Page, J. C. Warriner, J. S. Warriner, K. K. Neuman, D. E. George, C. R. Eyster, and F. C. Bidstrup. 2010. Adult survival, sex ratio, mating opportunity, and site fidelity in the Snowy Plover. Ibis forthcoming.

Székely, T., I. C. Cuthill, S. Yezerinac, R. Griffiths, and J. Kis. 2004. Brood sex ratio in the Kentish Plover. *Behavioral Ecology* 15: 58–62.

Székely, T., and J. D. Reynolds. 1995. Evolutionary transitions in parental care in shorebirds. *Proceedings of the Royal Society of London, Series B* 262: 57–64.

Székely, T., G. H. Thomas, and I. C. Cuthill. 2006. Sexual conflict, ecology, and breeding systems in shorebirds. *BioScience* 56: 801–808.

Thuman, K. A., and S. C. Griffith. 2005. Genetic similarity and the nonrandom distribution of paternity in a genetically highly polyandrous shorebird. *Animal Behaviour* 69: 765–770.

Trail, P. W. 1985. The intensity of sexual selection: Intersexual and interspecific comparisons require consistent measures. *American Naturalist* 126: 434–439.

van Rhijn, J. G. 1973. Behavioural dimorphism in male ruffs, *Philomachus pugnax* (L.). *Behaviour* 47: 153–229.

Wallander, J., D. Blomqvist, and J. T. Lifjeld. 2001. Genetic and social monogamy—does it occur without mate guarding in the Ringed Plover? *Ethology* 107: 561–572.

Walters, J. R. 1984. The evolution of parental behavior and clutch size in shorebirds. In *Shorebirds: Breeding behavior and populations,* ed. J. Burger and B. L. Olla, 243–287. New York: Plenum Press.

Warriner, J. S., J. C. Warriner, G. W. Page, and L. E. Stenzel. 1986. Mating system and reproductive success of a small population of polygamous Snowy Plovers. *Wilson Bulletin* 98: 15–37.

Westneat, D. F., and P. W. Sherman. 1997. Density and extra-pair fertilizations in birds: A comparative analysis. *Behavioral Ecology and Sociobiology* 41: 205–215.

Widemo, F. 1998. Alternative reproductive strategies in the ruff, *Philomachus pugnax*: A mixed ESS? *Behaviour* 56: 329–336.

Winkler, D. W., and J. R. Walters. 1983. The determination of clutch size in precocial birds. *Current Ornithology* 1: 33–68.

Yom-Tov, Y. 2001. An updated list and some comments on the occurrence of intraspecific nest parasitism in birds. *The Ibis* 143: 133–143.

Zharikov, Y., and E. Nol. 2000. Copulation behavior, mate guarding, and paternity in the Semipalmated Plover. *The Condor* 102: 231–235.

Breeding Biology

CONTENTS

SHOREBIRDS BREED ON all continents, including sub-Antarctic islands. Many species that breed in northern latitudes undertake amazing migrations between wintering and breeding sites, while others are permanent residents. Shorebirds exhibit tremendous variation in mating system, parental care, and breeding biology. Still, the vocalizations and displays used by breeding shorebirds appear to have changed little over millions of years. Moreover, shorebirds are rather constant in clutch size, and they hatch precocial chicks from large, yolk-rich eggs; their chicks develop quickly, especially in Arctic realms. Family groups remain together for varying lengths of time. Collectively, these attributes represent great diversity and constancy all wrapped up in one taxon. In this chapter, I characterize the salient features of breeding biology and connect them to important management and conservation issues. Detailed treatment of some management is left to later chapters, where I address habitat management and predator control.

PHILOPATRY, BREEDING SITE FIDELITY, AND DISPERSAL

The return of first-time breeders to natal sites (philopatry) and adults to previously occupied breeding locales (breeding site fidelity) has been a subject of many papers (Table 5.1) and one major review (Oring and Lank 1984). The general patterns from this body of work are that site fidelity varies strongly among species and, that in some species there are large differences between the sexes. This latter observation often has been interpreted in the context of a species' mating system (Greenwood 1980).

VARIATION AMONG SPECIES

Adult return rates are lowest for nonterritorial phalaropes, colonial-breeding recurvirostrids, lek-breeding calidrids, and probably other species that are dependent on ephemeral wetlands. These same groups appear to have lower levels of philopatry compared with species with mating systems based on defense of an all-purpose territory. To some extent, interspecific differences may be related to the predictability of breeding habitats. For example, in the interior of North America, the vagaries of weather occasionally lead to the drying of seasonal wetlands, which precludes breeding by American Avocets, Black-necked Stilts, and nonterritorial Wilson's Phalarope. Consequently, these species have some of the lowest return rates. By comparison, territorial Marbled Godwits, Willets, and Killdeer in these same regions, and even the same habitats, have higher return rates. Similar arguments may apply to the ephemeral nature of Arctic breeding habitats and the social systems of nonterritorial Red and Red-necked Phalaropes and some calidrine sandpipers (such as the Buff-breasted Sandpiper).

SEX DIFFERENCES

Males and females of many species also differ in site fidelity. In most species, males return at higher rates than females (Oring and Lank 1984) (Table 5.1). In a review of mating systems and site fidelity in birds and mammals, Greenwood (1980) predicted that the sex investing more in defense of resources on a territory (males of most calidridines) should return at higher rates than the limiting sex owing to advantages of territory ownership in competition for mates. By contrast, in nonterritorial species (such as phalaropes), the sex investing more in parental care (males) should benefit from site-fidelity owing to the advantages of familiarity in rearing young (Reynolds and Cooke 1988). In phalaropes, females compete vigorously for mates in roving flocks, and mate acquisition is less influenced by site familiarity than by direct female interactions (Colwell and Oring 1988b). Hence, female phalaropes would benefit from competing for mates wherever they may be encountered. These predictions are mostly supported by the literature (see Table 5.1). The exceptions appear to be colonial recurvirostrids and lek-breeding species.

ROLE OF EXPERIENCE

One important factor contributing to intraspecific variation in return rates is individual breeding experience. Adult males and females are more likely to return to breed at a site when they have successfully reared young. In many instances, this relationship has been examined by comparing return rates of individuals that hatched a clutch with those that failed, rather than evaluating fledging success. This pattern is especially true for first-time breeders. In yearlings, for instance, the fate of initial breeding attempts appears to dictate the tendency to return to a site. In the Spotted Sandpiper, there was no detectable sex difference in the return rate of adults that had bred locally in a previous year. However, successful breeders returned at significantly higher rates (63% for both sexes) than unsuccessful breeders (29% for males and 26% for females; Reed and Oring 1993). Clearly,

TABLE 5.1
Summary of sex differences in adult site fidelity and philopatry for selected shorebird species

SPECIES	% ADULT RETURN		% NATAL RETURN		MATING SYSTEM[a]	SOURCE
	MALES	FEMALES	MALES	FEMALES		
Black-necked Stilt	20	25	2	0	M/Colonial	b
American Avocet	20	50	0–2	0–3	M/Colonial	b, c, d
Snowy Plover	77	64	17	12	RDP	e, f
Semipalmated Plover	59	41	1	2	RDM	g
Killdeer	60	20	0–10	0	RDM	h, i
Pacific Golden-Plover	100	25	—	—	RDM	J
Eurasian Golden-Plover	78	77	—	—	RDM	k
Common Redshank	42–52; 84	35–45; 67	—	—	RDM	l, m
Spotted Sandpiper	63	63	18	11	RDPa	n
Black-tailed Godwit	78	69	0–2	0–1	RDM	o, p
Long-billed Curlew	71	69	9	2	RDM	q
Purple Sandpiper	60	54	—	—	RDM	r
Semipalmated Sandpiper	48	44	5	4	RDM	s
Western Sandpiper	54	37	4	1	RDM	t
Dunlin	90	65	11	10	RDM	m, u
Temminck's Stint	79	72	8	8	RMCP	v
Wilson's Phalarope	27	3	12	3	MAPa	w
Red-necked Phalarope	38–56	34–61	8–17	2–2	MAPa	x
Red Phalarope	9–16	3–10	0–2	0–0	MAPa	y
Buff-breasted Sandpiper	11	13	0	0	MDPy	z
Subantarctic Snipe	81	87	14	15		aa

Sources: b. Sordahl 1984; c. Robinson and Oring 1997; d. Plissner et al. 1999; e. Warriner et al. 1986; f. Colwell et al. 2007; g. Flynn et al. 1999; h. Lennington and Mace 1975; i. Colwell and Oring 1989; j. Johnson et al. 1993; k. Parr 1980; l. Thompson and Hale 1989; m. Jackson 1994; n. Reed and Oring 1993; o. Jonas 1979; p. Groen 1993; q. Redmond and Jenni 1982; r. Pierce 1989; s. Gratto et al. 1985; Gratto 1988; t. Holmes 1971; Ruthrauff and McCaffery 2005; u. Soikkeli 1967, 1970; v. Hildén 1975, 1978, 1979; w. Colwell et al. 1988; Colwell and Oring 1989; x. Reynolds and Cooke 1988, Schamel and Tracy 1991; y. Schamel and Tracy 1991; z. Lanctot and Laredo 1994; aa. Miskelly 1999.
[a]Mating system terminology follows Emlen and Oring (1977) and Oring (1982): RDM = resource defense (territoriality) monogamy; MAPa = mate-access polyandry; RDPa = resource-defense polyandry; MDPy = male-dominance (lek) polygyny; RMCP = rapid multiple-clutch polygamy.

the return of first-time breeders was strongly influenced by success or failure (Oring and Lank 1982).

The positive relationship between reproductive success and probability of return is of significance to management techniques used to increase local productivity of shorebird populations. Predator exclosures (cages of varying design that protect eggs from predators) are a nonlethal method commonly used to increase hatching success (see details in Chapter 12). Exclosures have been shown to be an effective short-term management tool, with elevated success often lasting a few breeding seasons (e.g., Neumann et al. 2004; Isaksson et al. 2007; Hardy and Colwell 2008). The long-term success of exclosures is often diminished, however, when individual predators, especially intelligent species such as corvids, learn that young shorebirds leave exclosures shortly after hatch. Under natural conditions, hatching success is probably a reliable indicator of the probability of fledging young, as evidenced by positive covariation between these two variables (e.g., Oring, Colwell, and Reed, 1991). However, exclosures deal with only half the predation problem; they fail to address predation on chicks. Moreover, exclosures may effectively decouple the information used by individuals to assess habitat quality from the decision to return to a site. Specifically, if the same predators of eggs take young, then exclosures falsely inform adults about habitat quality. Adults return at high rates owing to increased hatching success but are destined to fledge few, if any, young. This has been shown in a number of studies in which hatching success was increased through the use of exclosures but did not result in increased fledging success (e.g., Neuman et al. 2004). In effect, exclosures may contribute to a local area becoming a population sink, especially if the practice is applied to a large percentage of nests in an area. Consequently, exclosures are probably best viewed as a short-term management tool, which should be integrated with other forms of predator control, including habitat management.

Return rate (or site faithfulness) is often calculated as the percentage of adults of either sex that are site faithful. There are several important considerations when considering data reported in the literature. First, as a qualitative measure, return rate is a surrogate for dispersal distance, which is a continuous variable. Dispersal is often measured as the distance between territories or nests in successive years. Natal return rates show the tendency for young to return to a local breeding population. A better measure of natal movement, however, is the distance between an individual's natal nest (where it hatched) and the location of its first nest location. This distance is almost certain to vary among areas of relatively continuous suitable habitat versus those characterized by disjunct patches. Second, authors vary in their definition of "return." In some cases, a single observation of an individual where it was marked in a previous year qualifies as a return. This may be appropriate for nonterritorial phalaropes (e.g., Colwell et al. 1988), but typically site fidelity is defined as an individual breeding at a site where it has bred in a previous year. In reality, it may be meaningful to record a nonterritorial female phalarope as having returned if she was observed once in a location where she nested a previous year. By contrast, this may not be a useful measure of site fidelity for a territorial shorebird. Finally, interpretation of return rate as a measure of dispersal is confounded by sex differences in survival and by the interspecific variation in male and female tendencies to wander in search of mates. That is, differences in return rates between males and females may result from differences in survival rather than a sex bias in dispersal.

Measuring dispersal has important implications for demography. For instance, dispersal is integral to estimates of neighborhood size, which is essential to measures of effective population size (N_e; Koenig 1988). The difficulty of relocating marked individuals, especially yearlings dispersing for the first time, has made measures of dispersal difficult to come by (Table 5.2). Nevertheless, several recent papers have quan-

TABLE 5.2
Summary of average (±standard deviation) natal dispersal for selected shorebirds

SPECIES	DISPERSAL (KM)		SAMPLE SIZE		MIGRATORY STATUS	SOURCE
	MALES	FEMALES	MALES	FEMALES		
American Avocet	1.1±0.9	7.3±7.1	3	5	Migrant	a
Double-banded Plover	1.2	5.2	24	16	Migrant	b
Snowy Plover	16.0±29.2	18.2±33.6	26	20	Partial migrant	c
Black-tailed Godwit	2	1			Migrant	d
Semipalmated Sandpiper	5	4	18	17	Migrant	e
Subantarctic Snipe	0.1±0.1	0.2±0.1	8	9	Resident	f

Sources: a. Robinson and Oring 1997; b. Pierce 1989; c. Colwell et al. 2007a; d. Groen 1993; e. Gratto et al. 1985, Gratto 1988; f. Miskelly 1999.

tified natal and adult dispersal for several species of shorebird. Jackson (1994) marked hundreds of individuals of three species (Common Redshank, Dunlin, and Common Ringed Plover) breeding in the West Isles of Scotland. He relocated roughly equal numbers of males and females first marked there as chicks and showed that females dispersed farther than males. This corroborates the general pattern in birds. Along the Pacific Coast of North America, the Snowy Plover has been studied for better than 25 years in the Monterey Bay vicinity (Warriner et al. 1986; Stenzel et al. 1994, 2007). Over the course of a long breeding season, both males and females move widely in search of mates and alternative breeding sites. Some females moved hundreds of kilometers within a breeding season. In coastal northern California, near the northern extent of the species' breeding range, male and female Snowy Plovers dispersed similar distances from natal to breeding sites (Colwell, McAllister et al. 2007). The long distance dispersal of Snowy Plovers suggests that individuals breeding along the Pacific Coast are effectively one population. Recently, opponents of the Endangered Species

Act used a preliminary genetic analysis, showing that the coastal population was similar to interior birds, to challenge the conservation status of the listed population segment. A more comprehensive genetic analysis confirmed that Snowy Plovers across North America were genetically similar and best considered as one subspecies (Funk et al. 2007). Occasional movements of birds between the coastal population and interior sites in the western U.S. suggest that some gene flow exists. Nevertheless, the U.S. Fish and Wildlife Service upheld its decision to list the population segment as threatened based on other arguments.

SPRING ARRIVAL SCHEDULES

The timing of arrival of individual males and females to breeding areas varies with a species' mating system (Myers 1981). Males of most monogamous species establish themselves on territories prior to or simultaneously with the arrival of the first females. Males precede females to Arctic breeding grounds in several calidridine sandpipers (Holmes 1966; Myers 1981; Reynolds et al. 1986; Handel and Gill 2000). The sex differences in arrival are reversed in polyandrous species, even those not defending a territory. Female Spotted Sandpipers arrive earlier than most males, and females establish territories to which they attract multiple mates (Oring and Lank 1982). In phalaropes, sex ratios in early season flocks are strongly female-biased (Reynolds et al. 1986) and characterized by intense competition for mates (Colwell and Oring 1988b). These patterns are best explained by sexual selection acting on individuals of the competing sex for advantages in mating (Myers 1981; Oring and Lank 1982; Reynolds et al. 1986).

COURTSHIP BEHAVIOR

Like many birds, shorebirds engage in a rich array of behavior associated with defense of territory or mate, courtship, copulation, and nesting. Detailed accounts of these behaviors are provided by Bent (1962), Cramp and Simmons (1983), and del Hoyo et al. (1992), as well as the various species accounts of the *Birds of North America*. In chronological order, individuals compete for territories or mates directly, and court opposite-sex individuals as prospective partners. Once paired, mates initiate stereotypical displays associated with copulation. In most species, males and females begin choosing and constructing a nest, in which the female eventually lays her eggs. In a few shorebirds (such as the polygynous Buff-breasted Sandpiper and Ruff), the female alone selects the nest site.

VOCALIZATIONS

Shorebirds do not learn vocalizations in a manner akin to that described for many oscines, in which males memorize their song during a critical (learning) period and then later practice what they have memorized (Miller and Baker 2009). Consequently, the courtship and breeding vocalizations and displays of shorebirds are evolutionarily quite conservative; in other words, they have changed little over millions of years (Miller and Baker 2009). For instance, the nuptial displays of several pairs of tringine sandpipers involve repeated harmonic trills that

are given by males as they flutter in undulating flight over their territories. The acoustic spectrograms of the displays of Lesser Yellowlegs and Willet, two species that diverged approximately 8.5 million years ago, are quite similar. There is also great similarity in acoustic and flight displays of Green and Solitary Sandpipers, which diverged approximately 11.5 million years ago (Miller and Baker 2009). The similarities extend to alarm calls, as evidenced by parental alarm trills given by Black Turnstone, Red Knot, and Purple Sandpiper, species that diverged approximately 41 million years ago (Miller and Baker 2009). Miller and Baker (2009) suggest that homologies exist in courtship flight displays of some families (for instance, Charadriidae and Scolopacidae) that diverged about 93 million years ago!

The displays that lead to pairing and subsequent breeding are diverse and varied in complexity and duration. Shorebirds begin courting in late winter or early spring. Some tringines (such as Solitary and Green Sandpipers) perform very long flight displays even over kilometers while singing complex songs to attract mates (Oring 1968, 1973). The highly stereotyped precopulatory and postcopulatory displays of the American Avocet (Hamilton 1975), a short-distance Nearctic migrant, have been observed in late-winter flocks before migration and coincident with their prealternate molt (Evans and Harris 1997). This suggests that some individuals may pair in winter. Moreover, some species are known to copulate during migration (Rivers and Briskie 2003). Exceptions to this may be found in species that reside year-round on or near breeding areas (such as the Black Oystercatcher). However, courtship and pair formation primarily take place at breeding areas. This stands in contrast to dabbling ducks, which pair in the autumn or early winter (Baldassarre and Bolen 2006).

FLIGHT DISPLAYS
AND COURTSHIP BEHAVIOR

Once established on territories, males defend their turf and attract mates using flight displays and vocalizations, which involve long-distance communication. Miller (1984) provides an excellent overview of this facet of shorebird communication. The visual and acoustic features of these displays are adapted to the open, often windy habitats where many shorebirds breed. For example, the aerial displays of many tundra-breeding calidridine sandpipers are given by males as they hover over territories suspended by unrelenting Arctic winds. Male Semipalmated Sandpipers trill as they flutter above their tundra territories, effectively broadcasting their signal above the wind-produced distortion at ground level. Unpaired male Semipalmated Sandpipers spend 18% of their time in flight, most of which is aerial display. This represents a significant energy cost to individuals (Ashkenazie and Safriel 1979b) and may be related to the prevalence of reversed size dimorphism in shorebirds (Jehl and Murray 1986). These calidridine trills contrast with the booming display of male Pectoral Sandpipers as they chase prospective mates low over wet tundra. In Wilson's Snipe (and other snipe), winnowing (or bleating) is produced day and especially at night by males and occasionally females (Bent 1962; Tuck 1972). Displaying individuals dive, during which the outer rectrices vibrate (Tuck 1972).

Many temperate-breeding shorebirds also use flight displays that combine visual and acoustic signals. Male Willets perform an undulating territorial display flight, which emphasizes their contrasting white and dark wing markings, as they call incessantly ("pill-will-willet"). The distinctive flight display of the Northern Lapwing was analyzed in detail by Dabelsteen (1978) and summarized by Miller (1984). Territorial males perform a distinct sequence of flights as they course over their territory. Males initiate the sequence with "butterfly flight," which entails slow, deep wing-beats. This is followed by a zig-zag flight course where individuals revolve on their long axis. Later in the flight sequence, a humming is produced by vibrating outer primaries. There is strong sequential dependency to this display (Miller 1984).

Nevertheless, the lapwing song flight exhibits considerable variation, which may be related to factors intrinsic (endocrinology, stage of the breeding cycle) or extrinsic (date, weather) to the individual.

Other courtship behaviors are used by individuals to display at close range, whether these interactions involve intrasexual competition or intersexual mate attraction. These displays are often fascinating in their stereotypy, varying widely across taxa. The variability in courtship displays runs the gamut from lek-breeding sandpipers to territorial plovers and colonial-breeding avocets and stilts. Among lek-breeding species, such as the Ruff and Buff-breasted Sandpiper, males display on arenas in close proximity to one another; females visit the lek to choose a mate. Male Buff-breasted Sandpipers use wing-flashing displays, which emphasize the whitish underwing lining, to attract females; these displays have been observed during spring migration well south (Oklahoma) of the species' breeding grounds (Oring 1964), and they may even occur on wintering grounds. On the lek, in close proximity to females, male Buff-breasted Sandpipers perform an upright display in which the wings are extended outward in the direction of a female, which suggests an embracing posture (Fig. 5.1). This display has been popularized in several articles (Myers 1979; Lanctot 1998). Myers (1979) described the male's display as follows: "As the females approach, he first hulks over, ruffling his back feathers and starting a quickened tread. Abruptly he rears back, thrusting his head up and wings out, keeping his bill parallel to the ground while marching in place. Only now does he vocalize, a subtle tic-tic-tic timed to match the slow footsteps taken in place. As a crowning gesture he draws his neck in and throws his bill back, gazing catatonically toward the Arctic sky. The females crowd forward, inspecting minute details of his underwing."

Elaborate, complex, and ritualized courtship and copulatory behaviors are especially noteworthy in the stilts, avocets, and large plovers of the genus *Vanellus* (Miller 1984). A postcopulatory

FIGURE 5.1. A male Buff-breasted Sandpiper displaying to conspecifics near Barrow, Alaska. Photo credit: Noah S. Burrell.

wing-raised run accompanied by vocalizations is widespread in many species of the lapwings (Maclean 1972b; Miller 1984). In many recurvirostrids, a ritualized precopulatory sequence precedes a very quick copulation, followed by a wonderful postcopulatory display. Female American Avocets, for example, assume a posture in which the body is held parallel to the water with head and neck extended. As the sequence advances, males move from one side of their mate to the other, always from behind. Each time a male sidles next to his mate, he preens his breast feathers in an upright posture. As preening becomes more exaggerated, males sometimes splash the female with water. Eventually, the male jumps on the female's back, after which there is a quick juxtaposition of cloacae. Immediately after copulation, the pair cross bills, extend wings over each other's back, and walk briskly for a short distance through the shallow water until they part and resume feeding (Hamilton 1975). Very similar displays occur in stilts.

SCRAPING AND NEST BUILDING

Once paired, males and females of virtually all species of shorebird exhibit some form of a stereotyped courtship display known as SCRAPING, a term first used by Huxley and Montague (1926) to describe the nesting ceremony of the Black-tailed Godwit. Even the Solitary Sandpiper and Green Sandpiper scrape, although they usurp arboreal nests built by songbirds. Scraping occurs principally early in the breeding season as part of courtship, but it continues throughout the breeding season whenever birds renest after clutch failure. Scraping begins with a male and female entering suitable nesting habitat, whether this is tundra, prairie, or simple unvegetated substrates of beaches, salt pans, or cobble of braided rivers. In Wilson's Phalarope, females nearly always lead males either by walking from the wetland edge or by initiating looping flights that circle over grassy habitats near wetlands (Howe 1975). In other species, pairs simply walk short distances into suitable nesting habitats within a territory. In most species, males typically initiate the scraping ceremony by bowing with their bill directed downward and using their feet to scrape a shallow depression in the substrate. This may entail considerable effort in vegetated habitats, whereas in sandy substrates less exertion is required. The male often follows his mate into the scrape and further excavates the depression. Other courtship behaviors, including copulation, often occur during the interval when a male and female are scraping. Depending on the species and habitat where they nest, individuals may toss small twigs, pieces of vegetation, or pebbles as they sit in a scrape. These head-tossing motions are rather stereotypic of the scraping display.

Surprisingly, scraping behavior has not been well studied regarding the dual role that it plays in courtship and nest-site selection. Early on in pair formation, scraping probably functions as courtship behavior. The ritualized behavior may strengthen the pair-bond and synchronize the reproductive physiology of mates. Evidence of this comes from the observation that scraping may take place over intervals of days to weeks (in some temperate-breeding species with long breeding seasons) prior to females laying their first clutch. But the duration of scraping is shortened considerably when pairs renest. This suggests that scraping performed early in the season serves to solidify the pair-bond or synchronize the reproductive physiology of a male and female. However, as scraping continues, it eventually transitions into the process of nest-site selection and nest building. This stems from the observation that females eventually select a scrape in which to lay their first egg of a clutch.

Across all species, pairs may create several to many scrapes, but they gradually focus their attention on one where building continues and eggs are eventually laid. In the Snowy Plover, for example, both nests and scrapes occur in habitats that are more open and sparsely vegetated compared to random locations. However, scrapes have slightly (but significantly) more vegetative cover than nests, which suggests that some scrapes (probably those early in the courtship sequence) are not part of nest-site selection (Muir and Colwell 2010).

SAME-SEX BEHAVIOR

Same-sex sexual behavior has occasionally been reported in birds, including nine species of shorebird (MacFarlane et al. 2006). These behaviors specifically refer to those that include precopulatory displays that lead to mounting, as opposed to those that involve allo-preening and parental behaviors. Female-female sexual behavior occurred predominantly among socially monogamous species, and this was true of the few shorebird examples (two stilts and one oyster-catcher). Male-male sexual behavior was more frequent among polygynous species, including the Ruff.

BREEDING DENSITIES

Densities of nesting shorebirds vary greatly within a species' range. Spatial variation in breeding density may be influenced by a variety of factors, including social system and habitat availability. And breeding densities may vary substantially from year to year, owing to occasionally high mortality occurring during the nonbreeding season, especially when immigration is low.

In the high Arctic, some calidridine sandpipers (the Red Knot and Sanderling, for example) breed at low densities in polar deserts. Elsewhere in the Arctic, high densities of breeding shorebirds occur amid habitats with abundant wetlands. For example, in the Arctic National Wildlife Refuge, the highest densities of 14 species occurred in the wetland-rich coastal plain (Brown et al. 2007). In the Yukon-Kuskokwim Delta of Alaska, Western Sandpiper densities are among the highest recorded (approximately 200 pairs/km^2) for any breeding shorebird (Ruthrauff and McCaffery 2005). Across the northern Nearctic, Greater and Lesser Yellowlegs breed at low densities amid boreal forests, which may account for the few nests that have been found for these species. In the Great Basin region of the western United States, wetland habitat is patchily distributed, and great distances separate a few large wetland complexes (Haig et al. 1998). Accordingly, the seven species that breed there (American Avocet, Black-necked Stilt, Killdeer, Snowy Plover, Long-billed Curlew, Willet, and Wilson's Phalarope) vary in abundance across the landscape (Oring and Reed 1996). In the prairie pothole region of Canada, a diverse assemblage of eight species occur in habitats ranging from open alkali flats (Killdeer, Piping Plover, and American Avocet), bordering wet marshes (Wilson's Snipe and Wilson's Phalarope), and sparsely to densely vegetated uplands (Marbled Godwit, Willet, and Upland Sandpiper) (Colwell and Oring 1990). The nearest neighbor distances of conspecifics (inversely related to breeding density) are lowest for nonterritorial and colonial species and greatest for territorial species. In the southern Great Plains, four shorebirds (American Avocet, Black-necked Stilt, Killdeer, and Snowy Plover) breed in unpredictable habitats that fringe ephemeral playas and saline lakes, but these species did not nest around human-constructed wetlands (Conway et al. 2005). Some species have adapted to human-altered landscapes. Killdeer occasionally nest on driveways and on rooftops (e.g., Ankeny and Hopkins 1985) where substrates offer suitable habitat.

SELECTION OF A BREEDING SITE

Natural selection has strongly influenced the nesting habits of birds, including such intriguing life-history phenomena as clutch size, egg-laying intervals, and nest-site selection (Martin 1993). Shorebirds are no exception. Predation is probably the most important ecological factor that has shaped shorebird nesting biology, based on the strong impact predators have on individual reproductive success (e.g., Oring, Colwell, and Reed 1991; Oring, Reed et al. 1991) and the productivity of populations (Evans and Pienkowski 1984). But individuals must select a breeding site with habitat that is suitable and of high quality for other facets of their daily life, including foraging. In some cases, notably oystercatchers, shorebirds nest in habitats separate from foraging areas and regularly commute some distance to feed elsewhere. In the Furneaux Islands, Australia, two oystercatchers breed sympatrically. The Pied Oystercatcher only nests near their foraging areas in rocky

intertidal habitats, whereas the Sooty Oystercatcher sometimes nests in habitats more distant from their principal foraging areas (Lauro and Nol 1995). This is also true of the American Oystercatcher breeding on the Atlantic seaboard of North America, where individuals commute from a few meters to nearly 3 km daily between nesting and feeding areas (Nol 1989).

The habitat cues used by individuals to settle in an area can be viewed in a spatially hierarchical manner (Hutto 1985). Individuals must choose where to breed in a landscape of varying environmental features. Clearly, wetland density affects the proximity and availability of food for adults and chicks. Once an individual establishes itself in a particular breeding habitat, however, other habitat features influence where to nest. At finer spatial scales, the density of vegetation or suitability of substrates may be critical to establishing a nest site, and these features directly affect reproductive success. In all cases, selection of particular habitats can have important consequences for individual reproductive success. Consequently, differences in fitness coupled to heritability of behaviors may lead to the evolution of nesting habits.

Landscape-level habitat features, such as the presence, amount, and quality of wetland habitats, almost certainly influence the settlement of shorebirds. Evidence of this comes from the patchy distributions and varying densities of shorebirds on the breeding grounds. In some cases, researchers have demonstrated a positive correlation between species' densities and the density of wetlands at a coarse spatial scale. For example, densities of tundra-breeding shorebirds across the North Slope of Alaska vary greatly, with densities increasing downslope from drier tundra to the coastal plain near the Arctic Ocean. Highest breeding densities occur on the coastal plain, where wetlands occur in highest density (Brown et al. 2007). Similar patterns exist for temperate-breeding shorebirds. In the midcontinent of North America, the density of breeding Long-billed Curlews varies greatly across habitats that vary in the amount of grassland (Jones et al. 2008). Four species of shorebird (American Avocet, Black-necked Stilt, Killdeer, and Snowy Plover) breed in the Playa Lakes region of the southern Great Plains. These species were generally absent from constructed wetlands. The wetlands they used, however, tended to have more mudflats than the wetlands that lacked nesting birds (Conway et al. 2005).

Increasing evidence indicates that individuals of a variety of bird species use the presence of conspecifics as one of many habitat cues to select a potential breeding site (Ahlering and Faaborg 2006; Campomizzi et al. 2008). Conspecific attraction may most strongly influence the selection of breeding habitats by yearlings or older birds seeking to breed for the first time. Specifically, individuals searching for breeding opportunities may cue on conspecifics and settle in occupied areas; conversely, they may avoid unoccupied sites. Furthermore, the quality of a breeding site may be gauged by the presence of potential mates, incubating birds, adults tending chicks, and recently-fledged juveniles. Only one study has specifically examined social attraction to conspecifics in a breeding shorebird (Nelson 2007). Snowy Plovers breed at low population densities along the Pacific Coast of North America. In northern California, a small color-marked population was monitored for several years. At a landscape scale, naive plovers (immigrants that had never bred in the local population) were more likely to settle at sites that had greater numbers of experienced plovers (breeders that had returned from a previous year). At a finer spatial scale, along a 10 km stretch of ocean beach with the highest local breeding density, nests were patchily distributed, and individuals commonly initiated nests closer to conspecifics than expected by chance.

The benefits of breeding near conspecifics can be extended to include other species of bird, especially those that aggressively defend their nests. Some shorebirds nest in close proximity to other species, including other shorebirds. Benefits usually accrue to "timid" species that associate with "bold" ones (Drycz et al. 1981; Paulson and Erckmann 1985; Nguyen et al. 2006). For example, Common Snipes and Common Redshanks nesting near Black-tailed Godwits experienced lower nest predation (Drycz et al. 1981). Snowy Plovers nesting within 100 m of Least Tern (*Sternula antillarum*) nests had higher nesting success than plovers nesting farther from tern colonies (Powell 2001). Thompson and Thompson (1985) reexamined the response of Dunlin and Eurasian Golden-Plovers to disturbance by humans walking directly at them when they were in conspecific and mixed species flocks on breeding grounds. Individuals were assumed to be off-duty (on incubation recess) as the study was conducted in May. Dunlin took flight sooner when associating with plovers. This association between the two species has lead to the Dunlin being referred to as the "plovers' page," and it prevails during the nonbreeding season, too. It is unknown, however, how this association affects reproductive success. Moreover, to date, no study has specifically addressed whether individuals of any shorebird species *select* nest sites near the nest of an aggressive species by evaluating the timing of initiation and placement of nests between two different species.

NEST-SITE SELECTION

Bird nests have two principal functions that are directly related to fitness. Nests facilitate survival of eggs in the face of predation and other forms of nest failure, and the nest cup enhances the thermal environment necessary to develop embryos and hatch chicks (Hansell and Deeming 2002). To date, most research has emphasized the role of predation in the methods and approaches of studying shorebird nests. For instance, the placement of a nest in open habitats may afford some species (such as plovers) with an unobstructed view of the surrounding landscape. This facilitates early departure from a nest when a predator is detected. Conversely, dense vegetation at nests of sandpipers is often interpreted to increase the crypsis of incubating adults and eggs. Both observations articulate a relationship between habitat and nest survival.

Only a few recent studies have evaluated the thermoregulatory function of a shorebird nest, and none has specifically addressed whether nests are established in habitats that enhance the thermal environment of the eggs or tending adult. Moreover, the thermal environment may have subtle but significant effects on incubation period, which may affect nest survival. There are, however, numerous logical candidates for such a study, including the burrow-nesting Crab Plover, which nests in hot, arid regions, and Arctic-breeding sandpipers, some of which cover their eggs with vegetation when on incubation recess.

Once established at a breeding site, individuals use a variety of habitat cues to choose a suitable site where they establish a nest. Nest-site selection refers to the behavioral process whereby individuals use a suite of environmental variables to establish a nest. The expression of these behaviors is often gauged by measuring habitat in the vicinity of nests and comparing features with other sites that are unoccupied or randomly chosen by a researcher. Additionally, the quality of habitat at nests may be assessed by evaluating relationships between habitat variables and reproductive success, as measured by hatching success or estimates of nest survival.

The importance of predation is especially apparent in the behaviors of individuals when they select a nest site. Virtually all shorebirds construct simple nests on the ground. There are several exceptions to this rule: the Crab Plover digs a nest burrow, and three tringine (Solitary, Green, and Wood Sandpipers) use old passerine (often thrush or sometimes jay) nests at varying heights in trees in boreal forests. There are distinct differences in the nesting habitats of shorebirds that fall along familiar taxonomic lines. The recent consensus on shorebird phylogeny (van Tuinen et al. 2004) identified two distinct clades: one consisting of sandpipers and their allies, and the other of plovers, oystercatchers, thick-knees, Ibisbill, Crab Plover, and avocets and stilts. These two groups differ in their nesting habits, including the tendency to nest in vegetated versus open habitats and the manner in which adults respond to predators near the nest (Gochfeld 1984). Sandpipers and their allies breed in vegetated habitats where camouflaged plumages of adults and cryptic eggs hide nests from predators. The exception to this appears to be the Plains-wanderer, which nests in sparsely vegetated habitats. By contrast, plovers and their allies nest in the open, detect predators at a greater distance, and rely on egg crypsis to enhance clutch survival. Moreover, these differences between concealed and open nesters correlate with behavioral differences in how individuals respond to predators. Plovers typically leave the nest at a greater distance when they detect predators, whereas sandpipers continue to incubate upon close approach by predators (or humans). Accordingly, most studies of shorebirds (and other ground-nesting birds) emphasize habitat characteristics as they relate to concealment versus early detection. Researchers often relate these habitat features to clutch failure or nest survival, assuming that predation of eggs is the principal selective force shaping facets of nest-site selection behavior (Ricklefs 1969; Martin 1993). This is reasonable given the often high rates of clutch loss to predation (Evans and Pienkowski 1984).

For some species (plovers, oystercatchers, avocets and stilts, and thick-knees), the openness of habitat is an important feature influencing selection of a nest site. This is based on the observation that the nesting strategy of these species relies on early detection of predators and the departure of incubating adults from the nest such that eggs camouflaged by surrounding substrates or vegetation will survive. For example, Eurasian Golden-Plovers show strong selection for nesting on flat ground (avoiding slopes), and nest survival was higher there than on slopes (Whittingham et al. 2002). This preference for flat habitats may allow an incubating bird to detect ground-hunting predators earlier, thus reducing predation risk. In fact, golden-plovers tend to leave their nests at great dis-

FIGURE 5.2. Snowy Plovers nesting in two distinct habitats in coastal northern California: (A) gravel bars of the Eel River and (B) sandy beaches fronting the Pacific Ocean. On the river, egg crypsis amid coarse substrates leads to high hatching success. On beaches, eggs contrast with fine substrates, and the soft sand often reveals the tracks of plovers coming and going from the nest during incubation. Consequently, beach nests survive poorly compared with river nests. Similar survival advantages accrue to chicks reared on gravel bars in comparison with those raised on beaches. Photos: Mark Colwell and Jordan Muir.

tances (several hundred meters) when predators approach (Byrkjedal and Thompson 1998). Other plovers (such as the Semipalmated Plover, Piping Plover, Wilson's Plover, and Snowy Plover) also prefer to nest in habitats that offer unobstructed views from the nest (Burger 1987; Bergstrom 1988; Nguyen et al. 2003; Powell 2001; Muir and Colwell 2010).

When open-nesting species leave the nest in response to danger posed by predators, they too rely on habitat features to provide a background to hide cryptically patterned eggs. Consequently, the characteristics of vegetation and substrates may be a feature of habitat that these species select in a nest site. In some habitats, Snowy Plovers select nest sites with heterogeneous substrates and consisting of rocks approximately the size of eggs; in finer, sandy substrates, the eggs contrast with substrates (Fig. 5.2). Moreover, nests survive well in more heterogeneous habitats of coarse gravel and rock where eggs match the size of substrates. By contrast, nests survive poorly in sandy habitats where eggs stand out against fine, homogeneous substrates (Colwell et al. 2005). Other

studies indicate that nesting vegetation and substrate can influence nest survival. In Arctic-breeding plovers (*Pluvialis*), eggs are well camouflaged amid lichen but less well concealed against other types of vegetation. This crypsis is evident in the higher survival rates of nests in lichen vegetation (Byrkjedal and Thompson 1998). Similarly, Stone-Curlew eggs survived better when the base color of the eggs matched the background color of substrates (Solis and de Lope 1995).

Only a few studies have examined the characteristics of nest sites of sandpipers. In the prairies of Saskatchewan, eight species breed in mostly ephemeral wetlands and adjoining upland habitats (Colwell and Oring 1990). These species differ in types of cover and vegetation structure at nests, with some species nesting in unvegetated habitats of alkali flats (Piping Plover, Killdeer, and American Avocet) and others nesting in upland habitats at varying distances from wetlands (Wilson's Phalarope, Marbled Godwit, Willet, Wilson's Snipe, and Upland Sandpiper). Several species select nest sites that differ from random locations, suggesting preference for vegetation that offers greater concealment of eggs and adults. However, there was no relationship between the fate of Wilson's Phalarope clutches and any measures of either vegetation height or density that may be related to concealment of adults or eggs (Colwell 1992).

Surprisingly few studies have quantified habitat at nests of Arctic-breeding shorebirds, perhaps because suitable nest sites do not appear limited by availability of habitat (Smith et al. 2007). Nevertheless, individuals may select particular features of the nest site, given the importance of predation and role that vegetation plays in concealing incubating adults and eggs. Smith et al. (2007) quantified habitat at nests as well as food and parental behavior to assess their relative roles in determining nesting success of five species (Black-bellied Plover, Semipalmated Plover, Ruddy Turnstone, White-rumped Sandpiper, and Red Phalarope) breeding in the Canadian Arctic. There was little evidence that the structure of vegetation at nest sites influenced nesting success, nor did these species prefer to nest in habitats where food was more available. However, there was a tendency for incubation behavior to influence nesting success. Species taking fewer incubation recesses tended to have higher success. The strongest factor affecting nesting success was annual variation in predation pressure, which was related to alternative prey (lemmings).

Although studies of habitat selection coupled with nesting success have enhanced our understanding of breeding biology and factors affecting productivity, several critical issues remain unresolved. First, the behavioral process of habitat selection can be evaluated at a variety of spatial scales (Johnson 1980). The few studies of nest-site selection in shorebirds have addressed third-order selection. That is, they compare habitat at nests with that available in the immediate vicinity within the home range of an individual. Few studies have examined selection at larger spatial scales, which is valu-

able for habitat management. For any species faced with lost or degraded breeding habitat, it is important to demonstrate that individuals use (that is, select or prefer) certain habitats and avoid others. This addresses the issue of habitat limiting a population. If suitable breeding habitat is unoccupied within a species' range, then habitat cannot be the most important factor limiting the population. On the other hand, if breeding habitat is saturated, then all suitable areas are occupied. This should be evaluated across the range of a species, and the argument applies regardless of whether a population is common or rare. Moreover, resolving this issue is a critical foundation for habitat restoration as a mechanism for increasing population size. Studying higher-order habitat selection is germane to conservation efforts for several temperate-breeding shorebirds in which habitats have been compromised by human activities. It also may apply to remote areas of the Arctic where human activities increasingly degrade habitat.

A second unresolved issue surrounding shorebird selection of nesting habitats concerns the role that individual behavior and learning plays in the outcome of breeding. Individuals of varying age and breeding experience may differ in their response to approaching predators. The role of experience could be assessed by comparing yearlings that have never bred with older experienced birds. This approach requires a marked population studied over sufficient time to track changes in behavior of individuals. These age-related behaviors may be as important in determining nest fate as the physical features of the nest site. Embedded in this behavioral process is the role that scraping behavior plays in the selection of a nest site. Courtship scrapes of the Snowy Plover differ subtly from nests and random locations (Muir and Colwell 2010). Studying the habitat at multiple nest scrapes of pairs as they progress through courtship to egg laying would provide valuable insight into the process of nest-site selection.

EGGS

MORPHOLOGY

The shape, color, and size of shorebird eggs vary greatly among and even within species. Lack (1968) suggested that the pyriform shape evolved such that the four eggs typical of shorebird clutches fit well together, with the pointed ends of the eggs directed toward the middle of the nest. This may enhance incubation of the clutch and concealment of the eggs from visual predators. Lack (1968) did not elaborate on these arguments. More recently, Barta and Székely (1997) modeled egg shape in relation to clutch size and concluded that pyriform was the most efficient shape for the transfer of heat from incubating adult to embryo in the four-egg clutches typical of shorebirds.

Shorebird eggs also vary in color. The most plausible explanation for the varied background coloration and markings on the shell is that these patterns enhance egg crypsis and reduce risks of predation. This argument may apply best to species (such as plovers, thick-knees, and oystercatchers) most reliant on their eggs blending with background substrates to avoid detection by visual predators. In support of this, Lack (1968) cited variation in the color of Yellow-wattled Lapwing eggs in two soils. Lapwings typically have ochre-colored eggs that are blotched dark brown, but when they nest on reddish, laterite soils their eggs are colored red and blotched reddish brown. No study, however, has examined survival of clutches in habitats of varying backgrounds that afford different crypsis to eggs.

The size (mass in grams relative to female mass; Fig. 5.3) of shorebird eggs has been the subject of considerable interest, beginning with Lack's (1968) interspecific comparisons of shorebirds and other avian taxa. Compared with altricial species, shorebirds lay disproportionately large eggs that are energetically expensive to produce. Ricklefs (1974) estimated that daily egg laying in shorebirds required an increase in energy of 22% to 44%, compared with 7% to 13% in passerines. Consequently, compared with altricial species, the single egg of the Crab Plover is disproportionately large, constituting nearly one quarter the weight of the 400 g female (Lack 1968). Likewise, eggs of most plovers, which lay three- or four-egg clutches, are disproportionately large. Ricklefs (1984) examined variation in egg dimensions and neonate size within and among nine species of shorebird. Not surprisingly, there was great variation in egg size among species. Interestingly, recent work with stable isotopes examining the composition of eggs from 10 species of Arctic-breeding shorebirds indicates conclusively that these species are not "capital breeders" (Klaassen et al. 2001). In other words, females rely on energy acquired from feeding in Arctic habitats soon after their spring arrival to form eggs (such that they are "income breeders"). Evolutionarily, this makes sense given the migrations of these species and energy costs that would be incurred by females carrying excessive mass over long distances.

An interesting debate has continued for several decades regarding the relationships between egg size, female body mass, and mating systems. Specifically, several papers have examined whether socially polyandrous species or those laying multiple clutches in rapid sequence lay disproportionately small eggs compared with other shorebirds. The jacanas are often cited to illustrate the point that females that lay many eggs in rapid succession produce smaller eggs. Female Northern Jacanas are simultaneously polyandrous and lay frequent replacement clutches. In total, their four-egg clutch is a mere 23% of female mass, which is unusually small among shorebirds (Ross 1979). Ross (1979) analyzed 58 species of plover and sandpiper and concluded that egg size in polyandrous species was smaller than expected. His analysis, however, was confounded by phylogeny. The statistics treated species as independent of one another, ignoring a common evolutionary history. Subsequent analyses have controlled

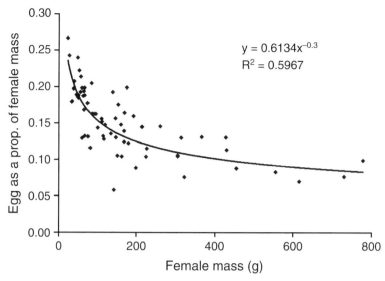

FIGURE 5.3. The relationship between female body mass and egg mass (as a proportion of female mass) for 71 species of shorebird. From Colwell (2006).

for phylogeny in comparative analyses. Ward (2000) determined that there was no evidence that polyandrous species laid smaller eggs than monogamous shorebirds. However, Liker et al. (2001) concluded that selection for reducing laying costs was the most parsimonious explanation for small egg size in polyandrous species. Finally, Lislevand and Thomas (2006) concluded that the relatively small eggs of polyandrous species were best explained by limitation in male incubation ability. In conclusion, polyandrous species lay small eggs, but the jury is still out regarding the relative strengths of various selection pressures shaping egg size in shorebirds.

Several studies have examined intraspecific variation in shorebird egg size (Väisänen et al. 1972; Väisänen 1977; Ricklefs 1984; Galbraith 1988; Grønstøl 1997; Nol et al. 1997; Dittmann and Hötker 2001; Lislevand et al. 2005). Similar to reports for birds in general (for a review, see Christians 2002), most studies have concluded that much of the variation in egg size occurs *among* females in a population rather than *within* individuals. In other words, egg size (mass or volume) of individual females is

quite repeatable. For example, repeatability of egg size measurements for Semipalmated Plovers breeding near Churchill, Manitoba, was 0.67 (Nol et al. 1997). In the Pied Avocet, variation in egg size was greater among than within clutches (Dittmann and Hötker 2001). Avocet eggs within clutches were relatively uniform in size, and differences among clutches accounted for 70% to 80% of variation in egg size. In the Northern Lapwing, repeatability was 0.53 to 0.58, depending on the measure (length, width, and volume). In the polyandrous Spotted Sandpiper, mean egg mass varied among females from 8.4 to 10.4 g, and most variation was attributable to female identity. In other words, females consistently laid the same sized egg (Reed and Oring 1997).

Replacement clutches tend to be smaller than first clutches laid by females, a pattern typical of all birds. Interestingly, clutches of second mates of polygynous male lapwings tended to be smaller than those of first mates. This may stem from a seasonal decline in food availability that compromises the energy that second females can muster to produce eggs (Grønstøl 1997). Alternatively, second mates may be younger and less capable of converting energy into reproductive effort. Collectively, the evidence indicates that egg size is strongly heritable, although environmental factors occasionally influence female investment in eggs. An unresolved issue raised by Grønstøl (1997) concerns the relationship between structural features of females and the size of their eggs. Specifically, in domestic turkeys, vent diameter, oviduct diameter, and internal distance between pubic bones vary greatly among females and correlate with egg dimensions (Hocking 1993).

In several studies, the intraspecific variation in egg size has been directly related to female body size or condition. Egg size in the American Oystercatcher was positively correlated with female body size (Nol et al. 1984; Nol et al. 1997). A similar pattern was obtained in the Semipalmated Plover; moreover, the relationship persisted even in unusually cold years, when food available to laying females was in short supply (Nol et al. 1997). Similar findings were reported for the Northern Lapwing (Lislevand et al. 2005). The size of an egg may influence the survival of young. Galbraith (1988) showed that substantial intraspecific variation in egg mass had consequences for the survival of Northern Lapwing chicks. Chicks from larger eggs were heavier, and they survived better than those from smaller eggs. These findings suggest that habitat quality, as indexed by the availability of food, may act to influence female condition, laying intervals, egg size, and subsequent survival of young.

EGG LAYING

Shorebirds can be categorized into one of two groups based on the temporal patterns of egg laying (Colwell 2006). In most sandpipers, the minimum interval between laying of successive eggs in a clutch is often slightly greater than 24 hours. By contrast, many plovers, oystercatchers, and thick-knees rarely or never lay at daily intervals. Rather, females lay at intervals of 48 hours or longer. For both groups, how-

ever, this minimum interval is often lengthened during periods of inclement weather, suggesting that food or nutrients (such as calcium) may compromise a female's ability to form eggs. The difference between these two groups in minimum laying interval appears to be related to several constraints: time and risk of clutch failure and energy conservation. Species breeding at northerly latitudes lay at shorter intervals than temperate or tropical species, which suggests a time constraint on breeding. In other words, the short northern breeding season has selected for short laying intervals. However, phylogenetic factors confound this interpretation, as most northern-breeding species are sandpipers, whereas plovers, oystercatchers, and thick-knees are distributed in temperate and tropical environs. There is a hint that species nesting in open habitats lay at longer intervals than those nesting in vegetation. This latter relationship suggests that the egg-laying interval is longer in open-nesting species as a hedge against total clutch loss. Long laying intervals may allow an individual to sample the risk of clutch failure with fewer eggs, thus reducing energy investment (Colwell 2006).

Most authors have considered shorebirds to be determinate layers. Females lay a fixed number of eggs in a clutch (Table 5.3), and, more specifically, if an egg is removed during the laying process, females will not replace that lost egg when completing the clutch. The result of a determinate pattern is a modal clutch size of typically four or three eggs, regardless of the removal of an egg. Several papers on plovers and anecdotal observations suggest that the characterization of shorebirds as determinate layers is more complex. In the Spur-winged Plover observational and experimental evidence shows that females lay in an indeterminate fashion when the first-laid egg disappears from an incomplete (one-egg) clutch. That is, females lay either three or four additional eggs, summing to an average clutch size of 4.5. But when the second, third, or fourth egg in a clutch disappears, females continue to lay in a determinate manner to finish the four-egg clutch (Yogev and Yom-Tov 1994, 1996). Similar findings have been reported for other plovers (Rinkel 1940; Klomp 1951) and the Spotted Sandpiper (L. W. Oring, personal communication). This variable laying pattern may also relate to the risk of clutch failure (Colwell 2006). Shorebirds nesting in open habitats, where the risk of clutch loss is high, may occasionally lay a first egg of a clutch and lose the egg to a predator or natural disturbance (drifting sand, tidal overwash, or flooding). Females often initiate a new nest and continue to lay in the manner described above for an indeterminate layer and complete the clutch of three or four eggs, depending on the species.

REPLACEMENT CLUTCHES

Clutch failure is a common phenomenon for shorebirds. Consequently, females of many species lay replacement clutches when first attempts fail, especially when there is sufficient time in the breeding season. In socially polyandrous species (such as phalaropes and the Spotted Sandpiper), females occasionally lay second, third, or fourth clutches for successive mates. The frequency with which species lay replacement (or multiple) clutches varies with latitude (and length of breeding season). Species breeding in temperate latitudes commonly replace clutches. For example, along the Pacific Coast of North America, the Snowy Plover initiates clutches over 120+ days (Warriner et al. 1986), and females commonly lay five to six clutches and occasionally more in a season. By contrast, most Arctic-breeding shorebirds rarely or infrequently lay replacement clutches owing to the short breeding season. This latter notion was evaluated by Naves et al. (2008), who determined that approximately 1% of clutches of five species breeding near Barrow, Alaska, constituted replacement clutches. Despite this low frequency, several issues may cause researchers to underestimate the frequency of replacement clutches, including (1) size and shape of study plots and (2) species-specific length of breeding season. The extent to which replacement clutches occur in a population can have implications for population estimates and monitoring (Evans and Pienkowski 1984; Naves et al. 2008).

CLUTCH SIZE

Compared with other precocial birds, shorebirds exhibit very little variation in the number of eggs laid in a clutch (Table 5.3; Maclean 1972a). Most species lay either four or three eggs; a small number lay two-egg or rarely one-egg clutches. Interspecific variation in clutch size follows several distinct patterns. First, there are differences among phylogenetic lines. Virtually all sandpipers lay four eggs, although the Tuamotu Sandpiper lays a single egg! Members of the plover clade often lay fewer eggs. Second, species breeding at temperate and tropical latitudes tend to lay smaller clutches than closely related species breeding at northerly latitudes (Winkler and Walters 1983). For instance, the clutches of Arctic-breeding plovers such as the Common Ringed Plover and Semipalmated Plover consist of four eggs, whereas temperate counterparts such as the Snowy, Kentish, and Wilson's Plover lay three eggs. These differences may reflect selection for maximizing reproductive output under time constraints of a short Arctic breeding season compared with the relatively lengthy breeding season of species occupying southerly latitudes.

Winkler and Walters (1983) reviewed clutch size variation in shorebirds and determined that clutch sizes tended to be smaller in groups in which adults fed the young (Table 5.3). The oystercatchers provide an illustrative example (Safriel et al. 1996; Hockey 1996). All species are characterized as highly precocial (O'Connor 1984); mobile young are fed until well after fledging by both adult parents. All oystercatchers lay three eggs, with comparatively long laying intervals between successive eggs and substantial intraclutch variation in egg size. Young oystercatchers hatch asynchronously to varying degrees across species. Consequently, there are often substantial differences in the size of young from successive eggs, which may affect patterns

TABLE 5.3

Latitudinal summary of shorebird clutch-size variation for taxa that do and do not feed their young
After Winkler and Walters 1983

FAMILY OR SUBFAMILY	N	TROPICAL		SOUTHERN		NORTHERN	
		4 EGGS	<4 EGGS	4 EGGS	<4 EGGS	4 EGGS	<4 EGGS
Adults feed young							
Dromadidae	1	0	1	—	—	—	—
Chionididae	2	—	—	0	2	—	—
Pluvianellidae	1	—	—	0	1	—	—
Burhinidae	9	0	9	—	—	—	—
Haematopodidae	8	—	—	0	5	0	3
Glareolidae	16	0	13	0	2	0	1
Rostratulidae	2	1	0	0	1	—	—
Total	39	1	23	0	11	0	4
Adults don't feed young							
Thinocoridae	4	—	—	4	0	—	—
Ibidorhynchidae	1	—	—	—	—	1	0
Recurvirostridae	13	3	0	5	1	4	0
Vanellinae	24	11	7	2	0	4	0
Charadriinae	38	0	9	1	13	9	6
Jacanidae	8	8	0	—	—	—	—
Scolopacidae	80	1	7	0	2	68	2
Total	168	23	23	12	16	86	8

of behavioral dominance among chicks. Early-hatched chicks, especially those from larger eggs, are fed more and are dominant to their siblings. Adults bring food to mobile young throughout the prefledging period, and they continue to feed juveniles once they attain flight and move about territories.

ABNORMAL CLUTCH SIZES

Occasional reports of larger or smaller clutches within taxa probably can be explained by one of two phenomena. Unusually large clutches have been reported for many species, and these large clutches appear to be the result of two females laying in a common nest. These sorts of "errors" may be more common in colonial-nesting stilts and avocets, but the occurrence of seven- or eight-egg clutches has been reported for a number of sandpipers. The occurrence of supernormal clutches may be more prevalent under conditions of high population density, when females may be able to locate nests of conspecifics more easily, and when predation is frequent, resulting in the need to dump eggs (Colwell 1986).

Sometimes, however, it is clear that that unusually large clutches have resulted from continuous laying by a single female, as was the case for a 20-egg Killdeer clutch (Mundahl et al. 1981).

Many species accounts report average clutch sizes that are slightly less than the modal clutch size (e.g., Hockey 1996). These averages are influenced by clutches that are slightly smaller than normal (clutch sizes with one less egg than the modal clutch size for a species). These smaller clutches tend to occur later in the breeding season, when females may be limited by energy or nutrients (such as calcium) in their ability to form eggs. However, smaller than normal clutches may be recorded because biologists monitor nests at long intervals, which is insufficient to accurately record "true" clutch size.

EVOLUTION OF CLUTCH SIZE

Lack (1968) stimulated much of the contemporary research addressing variation in clutch size, and this topic continues to receive considerable treatment among ornithologists, especially shorebird biologists. For shorebirds, this question has two important components (Walters 1984). First, why are clutch sizes in most species sharply truncated at four eggs? Second, what explains reduced clutch size (three, two, and rarely one) in other species? Several important considerations are relevant to discussions of interspecific and intraspecific variation in shorebird clutch size. First, shorebirds have precocial, nidifugous young. Consequently, Lack's hypothesis, which addresses the ability of adults to provision their young, generally does not apply to shorebirds. There are, of course, exceptions to the rule: Crab Plovers and oystercatchers have extended parental care of young hatchings from their one-egg and three-egg clutches, after which adults feed and teach the young how to forage. This may be important in understanding the variation in clutch sizes among shorebirds. Smaller clutch sizes tend to be more common in species that provision their young (Winkler and Walters 1983; Walters 1984).

In the vast majority of shorebirds, however, adults do not provision the young. Accordingly, the question of why the clutch size of shorebirds does not exceed four must address other hypotheses, including (1) energy or nutrient limitations on a female's ability to form eggs; (2) the physical ability of adults to effectively incubate eggs; (3) the risk of clutch loss to predators associated with clutches of different size, which require varying lengths of time to lay and incubate; and (4) parental ability to effectively rear young.

Several lines of evidence can be mustered to reject the notion that energy or nutrients (calcium; Patten 2007) limit clutch size. Given that shorebird eggs are large relative to the body size of most species (Ross 1979), it stands to reason that females may be constrained in their ability to form eggs. Some observations, however, argue against this explanation. In many species, females readily lay replacement clutches soon after failure of an initial clutch. In others (for instance, Temminck's Stint, Sanderling, and Mountain Plover), females lay two clutches in rapid succession, each cared for by a single adult. Clearly, the rapid production of multiple clutches laid in rapid succession contradicts the hypothesis that energy or calcium availability constrains clutch size.

A second possibility is that clutch size is limited by an adult's ability to cover and effectively incubate eggs. Occasional instances of five-egg clutches have been reported, and adults have successfully hatched young from all eggs. Ornithologists (e.g., Hills 1983; Székely et al. 1994; Yogev et al. 1996; Sandercock 1997) have experimentally tested this idea by adding one or two eggs to clutches with mixed results. In Wilson's Phalarope, experimentally enlarged clutches of five eggs typically hatched, whereas six-egg clutches were abandoned (Delehanty and Oring 1993) or failed to hatch all eggs. The large size of shorebird eggs is relevant to this hypothesis because the addition of one (or two) large egg(s) may compromise fitness via subtle effects on the adults' ability to effectively incubate, which may compromise egg viability and increase hatching asynchrony (Arnold 1999).

A third hypothesis posits that nest predation, which is the most important cause of reproductive failure in shorebirds, limits clutch size. Larger clutches require more time to complete, and this may lengthen the incubation period. Consequently, even the addition of one egg to a clutch can have biologically significant effects on the risk of loss to predators. Recently, Arnold (1999) made cogent arguments in favor of the combined and subtle effects of clutch size on the risk of nest predation, hatchability of eggs, and hatching asynchrony.

A final hypothesis is that parental ability to tend chicks constrains clutch size to four (Walters 1984). In effect, this hypothesis argues that broods of one or more additional chicks are difficult to defend from predators owing to the spatial geometry of the brood. There has been only one test of this idea. Safriel (1975) experimentally increased Semipalmated Sandpiper broods to five. Enlarged broods fledged fewer young than normal (control) broods of four.

INCUBATION

Incubation is a vital component of reproduction in all birds. It is defined by three elements: (1) the application of heat to develop embryos, (2) the maintenance of relatively high humidity to avoid desiccation, and (3) the behavior of egg turning (Deeming 2002). In shorebirds, where male–female social relationships span the full range of avian mating systems, the roles of the sexes in incubation are highly variable. In most monogamous shorebirds, males and females share incubation duties to some extent, although there is a tendency for females to provide less parental care as time progresses, especially in brood rearing (Jönsson and Alerstam 1990; Székely et al. 2006). In highly polygynous species such as Buff-breasted Sandpiper, Ruff, and Pectoral Sandpiper, females incubate alone. The opposite is true of most polyandrous

taxa. Males of some calidridine sandpipers, phalaropes, jacanas, and painted snipes provide all the parental care. These differences in the roles of the sexes have consequences for the expression of incubation patterns across species, including incubation schedules (total time on and off the nest, number and length of incubation recesses, and the daily timing of incubation bouts by males and females), incubation lengths (days between laying of the last egg in a clutch and hatch), and the condition (mass) of incubating adults. It is, therefore, not surprising that there is considerable interest in incubation in shorebirds, especially as it relates to the conflicting demands of the incubating adult and developing embryos (Tulp and Schekkerman 2006).

INCUBATION TIMING

For most shorebirds, adults generally begin regular incubation once the last egg of a clutch is laid. In the Long-billed Curlew, incubation begins with the laying of the penultimate egg (Hartman and Oring 2006), a pattern reported for many other avian species, not just shorebirds. In reality, however, adults occasionally visit the nest before clutch completion, and these visits involve sometimes lengthy bouts of sitting on the eggs, especially during cold weather. In the Spotted Sandpiper, both females and especially males incubate occasionally during the egg-laying period; males incubate on average about 14% of the daylight period (Maxson and Oring 1980). During the laying stage, Wilson's Plovers visited two nests sporadically, tending single eggs for 13% and 16% of daylight hours; this increased to 50% for a two-egg clutch (Bergstrom 1988). A similar increase in nest attendance with the laying of eggs was reported for Arctic-breeding Black-bellied Plovers (Byrkjedal and Thompson 1998). Common Greenshanks occasionally visit nests during the laying stage. During these visits, adults "tend to sit loosely, like lapwings, tilting forward and probably not placing the eggs against their brood patches" (Nethersole-Thompson and Nethersole-Thompson 1979). Whether warmth is applied to the eggs during these visits to partial clutches remains unknown.

Incubation constancy refers to the temporal pattern of incubation whereby adults tend eggs. Specifically, it refers to the percentage of time and frequency of on and off bouts in the incubation schedule of a species. Incubation constancy has been studied in a few shorebirds, using both direct observation and temperature probes placed in the nest to record the time that adults spend incubating. In the first few days after clutch completion, incubation constancy is often low and irregular, especially for females that may recoup energy invested in egg laying. Adults often are off the nest for extended periods of time. In general, incubation constancy increases to 80% to 95%, depending on whether a species is characterized by uniparental or biparental incubation. In biparental species, such as the Dunlin and Semipalmated Sandpiper, incubation constancy exceeds 97% (Norton 1972; Cresswell et al. 2003). By contrast, incubation constancy is lower for uniparental species such as the Pectoral Sandpiper, the females of which incubate on average 82% of the time. In the polyandrous Spotted Sandpiper, females may share incubation with their mate, but males often incubate alone. Incubation constancy (daylight hours) was 84% under biparental care versus 75% for clutches tended only by males (Maxson and Oring 1980).

Environmental factors, chiefly weather, may greatly affect patterns of incubation. Other factors that may contribute to variation in incubation behavior include time of day and the condition of the tending parent (Tulp and Schekkerman 2006). During the constant daylight of the Arctic summer, female Pectoral Sandpipers still had strong daily patterns to their incubation schedules. They had the fewest incubation recesses from 20:00 to 04:00. By contrast, females tended to increase the number of trips off the nest during the middle of the day (Cresswell et al. 2004). Cresswell et al. (2003, 2004) manipulated temperature of

the nest cups of both Semipalmated Sandpipers and Pectoral Sandpipers by use of added insulation and heating, respectively. In the biparental Semipalmated Sandpiper, adults increased the length of their incubation bouts by 10% (above 11.1 hours) in insulated nests, perhaps because they did not need to feed as much. In the uniparental Pectoral Sandpiper, females incubating clutches in artificially heated nests increased their total time at the nest by 3.6% (or 52 minutes). These observations suggest that the energetics of incubation and condition of the incubating adult determined the length of incubation bout, rather than the schedule of the nontending adult who was feeding at the time (in biparental species).

The incubation schedules of adults may also be influenced by the stage of incubation, with adults tending to be less attentive to nests early than late in incubation. This pattern may arise owing to the depleted energy and nutrient reserves of adult females that have recently completed the clutch. However, embryo development may also influence adult incubation patterns. Early incubation is often characterized by prolonged bouts in which adults are absent from the nest. By contrast, later in incubation adults extend incubation bouts. Female Snowy Plovers incubated less than 85% during mornings of the first 10 days of incubation. Incubation constancy increased abruptly 10 days into the 28-day incubation period, and remained at approximately 95% (Hoffmann 2005). In the Semipalmated Sandpiper, incubation bouts increased as the time of hatching approached, and this was independent of variation in weather (Cresswell et al. 2003). This variation in incubation constancy and nest attentiveness probably stems from several factors. Early in incubation, females may need to be off the nest more to recoup the energy and nutrients invested in eggs. At the same time, small embryos are more cold-tolerant, and adults require less energy to restart the development of cooled embryos when they are small compared with older, well-developed embryos (Turner 2002).

LENGTH OF INCUBATION

Incubation period, or the interval between the laying of the last egg and the hatching of chicks, varies substantially among species (Lack 1968). Some small, Arctic-breeding sandpipers (such as Temminck's Stint and phalaropes) require, on average, about 21 days to hatch eggs. By contrast, the average incubation period of several large, temperate-breeding shorebirds (such as lapwings and oystercatchers) exceeds 30 days. To some extent, this variation in incubation period is related to egg mass, with the longer incubation periods found in species that lay larger eggs (Nol 1986). There is, however, considerable scatter to the latter relationship. Species (and families) that run and catch prey require up to four more days of incubation than those of similar egg mass that are fed by their parents or that feed mostly by pecking. Nol (1986) proposed that this variation could be better understood by considering the foraging methods and development of young shorebirds. Specifically, young shorebirds that search visually and actively chase prey soon after hatch (such as plovers) require longer incubation for the development of the neural circuitry (of the optic tectum) necessary to accomplish these tasks. Incubation periods are shorter in sandpipers, in which the newly hatched young peck at prey, and oystercatchers and snipes, in which the adults feed their young to varying degrees.

Incubation period also varies within species. In some species, the average incubation period represents a wide range of individual records (Johnsgard 1981; Cramp et al. 1983). Incubation in Pied Avocets, for example, averages 23 days, but the range spans 20 to 28 days. Similar numbers exist for the Common Ringed Plover (24-day average, range: 21 to 28 days), Snowy Plover (28-day average, range: 24 to 38 days), and Ruddy Turnstone (range: 21 to 27 days), to name a few. Other species tend to have less variable incubation periods. This variation may stem from incubation constancy provided by uniparental versus biparental incubation of eggs. Uniparental species typically exhibit more variability owing to the influence of weather on embryo development during intervals when the single tending adult is on incubation recess. By contrast, biparental incubators may exhibit less variability in incubation period owing to the constancy of egg temperature maintained by two adults that rarely leave the eggs uncovered (Cresswell et al. 2003). In the Common Greenshank, in which males and females share equally in incubation, clutches averaged 23.9 and 24.4 days incubation in two different study areas in Scotland; extremes were 22.9 and 26.1 days (Nethersole-Thompson and Nethersole-Thompson 1979). As a side note, these discrepancies may also stem from definitions (When does incubation begin and end?) as well as the accuracy of observations in the field.

To some extent, intraspecific variation in incubation period correlates predictably with clutch initiation date: nests initiated progressively later in the breeding season require less incubation time. For example, at the northern extent of their breeding range in Saskatchewan, female Wilson's Phalaropes initiate clutches over a 45-day interval in May and June. The average incubation period in this population was 23 days. However, male phalaropes incubated some early-season clutches for up to 27 days, whereas late-season clutches took as little as 18 days to hatch (Colwell and Oring 1988a). Phalaropes are uniparental incubators, and males may occasionally incubate during prolonged cold spells in May that require them to take long incubation recesses. By the end of the breeding season, however, much warmer conditions prevail, and males apparently take shorter incubation recesses. Similar seasonal patterns of an inverse relationship of incubation and date have been recorded for temperate-breeding shorebirds. This is often interpreted as a consequence of warmer conditions prevailing during late spring or early summer such that the embryos left untended during incubation recesses do not lose heat and the development of embryos does not slow to the extent that it does during cooler, early spring periods.

GENDER ROLES

In monogamous species in which males and females share incubation, the daily incubation patterns of the sexes are often different (Ticehurst 1931; Wallander 2003). In Common Greenshanks, males often relieve their mates for nighttime incubation (Nethersole-Thompson and Nethersole-Thompson 1979). Similar patterns exist in many plovers, with literature suggesting that females tend to incubate more during the day and males more at night (Wallander 2003). In reality, this apparently equal division of labor in incubation is not universal and varies within species. Male Common Ringed Plovers, for example, incubate at night slightly more than females. Male-only nocturnal incubation prevails in other plovers, however, including Wilson's Plover (Thibault and McNeil 1995), Killdeer (Warnock and Oring 1996), and Kentish Plover (Kosztolányi and Székely 2002).

The reasons for these relatively consistent day–night patterns of incubation by males and females may be related to a variety of factors (Kosztolányi and Székely 2002). First, females may incubate during the day such that they can recover energy reserves by feeding at night, especially shortly after egg laying. This assumes that prey are more available to foraging birds at night. Second, males may need to defend the breeding territory during the day, thus restricting their ability to assist their mate. Third, the brighter and darker plumages of males of some species (and muted colorations of females) may make incubation more (or less) risky during the day because of the contrast with the habitat surrounding the nest site. Finally, the sexes may differ in their abilities to visually detect predators during the day. No study has systematically addressed these explanations, although Kosztolányi and Székely (2002) indicated that there was little support for the first two in the Kentish Plover.

HATCHING

In the few days immediately preceding hatch, the nearly fully-developed embryos become increasingly active in the egg. At the same time, adult behaviors change abruptly. The audible "peeping" calls of young and their tapping on the inner surface of the shell can be heard when the eggs first star. These sounds increase in frequency and volume as the external shell is pipped. Close to the time of hatch, distraction behaviors of adults, including tail-dragging and broken-wing displays (Gochfeld 1984), increase in vigor. Adults on incubation recess may also spend more time in the vicinity of the nest, acting as sentinels for danger (Nethersole-Thompson and Nethersole-Thompson 1979). The process of hatching, from first tapping through pipping to the hatching of the last egg may take considerable time. Byrkjedal and Thompson (1998) reported that the hatching interval spanned 5 to 9 days in the Black-bellied Plover and golden-plovers. In other instances, however, hatching proceeds rather quickly.

Because shorebirds generally begin regular incubation after the female lays the penultimate or last egg, the development of embryos in early-laid eggs is initiated at the same time as those in later eggs. Consequently, siblings tend to hatch over a short interval of a few hours to days. The comparatively synchronized pattern of hatching is advantageous given that adults lead nidifugous young away from the nest site soon after hatch. However, the synchrony with which young shorebirds hatch is variable. Tuck (1972) reported that hatching intervals between successive eggs in the four-egg clutch of Wilson's Snipe averaged 1.75, 2.25, and 1.00 hours, respectively. In other species, young have been reported to hatch over several days (Byrkjedal and Thompson 1998). When this happens to an extreme, it is not uncommon for adults to leave the last-hatched chick in the nest, often with fatal results. In other circumstances of asynchronous hatch, adults occasionally split

FIGURE 5.4. Neonates of most shorebird species hatch in a well-developed state. These young Killdeer are fully feathered in natal down and ready to leave the nest shortly after they hatch. In this clutch, the chicks hatched somewhat asynchronously compared with the norm for shorebirds. Photo credit: Jordan Muir.

the brood, with the male and female parent rearing young separately.

In many charadriiform birds, adults remove eggshells from the nest shortly after chicks hatch. This behavior is widespread in shorebirds. Shorebirds both walk and fly from nests carrying eggshells from newly hatched young. Adults also remove eggs damaged by predators in a similar manner. Adults dispose of shells at considerable distances, sometimes over water. Sometimes adults eat their eggshells, as was reported for some Common Greenshanks (Nethersole-Thompson and Nethersole-Thompson 1979). Removal of shells has long been interpreted as an antipredator behavior, first suggested in 1891 by Ogilvie (cited by Byrkjedal and Thompson 1998) who studied Stone Curlews. This was later demonstrated experimentally by Tinbergen et al. (1963) working with Black-headed Gulls (*Chroicocephalus ridibundus*). Specifically, adults remove eggs owing to the presence of the contrasting white inner membranes of the shell that are revealed in a hatched egg. An additional advantage of shell removal is that it avoids the possibility that the

shell from a newly hatched egg might "cap" one that is in the process of hatching, resulting in chick death.

CHICK GROWTH AND DEVELOPMENT

Shorebird chicks lie at the precocial end of the continuum describing the state of development of newly hatched birds. In most shorebirds, chicks hatch fully feathered in natal down (Fig. 5.4), equipped with well-developed hind limbs that facilitate their nidifugous nature. In some species, chicks are capable of difficult sensory-motor actions related to feeding and prey acquisition, whereas young of other species are tended and fed by adults for the first weeks or months of life. This subtle variation in the characteristics of neonate shorebirds correlates with other facets of posthatch development, specifically with thermoregulatory capabilities and the tendency to forage independently of parents.

The natal down of neonate shorebirds provides some structural properties to conserve heat. The patterning of natal down also serves to camouflage chicks when hiding from predators. Young chicks are dependent to varying degrees on adults for thermoregulation, which is provided by the parents applying the brood patch to young. Early in life, adults occasionally brood chicks for lengthy intervals, especially during cool times of the day and during inclement weather. As they age, however, young become increasingly independent of parents. Yet there is considerable variation in the age at which young become capable of maintaining their body temperature independent of adult brooding. Homeothermy, or the capacity to generate heat in response to cold stress, depends on the relative size of the skeletal muscles, especially pectoral muscles, and the metabolic activity of the muscle tissue (Ricklefs and Williams 2003). This capacity increases as chicks age and their muscles mature. Across species, neonates of shorebirds vary in their capacity for homeothermy, and this correlates positively with body size. Chicks of small species are less capable of homeothermy and consequently become thermally independent on adults at an older age compared with larger species (Visser and Ricklefs 1993). Small chicks of any species, however, are burdened with the potential loss of heat owing to high surface area to volume ratios (Ricklefs and Williams 2003).

The age of homeothermy has consequences for the activity budgets and survival of young. Neonates of small-bodied shorebirds require more brooding, which compromises the time that they can spend foraging (Visser and Ricklefs 1993). Chicks of high Arctic shorebirds are especially tolerant of cold. Young Red Knots are capable of maintaining their body temperature despite prolonged periods of cold, which enables them to feed more during the long days of the Arctic summer. As a consequence, young Red Knots have accelerated growth and require less brooding by adults (Schekkermann et al. 2003).

The time spent brooding varies predictably with age: as chicks age, grow, and replace their natal down with a full coat of juvenile feathers, they become increasingly capable of homeothermy and less dependent on parents for warmth. In the warm subtropical climate of South Africa, young of the Kittlitz's Plover, Blacksmith Lapwing, and Crowned Lapwing brooded 26%, 4%, and 34%, respectively, during daylight hours of the first 10 days of life (Tjørve et al. 2008). Brooding decreased markedly as chicks grew, but even old chicks brooded during periods of cold. Adult Snowy Plovers brooded chicks for increasingly shorter durations over their 28-day prefledging period, but parents occasionally brooded nearly fledged chicks (Colwell, Hurley, et al. 2007). The dependence of young on adults for warmth has management implications. During periods of inclement weather, neonates of small species may be especially vulnerable to hypothermia and death. These problems may be exacerbated by human activity near breeding areas. Chronic disturbance, coupled with cool, damp weather, may have serious consequences for the survival of young. Hot conditions may challenge thermoregulatory abilities of shorebirds as well. Grant (1982) showed that breeding Black-necked Stilts frequently soaked their belly feathers with water and used the moisture to increase humidity for eggs and chicks.

When young shorebirds hatch, they possess well-developed hindlimb and pectoral muscles that facilitate their nidifugous nature. Additionally, these muscles enable chicks to maintain their body temperatures for short periods, independent of brooding by parents. Chicks of large species are more capable of thermogenesis using these muscles than those of smaller species. In neonates, the hindlimb muscles are particularly large, whereas the pectoral muscles are somewhat smaller in comparison with adult mass. As young mature and become capable of flight, the growth of pectoral muscles increases, especially in smaller species such as the Dunlin (Krijgsveld et al. 2001).

At hatch, young shorebirds possess residual yolk in their abdomen as an energy reserve, which may enable them to survive for days if isolated or abandoned by parents, especially during inclement weather. Newly hatched chicks of most species are capable of feeding themselves, but chicks of some species (such as the Crab Plover, some lapwings, oystercatchers, and thick-knees) are fed by adults to varying extents over the first few days to months of their lives. Newly hatched plovers, for instance, are capable of visually locating and chasing active prey within hours of hatch (Nol 1986).

Young shorebirds require varying lengths of time to fledge. The age at which the young of different species fledge correlates with two factors, one intrinsic to the species and the other extrinsic. A general positive relationship exists between a species' mass and the fledging age: smaller species require shorter intervals to reach independence compared with larger species. For example, small calidridine sandpipers (Little Stint, Western Sandpiper, Semipalmated Sandpiper, and Baird's Sandpiper) that breed in the Arctic fledge at ages ranging from 14 to 18 days (Holmes 1972; Norton 1973; Ashkenazie and Safriel 1979a; Ruthrauff and McCaffery 2005; Tjørve et al. 2008). By contrast, fledging occurs at nearly double this interval (35 to 40 days) in larger, temperate and tropical species, such as Crowned and Blacksmith Lapwings (Tjørve et al. 2008). At the far extreme of this continuum are young Spotted Thick-knees, which are capable of flight (fledge) at approximately 54 days (Tjørve et al. 2008). In Eurasian Oystercatchers, the young fledge anywhere from 27 to 52 days after they hatch. This variation is undoubtedly related to the ability of the adults to provision chicks and the size hierarchy that develops among siblings (Kersten and Brenninkmeijer 1995). Oystercatchers remain dependent on adults for food for up to 3 months. Juvenile Crab Plovers remain with and are fed by adults, presumably their parents, throughout their first winter.

The short prefledging periods of shorebirds are accompanied by rapid growth and high energy expenditures for chicks. Compared with other precocial birds, shorebird chicks expend more energy and grow faster than predicted (Schekkermann et al. 1998, 2003; Tjørve et al. 2008). This is particularly the case for Arctic species, where the values are among the highest reported for birds. For example, Red Knot chicks in the high Arctic expend energy at a rate that is 89% above that predicted by an allometric relationship (Schekkermann et al. 2003). In the Curlew Sandpiper, the growth rate of chicks in tundra habitats of the Taimyr Peninsula was among the highest reported for birds (Schekkermann et al. 1998). Similar reports exist for other northern-breeding species, such as the Western Sandpiper (Ruthrauff and McCaffery 2005). These findings for sandpipers contrast with recent studies of tropical and subtropical breeding plovers, in which chick growth and energy expenditure is considerably lower. Tjørve et al. (2008) reported growth rates and energy expenditures for chicks of Kittlitz's Plover, Blacksmith Lapwing, and Crowned Lapwing that breed in South Africa. These three species had slower growth, lower energy expenditures, and longer prefledging periods than the temperate and Arctic species. Tjørve et al. (2008) provided a comparative perspective on latitudinal trends in growth and energy use by shorebird chicks, and reported positive correlations with latitude: shorebird growth rates increased 1.8% and energy expenditure increased 2.5% with each degree of latitude (Tjørve et al. 2008). A similar pattern emerges when chicks are examined in two populations of the same species at different latitudes. Pied Avocet chicks reared in Germany grew faster and used more energy than conspecifics in Spain, where food was less available (Joest 2003). The common interpretation of this latitudinal pattern in growth and energy expenditure is that it is related to food availability (Schekkermann et al. 2003; Tjørve et al. 2008). Specifically, chicks of northern-breeding species have more food available to them in the brief yet highly productive breeding season, compared with young reared in southerly latitudes. Further, support for the relationship between productivity and growth comes from observations that chicks grow slowly during cold, inclement weather. For instance, growth of Red Knot chicks slowed markedly during periods of cool weather and reduced food availability (Schekkermann et al. 2003). During cold periods, shorebird chicks sometimes appear to enter a state of torpor and may be "revived" by warming.

CONSERVATION IMPLICATIONS

The breeding biology of shorebirds is intimately tied to many aspects of management and conservation, as illustrated by the following examples. First, philopatry (or natal dispersal) is a critical facet of formulas to estimate effective population size, which is especially important in small populations. Dispersal also informs conservationists about the extent and history of gene flow within and among populations that may differ in measures of protection afforded by local, national, or international entities. An example of this is the population segment of the Western Snowy Plover that is protected under the U.S. Endangered Species Act despite the absence of genetic differences with populations elsewhere in North America (Funk et al. 2007).

Management of breeding habitats often relies on information on habitat availability and quality. Virtually all shorebirds can be characterized as having habitat preferences for their nests: sandpipers nest in vegetated habitats that afford crypsis to eggs and incubating adults, whereas plovers nest in sparsely vegetated areas and rely on camouflaged eggs to avoid clutch loss to predators. Accordingly, efforts to restore and enhance habitats to create features attractive to breeding birds require knowledge of habitat selection. It is critical that this ecological knowledge be conducted at the appropriate spatial scale for inference to be applied to restoration efforts. For example, the creation of open, sparsely vegetated habitats is widely embraced as a means to increase the attractiveness of habitat to several species of threatened plover. Most analyses of nest-site selection, however, have been conducted at fine spatial scales limited to the immediate vicinity of the nest (third-order selection of Johnson 1980). For restoration efforts, the appropriate scale of nest-site selection is a coarser spatial scale, one addressing whether plovers select open habitats (second-order selection).

Conservation efforts directed at shorebirds benefit directly from knowledge of breeding biology when agricultural land practices can be altered with minor impact on humans. In Europe, for instance, flooding of wet meadows is thought to benefit reproductive success by providing habitats with more food (Eglington et al. 2009). In these same habitats, the timing of agricultural practices may be altered to enhance the likelihood that shorebirds can complete nesting successfully before grazing or harvesting of hay begins, for instance (Durant et al. 2008).

A final example of applied ecology comes from relationships between weather, food availability, and various facets of breeding biology. Given that inclement weather can have effects on everything from egg-laying intervals to egg size and chick growth rates, it is not a far stretch to extend these relationships to situations in which humans can exacerbate these already difficult breeding conditions. For example, in coastal habitats where human recreational activity is high, buffer zones may be required to reduce the negative effects of human disturbance during the breeding season.

LITERATURE CITED

Ahlering, M. A., and J. Faaborg. 2006. Avian habitat management meets conspecific attraction: If you build it, will they come? *The Auk* 123: 301–312.

Ankeny, C. D., and J. Hopkins. 1985. Habitat selection by roof-nesting Killdeer. *Journal of Field Ornithology* 56: 284–286.

Arnold, T. W. 1999. What limits clutch size in waders? *Journal of Avian Biology* 30: 216–220.

Ashkenazie, S., and U. N. Safriel. 1979a. Breeding cycle and behavior of the Semipalmated Sandpiper at Barrow, Alaska. *The Auk* 96: 56–67.

———. 1979b. Time-energy budget of the Semipalmated Sandpiper *Calidris pusilla* at Barrow, Alaska. *Ecology* 60: 783–799.

Baldassarre, G. A., and E. G. Bolen. 2006. *Waterfowl ecology and management.* 2nd ed. New York: John Wiley and Sons.

Barta, Z., and T. Székely. 1997. The optimal shape of avian eggs. *Functional Ecology* 11: 656–652.

Bent, A. C. 1962. *Life histories of North American shorebirds.* New York: Dover.

Bergstrom, P. W. 1988. Breeding biology of Wilson's Plovers. *Wilson Bulletin* 100: 25–35.

Brown, S., J. Bart, R. B. Lanctot, J. A. Johnson, S. Kendall, D. Payer, and J. Johnson. 2007. Shorebird abundance and distribution on the coastal plain of the Arctic National Wildlife Refuge. *The Condor* 109: 1–14.

Burger, J. 1987. Physical and social determinants of nest-site selection in Piping Plover in New Jersey. *The Condor* 89: 811–818.

Byrkjedal, I., and D. B. A. Thompson. 1998. *Tundra plovers: The Eurasian, Pacific and American Golden Plovers and Grey Plover.* London: T & A.D. Poyser.

Campomizzi, A. J., J. A. Butcher, S. L. Farrell, A. G. Snelgrove, B. A. Collier, K. J. Gutzwiller, M. L. Morrison, and R. N. Wilkins. 2008. Conspecific attraction is a missing component of wildlife habitat modeling. *Journal of Wildlife Management* 72: 331–336.

Christians, J. K. 2002. Avian egg size: Variation within species and inflexibility within individuals. *Biological Review* 77: 1–26.

Colwell, M. A. 1986. Intraspecific brood parasitism in three species of prairie-breeding shorebird. *Wilson Bulletin* 98: 473–475.

———. 1992. Wilson's phalarope nest success is not influenced by vegetation concealment. *The Condor* 94: 767–772.

———. 2006. Minimum egg-laying intervals in shorebirds. *Wader Study Group Bulletin* 111: 50–59.

Colwell, M. A., S. J. Hurley, J. N. Hall, and S. J. Dinsmore. 2007. Age-related survival and behavior of Snowy Plover chicks. *The Condor* 109: 638–647.

Colwell, M. A., S. E. McAllister, C. B. Millett, A. N. Transou, S. M. Mullin, Z. J. Nelson, C. A. Wilson, and R. R. LeValley. 2007. Natal philopatry and dispersal of the Western Snowy Plover. *Wilson Journal of Ornithology* 119: 378–385.

Colwell, M. A., C. B. Millett, J. J. Meyer, J. N. Hall, S. J. Hurley, S. E. McAllister, A. N. Transou, and R. R. LeValley. 2005. Snowy Plover reproductive success in beach and river habitats. *Journal of Field Ornithology* 76: 373–382.

Colwell, M. A., and L. W. Oring. 1988a. Breeding biology of Wilson's Phalarope in southcentral Saskatchewan. *Wilson Bulletin* 100: 567–582.

———. 1988b. Sex ratios and intrasexual competition for mates in a sex-role reversed shorebird, Wilson's phalarope (*Phalaropus tricolor*). *Behavioral Ecology and Sociobiology* 22: 165–173.

———. 1989. Adult and natal philopatry of prairie shorebirds: Sex and species differences. *Wader Study Group Bulletin* 55: 21–24.

———. 1990. Nest-site characteristics of prairie shorebirds. *Canadian Journal of Zoology* 68: 297–302.

Colwell, M. A., J. D. Reynolds, D. Schamel, and D. M. Tracy. 1988. Phalarope philopatry. In *Acta XIX congressus internationalis ornithologici,* ed. Henri Ouellet 585–593. Ottawa: National Museum of Natural Science/University of Ottawa Press.

Conway, W. C., L. M. Smith, and J. D. Ray. 2005. Shorebird habitat use and nest-site selection in the Playa Lakes Region. *Journal of Wildlife Management* 69: 174–184.

Cramp, S., and S. E. L. Simmons, eds. 1983. *Birds of the Western Palearctic.* Vol. 3. Oxford: Oxford University Press.

Cresswell, W., S. Holt, J. M. Reid, D. P. Whitfield, and R. J. Mellanby. 2003. Do energetic demands constrain incubation scheduling in a biparental species? *Behavioral Ecology* 14: 97–102.

Cresswell, W., S. Holt, J. M. Reid, D. P. Whitfield, R. J. Mellanby, D. Norton, and S. Waldron. 2004. The energetic costs of egg heating constrain incubation attendance but do not determine daily energy expenditure in the pectoral sandpiper. *Behavioral Ecology* 15: 498–507.

Dabelsteen, T. 1978. An analysis of the song-flight of the Lapwing (*Vanellus vanellus* L.) with respect to causation, evolution and adaptations to signal function. *Behaviour* 66: 136–178.

Deeming, D. C., ed. 2002. *Avian incubation.* Oxford: Oxford University Press.

Delehanty, D. J., and L. W. Oring. 1993. Effect of clutch size on incubation persistence in male Wilson's Phalarope (*Phalaropus tricolor*). *The Auk* 110: 521–528.

del Hoyo, J., A. Elliot, J. Sargatal, and N. J. Collar, eds. 1992. *Handbook of birds of the world.* Vol. 3. Barcelona: Lynx Edicion.

Dittmann, T., and H. Hötker. 2001. Intraspecific variation in the egg size of the Pied Avocet. *Waterbirds* 24: 83–88.

Drycz, A., J. Witkowski, and J. Okulewicz. 1981. Nesting of 'timid' waders in the vicinity of 'bold' ones. *The Ibis* 123: 542–545.

Durant, D., M. Tichit, E. Kerneis, and H. Fritz. 2008. Management of agricultural wet grasslands for breeding waders: Integrating ecological and livestock system perspectives—a review. *Biodiversity Conservation* 17: 2275–2295.

Eglington, S. M., J. A. Gill, M. A. Smart, W. J. Sutherland, A. R. Watkinson, and M. Bolton. 2009. Habitat management and patterns of predation of Northern Lapwings on wet grasslands: The influence of linear habitat structures at different spatial scales. *Biological Conservation* 142: 314–324.

Emlen, S. T., and L. W. Oring. 1977. Ecology, sexual selection, and the evolution of mating systems. *Science* 197: 215–223.

Evans, P. R., and M. W. Pienkowski. 1984. Population dynamics of shorebirds. In *Shorebirds: Breeding behavior and populations,* ed. J. Burger and B. L. Olla, 83–123. New York: Plenum Press.

Evans, T. J., and S. W. Harris. 1997. Status and habitat use by American Avocets wintering at Humboldt Bay, California. *The Condor* 96: 178–189.

Flynn L., E. Nol, and Y. Zharikov. 1999. Philopatry, nest-site tenacity and mate fidelity of Semipalmated Plovers. *Journal of Avian Biology* 30: 47–55.

Funk, W. C., T. D. Mullins, and S. M. Haig. 2007. Conservation genetics of Snowy Plovers (*Charadrius alexandrinus*) in the Western Hemisphere: Population structure and delineation of subspecies. *Conservation Genetics* 8: 1287–1309.

Galbraith, H. 1988. Effects of egg size and composition on the size, quality and survival of Lapwing *Vanellus vanellus* chicks. *Journal of Zoology (London)* 214: 383–398.

Gochfeld, M. 1984. Antipredator behavior: Aggressive and distraction displays of shorebirds. In *Shorebirds: Breeding behavior and populations,* ed. J. Burger and B. L. Olla, 289–377. New York: Plenum Press.

Grant, G. S. 1980. Avian incubation: Egg temperature, nest humidity, and behavioral thermoregulation in a hot environment. *Ornithological Monographs* 30: 1–75.

Gratto, C. L. 1988. Natal philopatry, site tenacity, and age of first breeding of the Semipalmated Sandpiper. *Wilson Bulletin* 100: 664–665.

Gratto, C. L., R. I. G. Morrison, and F. Cooke. 1985. Philopatry, site tenacity and mate fidelity in the Semipalmated Sandpiper. *The Auk* 102: 16–24.

Greenwood, P. J. 1980. Mating systems, philopatry and dispersal in birds and mammals. *Animal Behavior* 28: 1140–1162.

Groen, N. M. 1993. Breeding site tenacity and natal philopatry in the Black-tailed Godwit *Limosa l. limosa. Ardea* 81: 107–113.

Grønstøl, G. B. 1997. Correlates of egg-size variation in polygynously breeding Northern Lapwings. *The Auk* 114: 507–512.

Haig, S. M., D. W. Mehlman, and L. W. Oring. 1998. Avian movements and wetland connectivity in landscape conservation. *Conservation Biology* 12: 749–758.

Hamilton, R. B. 1975. Comparative behavior of the American Avocet and the Black-necked Stilt (Recurvirostridae). *Ornithological Monographs* 17: 1–98.

Handel, C. M., and R. E. Gill, Jr. 2000. Mate fidelity and breeding site tenacity in a monogamous sandpiper, Black Turnstone. *Animal Behaviour* 60: 471–481.

Hansell, M. H., and D. C. Deeming. 2002. Location, structure and function of incubation sites. In *Avian incubation: Behaviour, environment and evolution,* ed. D. C. Deeming, 8–27. Oxford: Oxford University Press.

Hardy, M. A., and M. A. Colwell. 2008. The impact of predator exclosures on Snowy Plover nesting success: A seven-year study. *Wader Study Group Bulletin* 115: 161–166.

Hartman, C. A., and L. W. Oring. 2006. An inexpensive method for remotely monitoring nest activity. *Journal of Field Ornithology* 77: 418–424.

Hildén, O. 1975. Breeding system of Temminck's Stint *Calidris temminckii. Ornis Fennica* 52: 117–146.

———. 1978. Population dynamics in Temminck's Stint *Calidris temminckii. Oikos* 30: 17–28.

———. 1979. Territoriality and site tenacity in Temminck's Stint *Calidris temminckii. Ornis Fennica* 56: 56–74.

Hills, S. 1983. Incubation capacity as a limiting factor of shorebird clutch size. Master's thesis, University of Washington.

Hockey, P. A. R. 1996. *Haematopus ostralegus* in perspective: Comparisons with other oystercatchers. In *The oystercatcher,* ed. J. D. Goss-Custard, 251–285. Oxford: Oxford University Press.

Hocking, P. M. 1993. Relationship between egg size, body weight and pelvic dimensions in turkeys. *Animal Production* 56: 145–150.

Hoffmann, A. 2005. Incubation behavior of female Western Snowy Plovers (*Charadrius alexandrinus nivosus*) on sandy beaches. Master's thesis, Humboldt State University.

Holmes, R. T. 1966. Breeding ecology and annual cycle adaptations of the Red-backed Sandpiper (*Calidris alpina*) in northern Alaska. *The Condor* 68: 3–46.

———. 1971. Density, habitat and the mating system of the Western Sandpiper (*Calidris mauri*). *Oecologia* 7: 191–208.

———. 1972. Ecological factors influencing the breeding season schedule of Western Sandpipers (*Calidris mauri*) in subarctic Alaska. *American Midland Naturalist* 87: 472–491.

Howe, M. A. 1975. Behavioral aspects of the pair bond in Wilson's Phalarope. *Wilson Bulletin* 87: 248–270.

Hutto, R. L. 1985. Habitat selection by nonbreeding, migratory land birds. In *Habitat selection in birds,* ed. M. L. Cody, 455–476. Orlando, FL: Academic Press.

Huxley, J. S., and F. A. Montague. 1926. Studies on the courtship and sexual life of birds. VI. The Black-tailed Godwit (*Limosa limosa*). *The Ibis* 12: 1–26.

Isaksson, D., J. Wallander, and M. Larsson. 2007. Managing predation on ground-nesting birds: The effectiveness of nest exclosures. *Biological Conservation* 136: 136–142.

Jackson, D. B. 1994. Breeding dispersal and site-fidelity in three monogamous waders in the Western Isles, U.K. *The Ibis* 136: 463–473.

Jehl, J. R., Jr., and B. G. Murray. 1986. The evolution of normal and reverse size dimorphism in shorebirds and other birds. *Current Ornithology* 3: 1–86.

Joest, R. 2003. Junge Säbelschnäbler (*Recurvirostra avosetta* L.) in unterschiedlichen Klimazonen: Physiologische und ethologische Anpassungen an ökologische Bedingungen in Norddeutschland und Südspanien. Ph.D. thesis, Christian-Albrechts-Universtät.

Johnsgard, P. A. 1981. *The plovers, sandpipers and snipes of the world.* Lincoln: University of Nebraska Press.

Johnson, D. H. 1980. The comparison of usage and availability measurements for evaluating resource preference. *Ecology* 61: 65–71.

Johnson, O. E., P. G. Connors, P. L. Bruner, and J. L. Maron. 1993. Breeding ground fidelity and mate retention in the Pacific Golden-Plover. *Wilson Bulletin* 105: 60–67.

Jonas, R. 1979. Brutbiologische Untersuchungen an einer population der Uferschnepfe *Limosa limosa. Vogelwelt* 100: 125–136.

Jones, S. L., C. S. Nations, S. D. Fellows, and L. L. McDonald. 2008. Breeding abundance and distribution of Long-billed Curlews (*Numenius americanus*) in North America. *Waterbirds* 31: 1–14.

Jönsson, P. E., and T. Alerstam. 1990. The adaptive significance of parental care role division and sexual size dimorphism in breeding shorebirds. *Biological Journal of the Linnaean Society* 41: 301–314.

Kersten, M., and A. Brenninkmeijer. 1995. Growth, fledging success and post-fledging survival of juvenile oystercatchers *Haematopus ostralegus. The Ibis* 137: 396–404.

Klaassen, M., Å. Lindström, H. Meltofte, and T. Piersma. 2001. Arctic waders are not capital breeders. *Nature* 413: 794.

Klomp, H. 1951. Over de achteruitgang van de Kievit, *Vanellus vanellus* (L.) in Nederland en gegevens over het legmechanisme en het eiproductievermogen. *Ardea* 39: 143–182.

Koenig, W. D. 1988. On determination of viable population size in birds and mammals. *Wildlife Society Bulletin* 16: 230–234.

Kosztolányi, A., and T. Székely. 2002. Using a transponder system to monitor incubation routines of Snowy Plovers. *Journal of Field Ornithology* 73: 199–205.

Krijgsveld, K. L., J. M. Olson, and R. E. Ricklefs. 2001. Catabolic capacity of the muscles of shorebird chicks: Maturation of function in relation to body size. *Physiological and Biochemical Zoology* 74: 250–260.

Lack, D. 1968. *Ecological adaptations for breeding in birds.* London: Chapman and Hall.

Lanctot, R. B. 1998. Sexual attitudes at northern latitudes. *Natural History* 107: 72–75.

Lanctot, R. B., and C. D. Laredo. 1994. Buff-breasted Sandpiper *Tryngites subruficollis.* In *The birds of North America,* ed. E. Poole and F. Gill, no. 91. Philadelphia/Washington, DC: Academy of Natural Sciences/American Ornithologists' Union.

Lauro, B., and E. Nol. 1995. Patterns of habitat use for Pied and Sooty Oystercatchers nesting at the Furneaux Islands Australia. *The Condor* 92: 920–934.

Lennington, S., and T. Mace. 1975. Mate fidelity and nesting site tenacity in the Killdeer. *The Auk* 92: 149–151.

Liker, A., J. D. Reynolds, and T. Székely. 2001. The evolution of egg size in socially polyandrous shorebirds. *Oikos* 95: 3–14.

Lislevand, T., I. Byrkjedal, T. Borge, and G.-P. Sætre. 2005. Egg size in relation to sex of embryo, brood sex ratios and laying sequence in Northern Lapwings (*Vanellus vanellus*). *Journal of Zoology (London)* 267: 81–87.

Lislevand, T., and G. H. Thomas. 2006. Limited male incubation ability and the evolution of egg size in shorebirds. *Biology Letters* 2: 206–208.

MacFarlane, G. R., S. P. Blomberg, G. Kaplan, and L. J. Rogers. 2006. Same-sex sexual behavior in birds: Expression is related to social mating system and state of development at hatching. *Behavioral Ecology* 18: 21–33.

Maclean, G. L. 1972a. Clutch size and evolution in the Charadrii. *The Auk* 89: 299–324.

———. 1972b. Problems of display postures in the Charadrii (Aves: Charadriiformes). *Zoologica Africana* 7: 57–74.

Martin, T. E. 1993. Nest predation and nest sites: New perspectives on old patterns. *BioScience* 43: 523–532.

Maxson, S. J., and L. W. Oring. 1980. Breeding season time and energy budgets of the polyandrous spotted sandpiper. *Behaviour* 74: 200–263.

Miller, E. H. 1984. Communication in breeding shorebirds. In *Shorebirds: Breeding behavior and populations,* ed. J. Burger and B. L. Olla, 169–241. New York: Plenum Press.

Miller, E. H., and A. J. Baker. 2009. Antiquity of shorebird acoustic displays. *The Auk* 126: 454–459.

Miskelly, C. M. 1999. Breeding ecology of Snares Island Snipe (*Coenocorypha aucklandica huegeli*) and Chatham Island Snipe (*C. pusilla*). *Notornis* 46: 57–71.

Muir, J. J., and M. A. Colwell. 2010. Snowy Plovers select open habitats for courtship scrapes and nests. *The Condor,* forthcoming.

Mundahl, J. T., O. L. Johnson, and M. L. Johnson. 1981. Observations at a twenty-egg Killdeer nest. *The Condor* 83: 180–182.

Myers, J. P. 1979. Leks, sex and Buff-breasted Sandpipers. *American Birds* 33: 823–825.

———. 1981. A test for three hypotheses for latitudinal segregation of the sexes in wintering birds. *Canadian Journal of Zoology* 59: 1527–1534.

Naves, L. C., R. B. Lanctot, A. R. Taylor, and N. P. Coutsoubos. 2008. How often do Arctic shorebirds lay replacement clutches? *Wader Study Group Bulletin* 115: 2–9.

Nelson, Z. J. 2007. Conspecific attraction in the breeding distribution of the Western Snowy Plover (*Charadrius alexandrinus nivosus*). Master's thesis, Humboldt State University.

Nethersole-Thompson, D., and M. Nethersole-Thompson. 1979. *Greenshanks*. Vermillion, SD: Buteo Books.

Neuman, K.K., G.W. Page, L.E. Stenzel, J.C., and J.S. Warriner. 2004. Effect of mammalian predator management on Snowy Plover breeding success. *Waterbirds* 27: 257–263.

Nguyen, L. P., K. F. Abraham, and E. Nol. 2006. Influence of Arctic Terns on survival of artificial and natural Semipalmated Plover nests. *Waterbirds* 29: 100–104.

Nguyen, L. P., E. Nol, and K. F. Abraham. 2003. Nest success and habitat selection of the Semipalmated Plover on Akimiski Island, Nunavat. *Wilson Bulletin* 115: 285–291.

Nol, E. 1986. Incubation period and foraging technique in shorebirds. *American Naturalist* 128: 115–119.

———. 1989. Food supply and reproductive performance of the American Oystercatcher in Virginia. *The Condor* 91: 429–435.

Nol, E., A. J. Baker, and M. D. Cadman. 1984. Clutch initiation dates, clutch size, and egg size of the American Oystercatcher in Virginia. *The Auk* 101: 855–867.

Nol, E., M. Sullivan Blanken, and L. Flynn. 1997. Sources of variation in clutch size, egg size, and clutch completion dates of Semipalmated Plovers in Churchill, Manitoba. *The Condor* 99: 389–396.

Norton, D. W. 1972. Incubation schedules of four species of calidridine sandpipers at Barrow, Alaska. *The Condor* 74: 164–176.

———. 1973. Ecological energetics of calidridine sandpipers breeding in northern Alaska. Ph.D. dissertation, University of Alaska.

O'Connor, R. J 1984. *The growth and development of birds*. New York: John Wiley and Sons.

Oring, L. W. 1964. Displays of the Buff-breasted Sandpiper at Norman, Oklahoma. *The Auk* 81: 83–86.

———. 1968. Vocalizations of the Green and Solitary Sandpipers. *Wilson Bulletin* 80: 395–420.

———. 1973. Solitary Sandpiper early reproductive behavior. *The Auk* 90: 652–663.

———. 1982. Avian mating systems. In *Avian Biology*, vol. 6, ed. D. S. Farner, J. R. King, and K. C. Parkes, 1–92. New York: Academic Press.

Oring, L. W., M. A. Colwell, and J. M. Reed. 1991. Lifetime reproductive success in the Spotted Sandpiper (*Actitis macularia*): Sex differences and variance components. *Behavioral Ecology and Sociobiology* 28: 425–432.

Oring, L. W., and D. B. Lank. 1982. Sexual selection, arrival times, philopatry and site fidelity in the polyandrous Spotted Sandpiper. *Behavioral Ecology and Sociobiology* 10: 185–191.

———. 1984. Breeding area fidelity, natal philopatry and the social systems of sandpipers. In *Shorebirds: Breeding behavior and populations*, ed. J. Burger and B. L. Olla, 125–147. New York: Plenum Press.

Oring, L. W., D. B. Lank, and S. J. Maxson. 1983. Population studies of the polyandrous Spotted Sandpiper. *The Auk* 100: 272–285.

Oring, L. W., and J. M. Reed. 1996. Shorebirds of the western Great Basin of North America: Overview and importance to continental populations. *International Wader Studies* 9: 6–12.

Oring, L. W., J. M. Reed, M. A. Colwell, D. B. Lank, and S. J. Maxson. 1991. Factors regulating annual mating success and reproductive success in spotted sandpipers (*Actitis macularia*). *Behavioral Ecology and Sociobiology* 28: 433–442.

Parr, J. 1980. Population study of Golden Plover *Pluvialis apricaria*, using marked birds. *Ornis Scandinavica* 11: 179–189.

Patten, M. A. 2007. Geographic variation in calcium and clutch size. *Journal of Avian Biology* 38: 637–643.

Paulson, D. R., and W. J. Erckmann. 1985. Buff-breasted Sandpipers nesting in association with Black-bellied Plovers. *The Condor* 87: 429–430.

Pierce, R. J. 1989. Breeding and social patterns of Banded Dotterels (*Charadrius bicinctus*) at Cass River. *Notornis* 36: 13–23.

Plissner, J. H., S. M. Haig, and L. W. Oring. 1999. Within- and between-year dispersal of American Avocets among multiple western Great Basin wetlands. *Wilson Bulletin* 111: 314–320.

Powell, A. N. 2001. Habitat characteristics and nest success of Snowy Plovers associated with California Least Tern colonies. *The Condor* 103: 785–792.

Redmond, R. L., and D. A. Jenni. 1982. Natal philopatry and breeding area fidelity of Long-billed Curlews (*Numenius americanus*): Patterns and evolutionary consequences. *Behavioral Ecology and Sociobiology* 10: 277–279.

Reed, J. M., and L. W. Oring. 1993. Philopatry, site fidelity, dispersal, and survival of Spotted Sandpipers. *The Auk* 110: 541–551.

———. 1997. Intra- and inter-clutch patterns in egg mass in the Spotted Sandpiper. *Journal of Field Ornithology* 68: 296–301.

Reynolds, J. D., and F. Cooke. 1988. The influence of mating systems on philopatry: A test with the polyandrous Red-necked Phalarope. *Animal Behavior* 36: 1788–1795.

Reynolds, J. D., M. A. Colwell, and F. Cooke. 1986. Sexual selection and spring arrival schedules of red-necked and Wilson's phalaropes. *Behavioral Ecology and Sociobiology* 18: 303–310.

Ricklefs, R. E. 1969. *An analysis of nesting mortality in birds*. Washington, DC: Smithsonian Institution Press.

———. 1974. Energetics of reproduction in birds. In *Symposium on avian energetics, Provincetown, Mass. 1973*, ed. R. E. Paynter, Jr., 152–292. [Cambridge, MA]: Nuttall Ornithological Club.

———. 1984. Egg dimensions and neonatal mass of shorebirds. *The Condor* 86: 7–11.

Ricklefs, R. E., and J. B. Williams. 2003. Metabolic responses of shorebird chicks to cold stress: Hysteresis of cooling and warming phases. *Journal of Experimental Biology* 206: 2883–2893.

Rinkel, G. L. 1940. Waarnemingen over het gedrag van de Kievit (*Vanellus vanellus* [L.]) gedurende de broedtijd. *Ardea* 29: 108–147.

Rivers, J. W., and J. V. Briskie. 2003. Lack of sperm production and sperm storage by Arctic-nesting shorebirds during spring migration. *The Ibis* 145: 61–66.

Robinson, J. A., and L. W. Oring. 1997. Natal and breeding dispersal in American Avocets. *The Auk* 114: 416–430.

Ross, H. A. 1979. Multiple clutches and shorebird egg and body weight. *American Naturalist* 113: 618–622.

Ruthrauff, D. R., and B. J. McCaffery. 2005. Survival of Western Sandpiper broods on the Yukon-Kuskokwim Delta, Alaska. *The Condor* 107: 597–604.

Safriel, U. N. 1975. On the significance of clutch size in nidifugous birds. *Ecology* 56: 703–708.

Safriel, U. N., B. J. Ens, and A. Kaiser. 1996. Rearing to independence. In *The oystercatcher*, ed. J. D. Goss-Custard, 219–250. Oxford: Oxford University Press.

Sandercock, B. K. 1997. Incubation capacity and clutch size determination in two calidridine sandpipers: A test of the four-egg threshold. *Oecologia* 110: 50–59.

Schamel, D., and D. M. Tracy. 1991. Breeding site fidelity and natal philopatry in the sex-role reversed Red and Red-necked Phalaropes. *Journal of Field Ornithology* 62: 390–398.

Schekkerman, H., I. Tulp, T. Piersma, and G. H. Visser. 2003. Mechanisms promoting higher growth rate in Arctic than in temperate shorebirds. *Oecologia* 134: 332–342.

Schekkerman, H., M. W. J. Van Roomen, and L. G. Underhill. 1998. Growth, behavior of broods and weather-related variation in breeding productivity of curlew sandpipers *Calidris ferruginea*. *Ardea* 86: 153–168.

Smith, P. A., H. G. Gilchrist, and J. N. M. Smith. 2007. Effects of nest habitat, food and parental behavior on shorebird nest success. *The Condor* 109: 15–31.

Soikkeli, M. 1967. Breeding cycle and population dynamics in the Dunlin (*Calidris alpina*). *Annales Zoologici Fennici* 4: 158–198.

———. 1970. Mortality and reproductive rates in a Finnish population of Dunlin, *Calidris alpina*. *Ornis Fennica* 47: 149–158.

Solis, J. C., and F. de Lope. 1995. Nest and egg crypsis in the ground-nesting Stone Curlew *Burhinus oedicnemus*. *Journal of Avian Biology* 26: 135–138.

Sordahl, T. 1984. Observations on breeding site fidelity and pair formation in American Avocets and Black-necked Stilts. *North American Bird Bander* 9: 8–11.

Stenzel, L. E., G. W. Page, J. C. Warriner, J. S. Warriner, D. E George, C. R. Eyster, B. A. Ramer, and K. K. Neuman. 2007. Survival and natal dispersal of juvenile Snowy Plovers in central coastal California. *The Auk* 124: 1023–1036.

Stenzel, L. E., J. C. Warriner, J. S. Warriner, K. S. Wilson, F. C. Bidstrup, and G. W. Page. 1994. Long-distance breeding dispersal of Snowy Plovers in western North America. *Journal of Animal Ecology* 63: 887–902.

Székely, T., I. Karsai, and T. D. Williams. 1994. Determination of clutch-size in the Kentish Plover *Charadrius alexandrinus*. *The Ibis* 136: 341–348.

Székely, T., G. H. Thomas, and I. C. Cuthill. 2006. Sexual conflict, ecology, and breeding systems in shorebirds. *BioScience* 56: 801–808.

Thibault, M., and R. McNeil 1995. Day- and nighttime parental investment by incubating Wilson's Plovers in a tropical environment. *Canadian Journal of Zoology* 73: 879–886.

Thompson, D. B. A., and M. L. P. Thompson. 1985. Early warning and mixed species association: The "Plover's Page" revisited. *The Ibis* 127: 559–562.

Thompson, P. S., and W. G. Hale. 1989. Breeding site fidelity and natal philopatry in the Redshank *Tringa totanus*. *The Ibis* 131: 214–224.

Ticehurst, C. B. 1931. The incubating sex amongst waders. *The Ibis* 3: 582–583.

Tinbergen, N., G. J. Broekhuysen, F. Feekes, J. C. W. Houghton, H. Kruuk, and E. Szulc. 1963. Egg shell removal by the Black-headed Gull *Larus ridibundus* L.: A behaviour component of camouflage. *Behaviour* 19: 74–117.

Tjørve, K. M. C., L. G. Underhill, and G. H. Visser. 2008. The energetic implications of precocial development for three shorebird species breeding in a warm environment. *The Ibis* 150: 125–138.

Tuck, L. M. 1972. *The snipes: A study of the genus Capella*. Canadian Wildlife Series Monograph No. 5. Ottawa: Canadian Wildlife Service.

Tulp, I., and H. Schekkerman. 2006. Time allocation between feeding and incubation in uniparental Arctic-breeding shorebirds: Energy reserves provide leeway in a tight schedule. *Journal of Avian Biology* 37: 207–218.

Turner, J. S. 2002. Maintenance of egg temperature. In *Avian incubation: Behaviour, environment and evolution*, ed. D. C. Deeming, 119–142. Oxford: Oxford University Press.

Väisänen, R. A. 1977. Geographic variation in the timing of breeding and egg size in eight European species of waders. *Annales Zoologici Fennici* 14: 1–25.

Väisänen, R. A., O. Hildén, M. Soikkeli, and S. Vuolanto. 1972. Egg dimension variation in five wader species: The role of heredity. *Ornis Fennica* 49: 25–44.

van Tuinen, M., D. Waterhouse, and G. J. Dyke. 2004. Avian molecular systematics on the rebound: A fresh look at modern shorebird phylogenetic relationships. *Journal of Avian Biology* 35: 191–194.

Visser, G. H., and R. E. Ricklefs. 1993. Temperature regulation in neonates of shorebirds. *The Auk* 110: 445–457.

Wallander, J. 2003. Sex roles during incubation in the Common Ringed Plover. *The Condor* 105: 378–381.

Walters, J. R. 1984. The evolution of parental behavior and clutch size in shorebirds. In *Shorebirds: Breeding behavior and populations*, ed. J. Burger and B. L. Olla, 243–287. New York: Plenum Press.

Ward, D. 2000. Do polyandrous shorebirds trade off egg size with egg number? *Journal of Avian Biology* 31: 473–478.

Warnock, N., and L. W. Oring. 1996. Nocturnal nest attendance of Killdeers: More than meets the eye. *The Auk* 113: 502–504.

Warriner, J. S., J. C. Warriner, G. W. Page, and L. E. Stenzel. 1986. Mating system and reproductive success of a small population of polygamous Snowy Plovers. *Wilson Bulletin* 98: 15–37.

Whittingham, M. J., S. M. Percival, and A. F. Brown. 2002. Nest-site selection by Golden-Plover: Why do shorebirds avoid nesting on slopes? *Journal Avian Biology* 33: 184–190.

Winkler, D., and J. R. Walters. 1983. The determination of clutch size in precocial birds. *Current Ornithology* 1: 33–68.

Yogev, A., A. Ar, and T. Yom-Tov. 1996. Determination of clutch size and the breeding biology of the Spur-winged Plover (*Vanellus spinosus*). *The Auk* 113: 68–73.

Yogev, A., and Yom-Tov, Y. 1994. The Spur-winged Plover (*Vanellus spinosus*) is a determinate egg layer. *The Condor* 96: 1109–1110.

———. 1996. Indeterminacy in a determinate layer: The Spur-winged Plover. *The Condor* 98: 858.

PART TWO

Nonbreeding Ecology and Demography

6

Migration

CONTENTS

M IGRATION IS THE PREDICTABLE seasonal movement of individuals between breeding grounds and wintering areas. It represents a critical segment of the annual cycle of shorebirds. Individuals spend considerable time preparing for and completing it; they often re-

quire immense energy stores to reach destinations; and there are substantial risks to undertaking migratory movements. Yet it must be a successful evolutionary strategy because over 60% of shorebirds migrate (Warnock et al. 2001). In fact, the migratory feats of some shorebirds rank among the most prodigious of all animals. Most species that breed in far northern latitudes migrate substantial distances to their wintering areas, which often lie south of the equator. The Bar-tailed Godwit, for instance, undertakes the greatest nonstop flight of any bird on its 11,000 kilometer trans-Pacific route from Alaska to New Zealand (Gill et al. 2009). Other shorebirds, notably temperate and tropical species, are more likely to be partial migrants with a mix of year-round residents and migrants in a single population. Migration is critical from the perspective of the population as well. Distinct populations move varying distances among suitable habitats from wetlands to grasslands. In some cases, individuals stage at single sites in enormous numbers that represent a large percentage of a species' world population. These staging areas offer rich food resources for individuals to recoup the energy reserves necessary for the next leg of their journey. At the same time, individuals must manage their weights to facilitate escape from predators. It is at these concentration points that populations are vulnerable to anthropogenic effects. In this chapter, I review the migratory habits of shorebirds, including the management of time, energy, and danger, use of the various flyways by shorebirds worldwide, and conservation implications derived from these patterns.

ORIGINS AND EVOLUTION

As with most behaviors, the tendency for individuals to migrate is subject to natural selection. As such, it has an evolutionary history that can be deciphered by examining the migratory habits of populations. Evidence to support the conjecture that natural selection has shaped shorebird migration comes from two main sources. First, comparative analyses based on well-established phylogenies provide contrasts in migratory habits of closely related species. Phylogenies allow us to reconstruct the evolutionary histories of groups and ask whether a migratory species, and hence migratory behavior, is a derived character. Second, in some populations, a mix of migratory and resident individuals exists, suggesting that contemporary ecological conditions continue to shape individual behaviors.

PHYLOGENETIC EVIDENCE

Recent advances in understanding the phylogeny of shorebirds have provided the framework for comparative analyses of closely related species that differ in migratory habits. Recall that the two clades, one representing plovers and the other sandpipers, differ in their geographic distributions. The origins of plovers lie in temperate and tropical regions, whereas the sandpipers are distinctly northern in their origins. Of the various shorebird groups, the plovers probably evolved in temperate and tropical latitudes (van Tuinen et al. 2004). Within the plovers, however, there are some species that are migrants and

other, closely related taxa that are nonmigratory. Using a phylogeny based on characters based on mitochondrial DNA analyses, Joseph et al. (1999) approached the question of the location of the "ancestral home" and the evolution of migratory behavior in 15 species of plover in the genus *Charadrius*. Accordingly, the ancestral condition in the plover clade is thought to be permanent residency, whereas migration is the derived character. Joseph et al. (1999) identified two major plover lineages both of which had origins in the southern hemisphere, specifically South America. Within these two lineages, they compared several species' pairs, one resident and the other migratory (Red-capped Plover and Kentish Plover; and Inland Dotterel and Oriental Plover, respectively). These two species' pairs are widely considered to be each other's closest relative based on morphological data (Bock 1958). Joseph et al. (1999) were interested in understanding how migration mapped on to the phylogeny of these plovers, and specifically whether inferences could be drawn regarding which species (in a pair) had evolved migratory behavior. With the Red-capped Plover and Kentish Plover, evidence suggested that the latter had evolved its migratory habits from a common ancestor with the former based in Australasia. Similarly, the Oriental Plover appears to have evolved a migratory habitat, colonizing eastern Asia, from ancestral stock derived from the Inland Dotterel in Australasia. Additionally, the Killdeer and Semipalmated Plover had evolved migratory nature from an ancestral plover centered in South America. However, Joseph et al. (1999) were cautious about their findings, urging that further analyses of phylogenetic relationships were warranted.

PARTIAL MIGRANTS

A second line of evidence that migration continues to evolve in local populations concerns partial migrants. These are species or populations that consist of a mix of individuals with varying migratory tendencies. Some undertake short-distance seasonal movements, whereas others remain as permanent year-round residents in breeding areas. This pattern applies to several coastal populations of plover worldwide, including the Banded Dotterel of New Zealand (Pierce 1989) and the Pacific Coast population of the Snowy Plover (Stenzel et al. 1994; Colwell et al. 2007). For the Snowy Plover, there is considerable variation in the movements of individuals. Some leave breeding areas at the end of the breeding season and migrate both north and south along the Pacific Coast; other plovers in the same population remain year-round in breeding areas. To date, no study has examined the genetics of this behavioral dichotomy by conducting a simple analysis correlating parent–offspring behavior. For instance, if the behavior is genetically determined, then offspring would be expected to adopt the migratory or resident tendencies of their parents. Some evidence, however, suggests that the migratory behavior is somewhat plastic and shaped by the environment experienced by young plovers. For instance, the tendency of newly independent Snowy Plovers to migrate or remain resident appears to be related to the timing of independence from parents (Colwell et al. 2007). Young that fledge early in the breeding season tend to become migrants, whereas chicks that fledge later in the year are more likely to become established as permanent residents very near their natal home. This may be related to the social environment that newly fledged plovers encounter at different times of year. Plovers are highly social, especially during the nonbreeding season. When juveniles fledge and go searching for conspecific flocks to join, they encounter different opportunities, depending on when they fledge. Young that fledge early (May and June) in the breeding season may develop their migratory ways because they wander more widely in the weeks of early summer as they search for a postbreeding flock. By contrast, late-fledging juveniles readily find that social environment in close proximity to where they were reared; hence, they become nonmigrants. It is likely that other temperate-breeding shorebirds such as the Black Oystercatcher, Spotted Sandpiper, Killdeer (Sanzenbacher and Haig 2002), American Avocet (Demers et al. 2008), and Black-necked Stilt (Hickey et al. 2007) are partial migrants, given that some individuals are resident year-round in breeding areas.

MIGRATION STRATEGIES

The varied migration patterns of shorebirds result from natural selection acting on individuals in their choice of pathways and stopover sites, the timing with which they come and go from various locations, and the way they manage fuel in the form of lipids and protein. Collectively, these individual choices manifest themselves in a species' migration strategy. Some species migrate long distances, with males preceding females to breeding grounds each spring. Others undertake shorter movements between staging areas, with females preceding males and juveniles on their southbound flights. There is also considerable intraspecific variation in migratory behavior, and it is natural selection acting on this variation among individuals that produces the strategies observed in nature. Three factors act via natural selection on individuals to shape a species' migratory pattern: time, energy, and danger (or risk of death) (Alerstam and Lindström 1990). From an evolutionary perspective, individuals must "manage" these three factors, with consequences for survival and reproductive success.

TIME

Time can best be understood as a constraint acting on the speed and schedule with which individuals move between nonbreeding and breeding areas. This is apparent in a simple seasonal contrast in the passage of birds to and from breeding areas each year. In the spring, the bulk of shorebird populations typically move north over a short window of time, often a few weeks to 1 or 2 months. During passage, individuals often coalesce in immense flocks at staging sites. For instance, large proportions of the Atlantic populations of the Red Knot, Sanderling, and Ruddy Turnstone stop each spring at Delaware Bay to feed on horseshoe crab eggs. By contrast and as with almost all birds, southbound movements are more protracted, often lasting several months (Morrison 1984), although there are always exceptions to the rule. Wilson's Phalarope stages at hypersaline wetlands of the Great Basin of North America (such as Great Salt Lake, Utah; Mono Lake, California; and Lake Abert, Oregon) for several weeks beginning in late June. Females precede males to staging areas owing to the parental care duties of males. Virtually all staging occurs over 2 months (July through August) (Jehl 1988).

The importance of time constraints is evident in the spring arrival schedule of males and females. These schedules have obvious fitness consequences stemming from access to territories and mates. Early arriving individuals have primacy in access to territories. This is particularly apparent in monogamous species in which males migrate earlier than females and establish territories from which they display to attract females. Males outnumber females in early season flocks of Dunlin arriving on tundra breeding areas in northern Alaska (Holmes 1966). In western Alaska, male Black Turnstones arrive significantly earlier than females to establish breeding territories (Handel and Gill 2000). Similar arguments apply to females of polyandrous species, such as the territorial Spotted Sandpiper and nonterritorial phalaropes (Oring and Lank 1982; Reynolds et al. 1987). Female Wilson's Phalaropes outnumber males in early season flocks, in which intense competition for mates takes place (Colwell and Oring 1988). Females successful in acquiring mates early in the breeding season have higher reproductive success because they are more likely to acquire second or third mates (Oring and Lank 1982; Colwell and Oring 1988).

In northern latitudes with short breeding seasons, early arrival may also affect breeding opportunities, regardless of gender. Individuals that arrive early and fail in a first nesting attempt may be more able to renest. There may,

however, be costs associated with early arrival. Morrison (1975, 1984) indicated that Red Knots arriving in early spring in the high Arctic of Canada may face food shortages owing to cold weather, with consequences for survival. There also may be negative fitness consequences of late arrival on breeding grounds if chicks from late nesting attempts are reared after the period of peak insect emergence (Meltofte et al. 2007).

The more protracted southbound passage of shorebirds stands in contrast to the spring pattern (Morrison 1984). This stems from the interactions between a species' mating system, parental care duties of the sexes, breeding success, migration distance, and the timing of departure for adult males, females, and juveniles (Myers 1981; García-Peña et al. 2009). In the Semipalmated Sandpiper, most adults precede juveniles south, and most early migrating adults are either failed breeders or females that left their mates to care for their young; males depart later than females. Juveniles are abundant in late summer flocks at staging areas in North America; by early autumn, the population is essentially resident on the nonbreeding grounds in northern South America (Morrison 1984). Lank et al. (2003) offered new insights into the migration patterns of some calidridine sandpipers by suggesting that the early departures and routes of adults stemmed from predation danger posed by later migrating Peregrine Falcons (*Falco peregrinus*).

ENERGY

The migrations of any organism require fuel. Individuals must manage their energy to successfully complete successive legs of the journey. Energy is the currency by which individuals power their migrations and negotiate the success of these movements. Birds store two principal fuels for migration: lipids and protein. Lipids are a more efficient fuel source because they provide twice the energy and metabolic water than protein (Klassen 1996; Gill 2007). Additionally, lipids can be synthesized from any type of digested food (Klassen 1996). Shorebirds store lipids subcutaneously and within the body cavity amid the organs. However, evidence from a variety of species (see Klassen 1996; Landys et al. 2000) indicates that birds also catabolize protein reserves, which yield substantial amounts of water that wards off dehydration during the strenuous exercise of long-distance flight. The fuel stores of migrant Bar-tailed Godwits consist of 35% nonfat components, which are probably protein and associated water (Lindström and Piersma 1993; Klassen 1996; Landys et al. 2000). In the Red Knot, premigratory fattening is accompanied by a dramatic increase in mass of the pectoral muscle (Vézina et al. 2007).

Migrants need to acquire and store sufficient fuel in order to successfully complete migration. Acquiring sufficient lipid reserves to undertake movements between successive staging areas is paramount for shorebirds. Adequate fuel is especially critical for long-distance, transoceanic migrants such as the Bar-tailed Godwit, some of which migrate nonstop from Alaska to New Zealand (Piersma and Gill 1998; Gill et al. 2005; Gill et al. 2009). Stored energy also may influence an individual's survival and reproductive success once it arrives at breeding areas. Individuals that arrive with sufficient energy reserves in northern latitudes may initiate clutches earlier, be more likely to replace failed clutches, and lay larger eggs, all of which increase reproductive success. Meltofte et al. (2007) reported that a few studies have found reduced egg volume during cold or late spring conditions and among late or replacement clutches (e.g., the Western Sandpiper: Sandercock et al. 1999). Nevertheless, late clutches of the Red-necked Phalarope and Western Sandpiper had increased egg volume (Schamel 2000; Schamel et al. 2004).

DANGER

Avian ecologists have traditionally emphasized the importance of laying down large fuel reserves to power the endurance flight that takes individuals between successive sites on their migration. There is, however, a cost to laying down large fuel stores, and this is where the element of danger influences an individual's migration strategy. The increased mass associated with stored lipids and protein can have consequences for an individual's ability to fly. Maximum fuel loads are limited by the morphological features of a species. In other words, a bird of a given size has a maximum mass (in lipids and protein) that it can carry and effectively get airborne. Smaller, but still hefty, gains in mass may render an individual less capable of evading predators or taking off at all (Hedenström and Alerstam 1992; Klaassen 1996; Ydenberg et al. 2002). Burns and Ydenberg (2002) showed that the speed of escape flights of Least and Western Sandpipers decreased with mass in association with heavier wing loading (ratio of body mass to wing surface area). Moreover, the relationship between wing loading and danger of predation by falcons was the most plausible explanation for the differences in mass of Western Sandpipers at two autumn staging sites in coastal British Columbia. Specifically, sandpipers at an unobstructed stopover site where they could detect falcons early maintained higher masses over a period of several decades compared with individuals stopping at a second site where the habitat obstructed early detection of falcons (Ydenberg et al. 2002).

PHYSIOLOGY OF MIGRATION

The extraordinary migratory feats of shorebirds are remarkable among vertebrates for several reasons (Jenni-Eiermann et al. 2002). First, the nonstop endurance flights of some species last upward of 200 hours. This is accomplished without intake of food or water, and birds thus must rely on stored energy in the form of various fuels to power flight and provide water to avoid dehydration. Additionally, shorebirds have high metabolic rates, approximately twice that of comparably sized passerines. When coupled with the extraordinarily high metabolic rates of prolonged exercise of these long-distance movements, some migrant shorebirds are truly performing at the upper limits of what is known for vertebrates.

Much like other migratory birds, shorebirds make use of three forms of fuel to power their movements: carbohydrates, lipids, and proteins (Jenni and Jenni-Eiermann 1998, 1999). Carbohydrates are stored as glycogen in the liver and amid skeletal muscle. Glycogen is readily available as an energy source and generally is not used to power endurance flight typical of migration; however, it is immediately available to power initial takeoff and evasive flight in response to predation or disturbance. As a result, glycogen probably constitutes the energy source that shorebirds draw upon at the very start of each migratory movement. In contrast to glycogen, lipids are stored as triglycerides, mainly as adipose tissue either subcutaneously or amid the organs. Lipids are generally thought to be the best fuel for several reasons (Jenni and Jenni-Eiermann 1998, 1999). First, lipids yield over eight times the chemical energy per gram of wet mass than protein or glycogen. Moreover, fat can be easily built from a variety of food types. Protein, however, is also an important fuel, especially among species that undertake long-distance endurance flights (such as the Bar-tailed Godwit and Red Knot). There is no specialized storage form for protein. Rather, it is built up in the skeletal muscles before each migratory movement. The advantages and disadvantages of these three energy sources depend on the migration strategy of a species. Lipids provide abundant energy, but when birds catabolize protein during endurance flight, it releases six times more water. This is critical to avoiding dehydration during the strenuous activity of migration. The amount of water needed to avoid dehydration depends on a variety of factors associated with the cost of flight, including bird mass and the altitude at which a bird flies.

Migrating shorebirds exhibit predictable seasonal changes in their behavior and physiology. These changes facilitate the accumulation of necessary energy reserves to complete their journey. On wintering grounds, before migration, the weights of individuals are often at the lowest of the year (Klaassen 1996). These lean weights are transitory as individuals begin to load up on fuel in the weeks preceding the initial migratory

movements. Lipid stores are deposited both subcutaneously and amid the organs within the peritoneal cavity. The deposition of these fuel reserves is enabled by accelerated intake rates and changes in the efficiency of the digestive system. The anatomical changes to the gut are accompanied by changes in the musculature associated with powered flight.

Shorebird migration has been characterized as a series of alternating episodes of endurance flight punctuated by residence at stopover sites (Stein et al. 2005). To fuel endurance flights of varying distance, individuals feed actively (**HYPERPHAGIA**) and preferentially on high-quality prey. At some staging sites, such as Delaware Bay, shorebirds capitalize on abundant food, albeit small packets of energy, in the eggs of horseshoe crabs. Growing evidence indicates that, in addition to hyperphagia, the digestive system and flight musculature are dynamic in form and function during these alternating episodes of flight activity and staging. In fact, there are predictable changes in an individual's internal anatomy and flight physiology that facilitate the deposition of lipid stores and build lean mass in organs associated with digestion and the musculature that powers flight. Migrants, especially those undertaking long-distance flights, may add more than 50% of their body mass in lipids (Piersma and Gill 1998).

Shorebirds are capable of incredible rates of food processing and energy assimilation. A study of 15 species held in captivity during their southbound migration estimated that shorebirds had the highest rates of energy assimilation recorded for vertebrates (Kvist and Lindström 2003). The average rate of energy assimilation of 12 of 15 species of shorebird exceeded six times the basal metabolic rate, and one species (Red Knot) topped 10 times the basal metabolic rate. As a result, on average, shorebirds studied in captivity increased their mass by approximately 10% per day. Shorebirds truly are the gluttons of the vertebrate world, and this attribute is critical to the time-constrained, long-distance migrations they undertake (Kvist and Lindström 2003).

Additional evidence indicates that accompanying these feats of massive energy assimilation are rather dramatic changes in the anatomy of the digestive system. Changes in the lean mass of migrants are accompanied by increased size of digestive organs during periods of hyperphagia. Similarly, the flight musculature increases in mass before and during migration periods. In fact, changes in lean mass of digestive organs (not associated with lipid storage) appear to be initiated during the period of migration as opposed to late in the wintering period, just before migration. For instance, during migration Western Sandpipers weigh 25% more than during the winter, and 40% of this gain stems from an increase in lean mass components. Most digestive organs (such as the proventriculus, gizzard, and small intestine) and flight muscles were 10% to 100% larger in migrating sandpipers than in conspecifics collected during the winter (Stein et al. 2005).

During migratory flight, catabolism of lean body mass components (digestive organs and flight muscles) occurs as a result of fasting and the need for amino acids and metabolic water, also derived from lipids (Guglielmo and Williams 2003). Metabolic water reduces the risk of dehydration during endurance flight. Bar-tailed Godwits arriving in spring at staging sites in the Dutch Wadden Sea maintained their water balance despite having just completed a 4,300 km nonstop journey from West Africa (Landys et al. 2000). It appears that migrating godwits catabolize muscle protein, which produces substantial amounts of free water.

At stopover sites, shorebirds rapidly increase the size and activity of the digestive system to reconstitute lean mass and store lipids for the next leg of the journey. The activity of enzymes associated with different functional segments of the digestive system also varies seasonally (Stein et al. 2005). During spring migration, Western Sandpipers have high levels of pancreatic lipase to breakdown lipids acquired from invertebrate prey (such as gravid female amphipods, *Corophium*: Stein et al. 2005). Migrants also have elevated levels of enzymes from the small intestine. The fueling strategy of the Western Sandpiper suits its comparatively short distance movements (mostly less than 1,000 km) between successive staging areas. This contrasts with several large sandpipers, notably the Bar-tailed Godwit and Red Knot, which undertake fewer, longer flights. In these species, immense energy stores are laid down in lipid and protein of flight muscle in the weeks before migration. Immediately prior to departure, guts atrophy as a weight conservation measure. Upon arrival at wintering areas, the digestive machinery enlarges and resumes normal function. In summary, the digestive anatomy and physiology of migrant shorebirds is in a state of constant flux associated with their migration strategies, the foods they eat, the time of year, and the need to refuel and to be ready for flight.

HOP, SKIP, AND JUMP

The outcome of individual behaviors to manage time, energy, and danger describes a species' migration pattern; at the level of the population, it is portrayed by the timing and routes of movements between breeding and nonbreeding areas. These patterns are highly variable among and within species. In some populations, individuals undertake tremendous nonstop movements, whereas others move short distances between successive staging areas. Piersma (1987) summarized shorebird migration strategies as falling along a continuum, based primarily on the distance traveled between successive stopover and staging sites (Fig. 6.1). Some shorebirds hop short distances between stopover areas and gain proportionately small fuel stores. Other species skip among sites, opportunistically accumulating moderate fuel loads. Still other species undertake truly prodigious migratory feats and jump long distances after accumulating large lipid reserves and while experiencing hypertrophy of flight muscles and atrophy of the digestive system. These interspecific differences arise owing to the ecological conditions

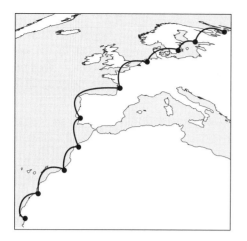

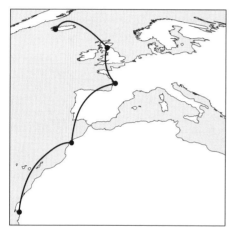

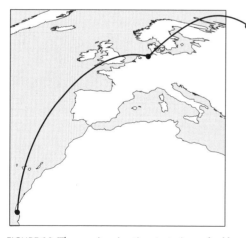

FIGURE 6.1. Three main migration strategies evolved by shorebirds in response to geographic barriers and the locations of staging areas. Strategies are categorized into (top) hop, (middle) skip, and (bottom) jump, although some species may exhibit a mix of these strategies. After Boere and Stroud (2006).

faced by individuals along their flyways, including the presence of geographical barriers and the availability of predictable food supplies at wetlands where birds refuel. The vagaries of weather confronting migrants may further add to the variation observed. A few well-chosen examples illustrate these strategies, and they highlight the variability inherent among individuals.

Migration strategies occasionally vary within species, sometimes by subspecies and other times with season. For instance, the *pacifica* subspecies of the Dunlin breeds in western Alaska and winters in estuaries along the Pacific Coast from Canada south through Mexico. During spring, Dunlins hop up the coast (Warnock et al. 2004), but in the fall they stage and molt on the Alaska peninsula before a single long-distance movement across the Gulf of Alaska to wintering areas along the eastern Pacific flyway (Warnock and Gill 1996).

The various subspecies of the Bar-tailed Godwit are prime examples of shorebirds that exhibit a jump migration strategy. The Alaskan subspecies (*baueri*) stands out among all birds because it undertakes the longest nonstop migration of any species, traveling 11,000 km from staging areas in Alaska to New Zealand wintering areas. Godwits travel this distance in approximately 6 to 10 days (Gill et al. 2005). The details of this migration have been worked out by a cadre of researchers who have (1) marked and resighted more than 10,000 individuals, (2) tracked movements using radio and satellite telemetry, (3) explored the changing anatomy and physiology of migrants, and (4) combined this information with knowledge of the timing of movements in relation to weather conditions.

After breeding in western Alaska, Bar-tailed Godwits move to coastal habitats along the Bering Sea, principally the Yukon-Kuskokwim Delta and estuaries along the Alaskan Peninsula. For 4 to 8 weeks, individuals forage in intertidal habitats, principally on small bivalves (*Macoma*), which are ingested whole and ground up by the gizzard. This rich food allows individual godwits to add up to 55% of their weight in fat, which approaches the maximum weight load that a bird can carry (Piersma and Gill 1998). Just prior to their southbound departure for New Zealand, two additional changes take place in godwit anatomy: their flight muscles increase in mass, and their digestive systems atrophy. The reduced size of the gut and liver is considered an adaptation to conserve weight in preparation for the long distance flight to come (Piersma and Gill 1998; Piersma et al. 1999; Landys-Cianelli et al. 2003; Battley and Piersma 2005). The southbound departure of godwits is timed to coincide with favorable tailwinds associated with moderate low pressure troughs south of the Aleutian Islands. These systems track north into the Gulf of Alaska, creating winds that carry godwits south over the Pacific. Godwits departing under these weather conditions probably encounter favorable (10 to 15 m s^{-1}) quartering or tailwinds for approximately 1,000 km after leaving their Alaskan staging sites. During passage over the leeward chain of the Hawaiian Islands and across the equator, godwits encounter light (2 to 8 m s^{-1}) crosswinds. As godwits near the final 1,000 km of their journey, tailwinds prevail in association with southeasterly tradewinds and austral westerlies (Gill et al. 2005). The northward migration of godwits is almost as impressive, with birds jumping at least 8,000 km on occasion. The migration strategies (and timing of prealternate molt) of individual Bar-tailed Godwits are quite predictable (Battley 2006). From one year to the next, godwits departed wintering quarters in New Zealand within a 1-week window. Among males, but not females, smaller individuals tended to depart earlier than larger birds.

At the other end of the distance continuum are many more species that exhibit short hop or a combination of hop and skip migration strategies. The Western Sandpiper is a good example of a coastal migrant that both hops and skips its way north each spring to breeding areas in western Alaska. This species has received considerable attention of late owing to the concerted efforts of biologists along the Pacific Coast of North America (e.g., Iverson et al.1996; Warnock and Bishop 1998; Guglielmo and Williams 2003;

Lank et al. 2003; O'Hara et al. 2005; Stein et al. 2005; Bishop et al. 2006; Williams et al. 2007). Western Sandpipers are probably a much more typical example of migration strategies of shorebirds because they concentrate their movements among coastal estuaries, although some individuals migrate through the interior. Consequently, Western Sandpipers are confronted with relatively minor geographic barriers to movement between successive stopover sites, especially compared with the Bar-tailed Godwit.

Western Sandpipers winter across a broad latitudinal range from British Columbia to Peru (Nebel et al. 2002; O'Hara et al. 2005). Each spring, individuals stop at multiple sites for several days, where they rest and refuel by adding lipids necessary for the next leg of the journey. Observers relocated radio-marked individuals at 10 estuaries and interior sites along a 4,200 km length of the Pacific Coast. Marked sandpipers varied greatly in their patterns of movement northward, with some birds traveling short distances (200 to 300 km) and others long distances (3,200 km) among sites (Warnock and Bishop 1998). The typical duration of stay at these staging sites was about 4 days, although some birds remained for longer than 10 days at San Francisco Bay, California. Highest levels of plasma triglycerides were observed north of San Francisco Bay, California, and they increase with latitude (Williams et al. 2007). Bishop et al. (2006) suggested that this heterogeneity of individual strategies was adapted to an unpredictable landscape. In particular, the vagaries of wet and dry years in the intermountain west of North America has made wetland habitats and food unpredictable from year to year, with consequences for individual fueling strategies and energy to power the next stage of migration.

The distinction between hop, skip, and jump strategies proposed by Piersma (1987) is blurred. Categories probably grade into one another owing to the effects of regional geography on the availability of high-quality stopover and staging sites. Butler et al. (2001) showed that the global distribution of nonbreeding shorebirds, including migration flyways and their staging sites, was concentrated in coastal areas associated with areas of upwelling, resulting in high productivity in near-shore oceanic waters. Piersma et al. (2005) went farther in their assessment of the relationship between staging areas and the nonbreeding distribution of Red Knots. They indicated that the flyways, staging sites, and wintering areas of the various subspecies were determined by the availability of coastal sites that offer large expanses of intertidal (preferably soft) substrates, where knots feed on hard-shelled prey. This observation is probably true of virtually all migratory shorebirds—the location and distribution of high-quality staging sites in association with geographical barriers (such as oceans, deserts, or extensive coastlines of rocky intertidal habitat) determine a species' migration system.

STOPOVER BEHAVIOR

It is obvious that an integral feature of a species' migration strategy is the manner in which individuals select and use stopover and staging sites. As individuals move along their migratory pathways, individuals stop for varying lengths of time to refuel. This stopover behavior has been well studied in a variety of species in coastal and interior habitats. Biologists commonly use several terms—turnover rate, stopover duration, and also length of stay—in discussing the individual behaviors of migrants at staging sites. Although related, these terms describe different phenomena: the first is related to the population, and the last two address individual behavior. To illustrate the meaning of *stopover duration,* consider the Western Sandpipers. Radio-marked birds staged at estuaries for 1 to 10 days, with most birds remaining less than 4 days at most sites (Iverson et al. 1996; Warnock and Bishop 1998; Bishop et al. 2006). One can describe this behavior by the stopover duration (length of time in days or hours) of individuals. Furthermore, biologists may be interested in the relationship between stopover duration and various measures of individual condition (e.g., mass and fuel stores: Holmgren et al. 1993). These relationships may shed light on the quality of a site.

For example, stopover duration should correlate negatively with the quality of foraging. In other words, individuals should remain for shorter periods at high-quality staging sites where they can accumulate lipids quicker; this assumes that faster passage and earlier arrival is beneficial. Dunn et al. (1988) were among the first to examine this relationship. They studied southbound Semipalmated Sandpipers staging on the Atlantic seaboard during the late summer. Individual sandpipers were faced with a 3,200 km nonstop flight to their wintering areas in South America. The stopover duration of individuals varied from a few days to several weeks. Stopover duration correlated negatively with fat load at this coastal site. Moreover, individuals staging early in the southbound migration typically staged longer than those migrating later in the fall (Dunn et al. 1988). Most studies, however, have found no relationship between fat and stopover duration at coastal or interior staging sites (Page and Middleton 1972; Lank 1983; Lyons and Haig 1995). Differences between coastal and inland sites may relate to weather conditions prompting initiation of migration by individuals confronted with short- versus long-distance flights.

Biologists often use *turnover rate* somewhat interchangeably with stopover duration. But, turnover rate (percentage change per unit time) is probably best understood and applied to the population of migrants using a particular staging area. Specifically, turnover rate expresses the percentage of individuals present one day that remain the next day. For example, daily turnover rate would be 100% if all Western Sandpipers staged for one day at a given estuary. In this case, if a biologist surveying for radio-marked individuals detected 10 on one day, all of those birds would have left the site the next day. This would be a high turnover rate (a high percentage change in population of marked birds per day). By comparison, a low turnover rate would occur when individuals remain (a longer stopover duration) for longer periods at the site.

The importance of turnover rate is probably best understood in the context of estimates of a species' abundance at a particular site (e.g., Bishop et al. 2000). These estimates can be of great value to understanding the relative importance of staging sites along a species' flyway; hence, they may improve conservation actions focused on wetland recognition under international treaty (for example, the Ramsar Convention) or other programs (such as the Western Hemisphere Shorebird Reserve Network). As it stands, most estimates of abundance used to assign relative importance to a site are based on one-time maximum counts, which grossly underestimate use. However, consider the value of knowledge of turnover rates in improving these estimates of species' use of any particular staging site along a flyway. If biologists conducted a series of surveys at 3-day intervals during the month-long peak of migration for the Western Sandpiper, and they combined these estimates with average turnover rate at a site, then they would almost certainly increase their estimates of the number of individuals staging at the site.

Several recent papers have attempted to improve estimates of shorebird use of a particular site or region by incorporating knowledge of length of stay. The Copper River Delta in south-central Alaska has long been recognized as an important staging site for spring shorebirds (Senner and Howe 1984). The estuary receives freshwater input from six major rivers, which create extensive tidal flats of soft sediments covering approximately 2.8 million hectares from Orca Inlet in the northwest to Controller Bay on the southeast. Early estimates of shorebird numbers (Isleib 1979; Senner et al. 1981; Senner and Howe 1984) suggested several million spring migrants used the area, most of which were Western Sandpipers and Dunlins. Bishop et al. (2000) used telemetry data showing that spring migrants spent roughly 2 to 3 days on the Copper River Delta to refine estimates of daily use by Western Sandpipers. First, they conducted aerial surveys daily to estimate the total shorebird abundance and then determined, using ground-based observations, the proportion of the total shorebirds that were Western Sandpipers. Next, they summed these daily counts over

the 21-day period of migration to arrive at a total spring count. Finally, they corrected this total count by the length of stay and the probability that a bird present on the delta would be detected (0.90; that is, 10% of sandpipers stayed for less than 1 day). If all birds remained for just 1 day, then their initial estimate was the total population size. Given that Western Sandpipers staged for slightly longer than this, they must have been counting some sandpipers multiple times. So the correction reduced the estimate by the product of length of stay and detection probability. In the end, Bishop et al. (2000) estimated that 1.2 to 4.1 million Western Sandpipers and 300,000 to 900,000 Dunlin used the Copper River Delta as a spring staging site. Recently, Skagen et al. (2008) incorporated a 7-day residency (stopover duration) based on empirical data from several calidridines to estimate the number of shorebirds using the prairie wetlands of North America. Their estimates of 7.3 million and 3.9 million spring and autumn migrants, respectively, provided a first attempt at quantifying the importance of these regions to Nearctic shorebirds.

In reality, a species' migration strategy represents the amalgam of behavioral decisions made by individuals in a population. The patterns of bird movement arise owing to the interaction between an individual's condition (such as mass and lipid stores) and the availability and quality of wetland habitats along their routes, including geography (Henningsson and Alerstam 2005a). In other words, migrants move along flyways with knowledge of their ever-changing condition and stop to replenish their fuel stores when necessary or the opportunity arises. An individual's decision to remain at a site or move to another wetland is probably strongly influenced by the nature of habitat present on the landscape. For instance, a shorebird may arrive at a staging site based on the presence of a flooded landscape that offers potential foraging habitats, but the decision to stay for any duration may be based on the quality of the wetland as gauged by foraging opportunities (intake rates). Lastly, experienced birds probably initiate movements using prior knowledge of the landscape, such as the location of subsequent staging sites. Young migrating north for the first time may simply follow experienced birds.

In some regions, availability of wetland habitats as staging sites varies greatly from year to year, as does the quality of food resources. In particular, the vagaries of weather in the form of widespread and prolonged drought may render large regions of the continental interior of North America unsuitable for years (Skagen et al. 2008). By contrast, coastal regions offer comparatively predictable habitats and food resources year after year. Consequently, the strategies used by migrating shorebirds may be strongly influenced by the interplay between spatial and temporal variability (or predictability) in the presence and quality of wetland habitats essential for feeding.

The timing of migratory movements along coasts coincides with the seasonal increase in productivity during the northern spring and summer. Along the Atlantic seaboard, spring passage of Ruddy Turnstones, Sanderlings, and Red Knots coincides with the increasing abundance of food provided by horseshoe crab eggs. The southbound migration of shorebirds begins in early summer and continues into the fall. Migrants capitalize on the abundance of invertebrates produced during the long, productive northern spring and summer. Many of the most productive "wintering" sites are situated at southerly latitudes coincident with the austral summer and a pulse of productivity (Hockey et al. 1992).

Migrants passing through the continental interior are more affected by unpredictable weather patterns and a wetland landscape in flux, which create ephemeral wetland conditions and less dependable food supplies. Extended droughts may render all but the most permanent wetlands useless as staging sites in some years. However, during wet intervals the region is awash in seasonal and permanent wetlands that offer abundant food. As a result, the occurrence of shorebirds at staging sites is probably more variable from year to year than at coastal wetlands (Skagen et al. 2008). Table 6.1 describes

differences in wetland conditions and shorebird behaviors across a continuum of habitat conditions that influence the quality and predictability of food. The patterns probably vary in the following way. At coastal sites where food resources are high quality, and food resources are more predictable in timing within the season and certainly among years, shorebirds such as the Red Knot exhibit high site fidelity and visit fewer locations. The same is generally true of a few species such as Wilson's Phalarope that stage at large lakes and depend on predictable nektonic food resources. By contrast, shorebirds such as the Pectoral Sandpiper that migrate through the continental interior use ephemeral wetlands that are strongly affected by the vagaries of weather. As a result, individual patterns of movement vary greatly in stopover duration within and among years and birds exhibit lower site fidelity. In both scenarios, individual length of stay probably depends on the quality of food resources, the interval since the last stopover, and the distance to the next staging or breeding site. Individuals in good condition will remain for shorter intervals at high-quality sites whereas those that have depleted energy reserves will remain for longer.

Evidence to support these differences comes from a comparison of stopover behaviors and variation in shorebird numbers. For example, during spring migration, Semipalmated and White-rumped Sandpipers staged at prairie wetlands for approximately twice the duration as coastal migrating Western Sandpipers (Skagen and Knopf 1994). These estimates, however, may be biased high owing to a purposeful attempt to mark early migrants in the prairies. Interior migrants are probably less site faithful to any particular staging area, more prone to wander across the landscape, and likely to stay for shorter durations. Pectoral Sandpipers were more transient in their use of prairie wetlands compared with what has been observed for coastal migrants such as the Western Sandpiper (Farmer and Parent 1997; Lehnen and Krementz 2005). Length of stay also has been shown to decrease with date in the Pectoral Sandpiper (Farmer and

TABLE 6.1

Comparison of foraging habitats, food resources, behavior of shorebirds staging at sites that vary across a continuum of predictability, and quality of food resources

	ESTUARY OR BEACH (E.G., SAN FRANCISCO BAY, BAY OF FUNDY, WADDEN SEA)	PERMANENT WETLAND OR LAKE (E.G., GREAT SALT LAKE, MONO LAKE, CASPIAN SEA)	SEASONAL WETLAND (E.G., PRAIRIE WETLANDS, PLAYA LAKES)
Foraging Habitat	Expansive tidal flat; constant, predictable	Limnetic zone, shore; constant, predictable	Shallow, wetlands; variable, unpredictable
Invertebrates	Diverse benthic invertebrates	Simple nektonic invertebrates	Simple macro invertebrates
Annual availability	High	High	Variable
Seasonal availability	High	High	Low
Daily availability	High	High	Moderate
Quality of food	High	High	Variable
Behavior of Individuals	Predictable	Predictable	Unpredictable
Annual site fidelity	High	High	Low
Movements	Few, infrequent	Few, infrequent	Many, frequent
Stopover duration[a]	Long	Long	Short
Typical species	Red Knot	Wilson's Phalarope	Pectoral Sandpiper
Population Use of Sites			
Variation in abundance	Consistently high	Consistently high	Highly variable
Turnover rate	Lower	Lower	Higher

[a] Dependent on distance to and from staging sites, need to build energy reserves, and the quality of food resources at the site.

Wiens 1999). High annual variation in shorebird abundance at individual sites in the prairies suggests that interior migrants are more opportunistic than their coastal counterparts (Skagen et al. 2008).

FLIGHT RANGE ESTIMATES

The observation that shorebirds, especially small ones such as the Semipalmated Sandpiper, undertake sometimes extraordinary nonstop transoceanic flights has contributed to a hefty literature addressing the energy requirements (lipid reserves and accumulated mass), flight speeds (including drag), and weather conditions (tail winds and wind drift) required for successful completion of a trip. Zwarts et al. (1990) provide an excellent summary of the components of accurate flight range estimates, based on the work pioneered by McNeil and Cadieux (1972), Summers and Waltner (1978), Davidson (1984), and Castro and Myers (1989). Flight range estimates have applied value because they detail the potential distances traveled by individuals between successive staging areas along a migratory route. As a result, this information is essential in conservation planning to protect these important staging sites. Moreover, quantifying the components of flight range estimates are essential to understanding how anthropogenic factors (for instance, disturbance and decreased intake rates) may compromise the quality of habitat at staging sites. In other words, if human disturbance limits intake rates, then individuals are less likely to lay down sufficient fuel stores to carry them successfully on their next leg of the journey and beyond.

A species' flight range is influenced by five elements (Zwarts et al. 1990): (1) body composition and mass at the time of departure from a staging site, (2) minimum body composition and mass necessary to survive at the next staging site after a migratory movement, (3) flight costs associated with body mass and wing shape, (4) flight speed, and (5) wind conditions encountered by individuals during migration. The amount of fuel burned is the difference between an individual's departure mass and subsequent arrival mass. The fuel required per kilometer of the journey is the product of flight costs and flight speed. And, finally, a species' flight range estimate can only be derived by including the wind conditions encountered during passage.

As previously pointed out, energy reserves exist in two forms in migratory shorebirds. Most energy is stored subcutaneously in the form of lipids, but reserves also exist in the form of protein in flight muscles. Both of these fuels yield metabolic water, although catabolized protein yields considerably more of it (Landys et al. 2000). This water is essential to keep a bird hydrated during strenuous migratory flight (Klaassen 1996). In the days immediately before migration, the flight muscles hypertrophy, adding mass without any noticeable increase in exercise by an individual (Lindström and Piersma 1993). In many long-distance migrants, mass may double prior to migration (Klaassen 1996; Piersma and Gill 1998). Winter mass is typically the lowest of the annual cycle. Just before migration, shorebirds begin to deposit immense lipid reserves, which may vary with an individual's distance from its breeding grounds. Jump migrants typically add more mass than found in species that hop or skip their way north. At the time of northward departure, the percentage increase in mass (above lean winter mass) associated with lipid reserves varies from 50% to greater than 100%. Individuals departing from the extreme south of the wintering range (e.g., South Africa: Zwarts et al. 1990) may lay down reserves that amount to 90% of their lean winter mass. By contrast, shorebirds departing the Banc d'Arguin in northwestern Africa depart with lipid reserves 30% to 60% above their lean winter mass (Zwarts et al. 1990). Klaassen (1996) pointed out that the relative differences in the composition of fuel stores as lipids and protein have consequences for the flight range of individuals. Moreover, these differences vary among species of shorebird that differ in size.

It is assumed that shorebirds using the long jump strategy (Piersma 1987) arrive at staging areas with virtually all of their fuel reserves depleted. Zwarts et al. (1990) suggested that the fat-free mass of newly arrived migrants may be 10% of their fat-free winter mass in the tropical winter. Other species that hop or skip their way along their flyways probably use less of their energy reserves in each movement between staging areas. Shorebirds continue to add mass at each staging site, typically adding up to 1% of their lean mass per day (Zwarts et al. 1990). Some species, however, have truly remarkable abilities to add mass. For example, fueling rates of fall-migrant Semipalmated Sandpipers in the Bay of Fundy are 4% to 10% d^{-1} (Hicklin 1993). Adult Wilson's Phalaropes staging at Mono Lake, California, before their nonstop migration to South America gain up to 2 g of fat per day and double their weights prior to departure (Jehl 1988).

Shorebirds power themselves on their migrations using flapping flight, which entails costs associated with carrying a fuel load of lipids and muscle protein. In some species, economies of weight are affected by the atrophy of digestive tissue (such as the stomach and liver) immediately before departure (Piersma and Gill 1998). Nevertheless, costs associated with flapping flight differ among species in a predictable allometric relationship: costs are greater for large compared with small species. This allometric relationship means that it is more costly for a large shorebird carrying a maximum fuel load to undertake migration compared with a small species (Klaassen 1996).

An individual has a maximum fuel load that it can carry, and this load limits a bird's ability to get off the ground. Observations suggest that some species approach this maximum mass just before departure on long migrations. Jehl (1988), for example, indicated that just before departure some Wilson's Phalaropes are so obese that they have difficulty lifting off the water and can be readily dip-netted from the surface of Mono Lake, California. Klassen (1996) argued that maximum fuel-carrying capacity decreases with mass such that larger species are predicted to have shorter flight ranges than small species. Thus, the maximum flight range of small species is greater than that of larger taxa because flapping flight is energetically more demanding

of a large bird carrying its maximum fuel load. These observations attest to the truly remarkable energy feats accomplished by Bar-tailed Godwits crossing the Pacific each autumn (Gill et al. 2009).

Several models exist to predict flight time in shorebirds based on morphology and fuel loads (Zwarts et al. 1990). Initial models made use solely of the mass arguments (Summers and Waltner 1978), whereas others added wing length to the formula (Castro and Myers 1989). Zwarts et al. (1990) adapted these models to include adjustments for mass composed of 60% fat and 40% protein. The outcomes of these models are projections of 35- to 65-hour flight times for small and large species, respectively (Zwarts et al. 1990). Recent evidence, however, indicates that Bar-tailed Godwits routinely exceed this range of values.

Flight speed also influences the performance of a migrant, and species of different size vary in their flight speeds (Whelam 1994). Small shorebirds fly faster than large ones, probably owing to the higher energy costs associated with flapping flight in larger birds (Klassen 1996). Consequently, potential flight ranges vary with species' mass. Unfortunately, there is limited reliable evidence upon which to judge the flight speeds of shorebirds. Zwarts et al. (1990), summarized data for 47 species that showed considerable variation, from 30 to 100 km h^{-1}. Most species' flight speeds are between 50 and 80 km h^{-1}. Flight range estimates often assume constant flight speeds.

A final environmental factor, wind, may influence an individual's flight range. Successful completion of a flight can be enhanced by tailwinds or arrested by headwinds (Butler et al. 1997). Evidence indicates that some shorebirds time their departures to coincide with favorable tailwinds that increase their flight range capabilities above that which would be possible under calm conditions. Once again, a stellar example comes from the Bar-tailed Godwit (Gill et al. 2005, 2009). Godwits leave staging areas on the Alaska Peninsula coincident with low pressure systems that provide direct or quartering tailwinds of sufficient velocity ($15\,\text{m s}^{-1}$) to power them for 1,000 to 1,500 km south. The importance of tailwinds in migration is not limited to species that undertake long-distance, nonstop migrations. There is evidence that shorebirds that undertake shorter movements between consecutive staging areas also benefit from wind-assisted flight. Each spring the Western Sandpiper is the most abundant migrant along the Pacific Flyway of North America. Butler et al. (1997) estimated the energy (fat) requirements of flight under calm conditions and wind-assisted flight for several scenarios of fat deposition. They compared their results with empirical data based on body mass of birds captured at staging sites along the flyway. They concluded that wind-assisted flight was necessary for successful completion of migration—that is, individuals arrived on breeding grounds with sufficient energy reserves for breeding. This was true even when fat deposition rates were relatively low ($0.4\,\text{g d}^{-1}$).

POPULATIONS AND FLYWAYS

The migrations of shorebirds take them from wintering grounds, where many individuals may spend as much as 8 to 10 months of the year, to breeding areas, which are often in remote, northerly regions. Conservationists have applied the flyway concept to migratory birds for decades. This concept was first used by Frederick Lincoln in the 1930s to manage the waterfowl harvest in North America (Lincoln 1935; Hochbaum 1956), where seasonal hunting regulations continue to be adjusted annually to match the varying size of waterfowl populations with different geographic origins. The utility of the flyway concept to conservation goals of other migratory birds has been addressed (Boere and Stroud 2006; Delany et al. 2009). The strength of the flyway concept lies in the recognition that effective management must be applied to distinct populations that move within geographically limited areas. Within these geographic areas, shorebirds often congregate at a limited number

of wetlands that are increasingly vulnerable to anthropogenic degradation and loss. This pattern has strong implications for the vulnerability of entire populations if habitat is lost (Myers et al. 1987).

FLYWAY CONCEPT

For any migratory species, a flyway encompasses the full range of breeding, stopping/staging, and nonbreeding areas occupied by a population during the annual cycle (Boere and Stroud 2006). For example, Wilson's Phalarope uses a flyway that connects breeding habitats in wetlands of the interior of western North America, summer staging sites at a few hypersaline lakes in the intermountain west, and nonbreeding areas in southern South America. In spring, the geographic extent of the population's movement is shifted eastward to include many prairie wetlands (Colwell and Jehl 1994). Under this definition, a species' flyway may vary greatly in association with population size, phylogeographic differentiation, and extent of breeding and nonbreeding range. For example, the small population of the Piping Plover that breeds along the Atlantic coast has a rather limited flyway connecting it to wintering areas along the Atlantic coast and into the Caribbean. By contrast, the abundant, Arctic-breeding Semipalmated Sandpiper is widely distributed across tundra habitats of the Nearctic, from which it migrates across a broad front spanning the midcontinent and Atlantic seaboard to wintering areas in South America (Morrison 1984).

A species' flyway consists of a series of locations where individuals stop for varying lengths of time to replenish fuel reserves for subsequent migratory movements. These locations are referred to as staging areas or stopover sites. The number of staging areas used by individuals that make up a flyway population of a species varies greatly in association with the physical barriers that confront migrants. Not surprisingly, wetlands are the most important habitats that serve as staging areas for shorebirds owing to the availability of rich invertebrate food re-

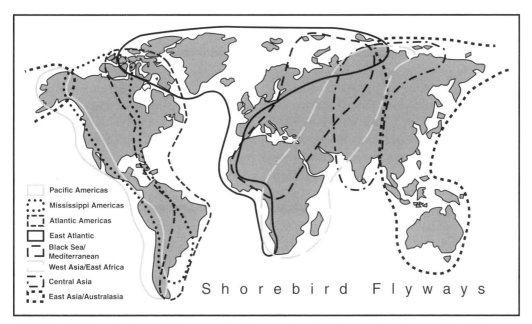

FIGURE 6.2. Eight shorebird flyways link northern breeding grounds with southern nonbreeding areas. A ninth flyway, the Patagonian (not shown), links the southern tip of South America with northern nonbreeding areas on that continent. From Boere and Stroud (2006).

sources. Worldwide, coastal estuaries in particular are overwhelmingly represented among staging sites (Butler et al. 2001). This probably stems from the predictable nature of tidal habitats and their abundant food. By contrast, seasonal wetlands of continental interiors are subject to the vagaries of weather, which render habitat and food less predictable. Consequently, shorebird use of interior wetlands is much more variable from year to year compared with coastal sites (Table 6.1). Exceptions to this rule include several hypersaline lakes in the Great Basin of western North America, where virtually the entire population of Wilson's Phalarope stage prior to migrating to South America. In accordance with the less predictable wetlands of the continental interior, Skagen and Knopf (1993) emphasized the opportunistic nature of shorebird migration in North America. They recommended that a conservation strategy for this region should encompass a large spatial scale with a system of protected areas (wetlands) widely dispersed across the landscape (Skagen et al. 2008).

In its application to conservation of migratory birds, the flyway concept differs slightly from the species-based definition previously provided (Delany et al. 2009). Traditionally, a flyway is a geographic area where multiple species co-occur, although these species may differ in their habitat preferences, migration strategies, and adherence to strict geographic boundaries of flyways (Boere and Stroud 2006). This is the flyway definition familiar to most avian ecologists, especially as it applies to hunted populations of waterfowl. For shorebirds, nine distinct flyways link populations to breeding and wintering areas (Fig. 6.2). In some cases, these geographic flyways have conservation value because they facilitate political cooperation of regional governments to manage species and protect habitats. The U.S. Migratory Bird Treaty Act (1918) is one example of such cooperation among the United States, Canada, Japan, Mexico, and Russia in managing migratory birds. Other international treaties of more recent origin apply to other flyways of the world. For example, the Australian government signed treaties with

Japan (Japan-Australia Migratory Bird Agreement; JAMBA; 1974), China (CAMBA; 1986), and Korea (ROKAMBA; 2002) to facilitate flyway conservation of waterbirds, many of which are shorebirds. Whereas these international agreements draw attention to the conservation status of migratory shorebirds worldwide, they are often limited in their utility for protecting habitats. For example, the government of South Korea aggressively has continued to convert the Saemangeum estuary on the Yellow Sea, mostly for rice cultivation. This estuary is a critical stopover site for migrant shorebirds breeding in the eastern Palearctic and western Alaska (Barter 2002).

FLYWAY DYNAMICS

It is important to recognize that contemporary flyways that are the focus of conservation efforts for shorebird populations are the outcome of rather recent geological and climatological events (Henningsson and Alerstam 2005b). Moreover, future events will continue to alter the flyways we know today, hence shaping our approach to conservation. The climatic upheaval of the Pleistocene, especially the Wisconsinan/Weichselian glacial spanning 115,000 to 10,000 years ago, produced dramatic changes to shorebird habitats, and these changes affected areas along the flyways throughout the annual cycle. Breeding habitats of northern latitudes expanded and contracted in association with ice sheets and associated glacial refugia. During the nonbreeding season, changing sea levels altered the availability of intertidal habitats as foraging areas for shorebirds. Sea levels during the last glacial maximum were estimated to be 130 m lower than today. These changes in climate and habitat must have had profound effects on the availability of food for shorebirds. Consequently, the distributions of shorebirds certainly have changed dramatically across all latitudes in the not-so-distant past.

In order to plan for future conservation in the face of climate change, it is useful to attempt to reconstruct and understand the paleohistory of flyways. This is no small task as it requires detailed phylogeographic data and understanding of effects of changing climate on habitats. Buehler et al. (2006) provide just such an assessment of the effects of varying Pleistocene climate on the phylogenetic divergence and range expansion of the six Red Knot subspecies, all within the context of the flyways used by the subspecies. Unlike most shorebirds, the Red Knot has apparently undergone a relatively recent divergence. The nominate form (*Calidris canutus canutus*), which breeds in the Siberian Arctic, diverged from an ancestral Palearctic population about 20,000 years ago; it winters in western Africa. The five other subspecies are of more recent origins. The three Nearctic subspecies (west to east: *roselaari, rufa, islandica*) diverged about 12,000 years ago, and winter in the Americas or western Europe (*islandica*). The other two Palearctic subspecies (east to west: *piersmai* and *rogersi*) diverged about 6,500 years ago. Today, they move along separate routes within the east Australasian flyway to winter in Australia. Buehler et al. (2006) coupled this phylogeny with changes in two critical habitat features that influence contemporary Red Knot distributions: polar desert breeding habitats (with less than 25% vegetative cover) and expansive intertidal flats of soft sediments that support large bivalve populations in wintering areas. They suggested that Nearctic subspecies stem from recent warming (12,000 to 14,000 years ago) and range expansion from the eastern Palearctic. Furthermore, two of these populations established new flyways in the Americas along ice-free corridors between major ice sheets. The *islandica* subspecies appears to have evolved a flyway connection of rather recent origins with wintering areas in western Europe. However, this interpretation contradicts traditional views of the affinities of Red Knot subspecies based on morphology. Specifically, *islandica* has been viewed as closely related to *canutus* owing to similarities in size, phenology of molt, and timing of movements through the western Palearctic. Additional divergence occurred in the eastern Palearctic forms.

The flyways and wintering areas of Red Knots worldwide are tightly linked to the presence of expansive soft sediment tidal flats (Piersma et al. 2005). These foraging habitats would have varied in their extent during recent times. Lastly, Buehler et al. (2006) postulated a relatively recent expansion of *rufa* to use spring staging areas along the Atlantic Seaboard. Specifically, they suggest that horseshoe crabs began to flourish only recently along the eastern shore of North America. They posited that staging habitats at places like Delaware Bay, where the eggs of spawning crabs provide rich and abundant food for northward migrants, are probably of relatively recent origins.

CONTEMPORARY PATTERNS

Individual shorebird populations are rather restricted in their use of flyways as they wend their ways between breeding and nonbreeding habitats. For instance, the six subspecies of Red Knot traverse distinct routes between their high Arctic breeding areas and wintering areas (Piersma 2007). Collectively, however, species concentrate in nine global flyways (Piersma and Lindström 2004; Boere and Stroud 2006). Excellent regional summaries are provided by Morrison (1984), Pienkowski and Evans (1984), and the numerous chapters in Evans et al. (1984) and Boere et al. (2007). What follows is an overview of these flyways, recognizing that many species, and certainly populations, vary in their distribution within and across flyways. As others have done, I approach flyways from a boreal perspective, because the principal migratory movements of shorebirds are north to south from breeding to wintering areas. Lastly, I use the term "winter" in reference to northern latitudes because transequatorial migrants reside in southern nonbreeding habitats during the austral "summer."

The New World (North, Central, and South America) consists of a distinct group of breeding species with limited movement outside this realm (Morrison 1984; Boland 1988). Most (56) of the 77 species breeding in the New World also winter in this region. An additional 20 species either breed or winter in the New World. Various species (Bristle-thighed Curlew) or subspecies (such as the Common Ringed Plover, Bar-tailed Godwit, Red Knot, Sanderling, and *arcticola* subspecies of Dunlin) breed in the Nearctic tundra and winter elsewhere. There are no species that winter in the New World and breed exclusively in the Palearctic (Boland 1988). However, a segment of one population of Red Knot (*C. c. roselaari*) migrates from the eastern Siberian Arctic (Wrangel Island, Russia) to winter on the Baja California peninsula of Mexico (Piersma 2007).

A leap-frog pattern of distribution exists for many shorebirds in the Nearctic (Boland (1990). Many taxa breeding at northerly latitudes overfly their temperate counterparts to occupy southerly wintering areas. This pattern occurs in 5 of 12 species with subspecies, 4 of 10 genera (such as *Numenius, Limosa, Arenaria,* and *Calidris*), 5 of 9 tribes (same as previous), and 2 of 4 families (Scolopacidae and Charadriidae). The cause of this leap-frog distributional pattern remains unknown, but with it one can effectively predict the distance that a species of Nearctic shorebird migrates based on its breeding latitude and taxonomy. Northerly breeding sandpipers migrate farther than temperate breeding plovers, oystercatchers, stilts, and avocets (Boland 1988).

In the Americas, shorebirds migrate between northern breeding grounds and southerly wintering areas along three main geographic flyways (Morrison 1984). Recently, these regions have been further subdivided for conservation purposes (Brown et al. 2001). The PACIFIC AMERICAS FLYWAY links populations breeding across northern and western regions of the New World, with wintering areas spanning a broad latitudinal range from Alaska south through Tierra del Fuego. A second pathway, the CENTRAL/MISSISSIPPI AMERICAS FLYWAY, links populations breeding across northerly latitudes of Canada with Central and South America. Use of this flyway is limited almost exclusively to the migration and breeding periods, owing to the cold winters in the midcontinent of North America. The exception to this may be the Playa Lakes region of the southern Great Plains. However, in general, few species winter in the interior of North America. The distribution of species in wetlands of the midcontinent differs from those in coastal regions (Skagen and Knopf 1993). Long-distance migrants, typically small, Arctic-breeding sandpipers, migrate through the central plains, whereas short-distance migrants, typically larger, temperate-breeding species, frequent wetlands in the intermountain west. The third New World flyway, the ATLANTIC AMERICAS FLYWAY, is used by northern populations moving to wintering areas across a broad range of latitudes along the Atlantic Coast of the Americas.

Most of the earth's landmass lies in the Northern Hemisphere, creating a pattern of northern breeding areas linked to southern wintering grounds. In South America, however, a small number of shorebird species reverse this movement and move north along the PATAGONIAN FLYWAY to winter in midlatitudes. Examples include the Two-banded Plover and Magellanic Plover.

In the Palearctic, shorebirds migrate along five flyways (Boere and Stroud 2007). Details on the patterns of species movements are best known for the western flyway, the EAST ATLANTIC FLYWAY, where focused research has been conducted for decades in a variety of countries. By contrast, details of shorebird movements along midcontinent and eastern flyways are less well known, owing to nonexistent or recent development of research programs focused on monitoring shorebird populations (Straw et al. 2006).

Shorebirds moving along the EAST ATLANTIC FLYWAY originate from Arctic breeding habitats spanning a broad longitudinal range from northeast Canada to central Siberia. Shorebirds using this flyway concentrate at coastal sites where expansive intertidal flats afford rich foraging opportunities. In some cases, populations or subspecies migrate through this region on schedules that differ slightly. For example, the *islandica* and *canutus* subspecies of the Red Knot co-occur to some extent during the southbound and northbound migrations when they stage in the Dutch Wadden Sea. However, the subspecies take quite different routes to their breeding areas in northeast Canada and western Siberia, respectively (Piersma 2007). Shorebirds using the East Atlantic Flyway winter across a broad latitudinal range, from the higher latitudes of Norway, where Purple Sandpipers winter at 65°, to the southerly extent of the African continent. Important staging areas include the Wadden Sea (of Denmark, Germany, and the Netherlands), Banc d'Arguin, Mauritania, The Wash and Severn estuaries of England, and various estuaries of South Africa (see chapters in Evans et al. 1984).

Three flyways link the central Palearctic region with wintering areas in Africa and southwestern Asia. However, surprisingly little has been published on shorebird populations using these three flyways. The most western of these three flyways is the BLACK SEA/MEDITERRANEAN FLYWAY, which links breeding areas in western and northern Siberia with North African wintering grounds. Shorebirds moving along this flyway stage at a small number of sites in the Black and Mediterranean Seas. Two additional flyways, the WEST ASIA/EAST AFRICA FLYWAY and the CENTRAL ASIA FLYWAY extend from the Siberian Arctic to eastern Africa and southwestern Asia, respectively. These regions are characterized by arid habitats, and little is published regarding shorebird use of this flyway (Mundkur 2006).

The EAST ASIA/AUSTRALASIAN FLYWAY extends broadly along the western Pacific region, linking Arctic breeding areas in the Russian Far East and western Alaska with wintering grounds extending from southeast Asia to Australia and New Zealand. Approximately 60 species use this flyway.

CONSERVATION IMPLICATIONS

In North America, Myers et al. (1987) first drew attention to the critical role that migratory staging areas held for shorebirds. They emphasized that shorebirds often concentrated at a few critical wetlands during their annual migrations. Loss or degradation of any one site could have serious consequences for populations. In some cases, they were prophetic in their statements. The Red Knot has been listed as endangered owing to a precipitous decline in its Nearctic population stemming from the loss of food resources at a Delaware Bay and perhaps other sites along the Atlantic Americas Flyway (Niles et al. 2008). Similar projections portend dire consequences for populations of many shorebirds migrating through the Australasian Flyway stemming from extensive conversion of coastal estuaries.

The importance of flyways and staging areas in the annual cycle of shorebirds has been a cornerstone of habitat conservation efforts for decades (e.g., Delany et al. 2009). The Ramsar Convention (1971) was established as an international effort to recognize and protect wetlands worldwide owing to their biological significance. One criterion used to establish the significance of a wetland is based on an assessment of waterbird use. Specifically, sites can be nominated based on the total number (20,000) or percentage (1%) of a flyway population that uses a site. Some of the earliest designated wetlands were estuaries in Europe, where thousands of shorebirds wintered or staged during migration. In England, The Wash was designated as a Ramsar site in 1988. The nomination was based, in part, on the significance of this estuary to seven shorebirds (Common Redshank, Eurasian Curlew, Eurasian Oystercatcher, Red Knot, Sanderling, Dunlin, and Black-bellied Plover) during migration periods. Additional data indicate that The Wash is an important wintering site for several shorebirds. It is likely, however, that the importance of estuaries to species' populations (that is, flyway criteria used to recognize the importance of a site) will vary over time with both changing population size and climate change. Moser (1988) illustrated the effects on flyway criteria of long-term changes in population size

and shifting distributions of wintering Black-bellied Plovers at 45 estuaries around the United Kingdom. During the mid-twentieth century, most plovers wintered at a limited number of estuaries, and population size was small. But as the wintering population increased, some of these estuaries reached carrying capacity, as evidenced by stable numbers. The carrying capacity of any estuary was mediated by the territorial behavior of nonbreeding plovers. Over time, "new" plovers spilled over into estuaries that had previously had fewer birds. The result was that estuaries that were earlier designated as "important" owing to the large percentage of the wintering population subsequently decreased in importance (as gauged by percentage of flyway population) as the population increased.

The effects of changing climate on the redistribution of wintering shorebirds may further exacerbate attempts to designate estuaries as important using the flyway criterion. Austin and Rehfisch (2005) showed that the winter distributions of shorebirds in the United Kingdom have shifted from the east to southwest during cold winters over the 30-year interval from 1969 to 1998. With warming climate, winter populations have increased in estuaries along the eastern coast and decreased across sites in the southwest of England. This redistribution of wintering populations occurred in eight of nine species. The only exception was the Eurasian Curlew, the species most likely not to suffer effects of climate owing to its large body size. There are additional conservation implications derived from this redistribution of wintering shorebirds. Austin and Rehfisch (2005) pointed out that the abundance of most species has declined substantially at many estuaries in southwestern England. Dunlin numbers, for example, have decreased to approximately 14,000 on the Severn Estuary, which is below the international population threshold for designating a site as a Special Protection Area in the United Kingdom (Stroud et al. 2001).

EFFECTS OF GLOBAL WARMING ON SHOREBIRDS

The migrations or shorebirds are wonderfully timed to coincide with the availability of food resources at staging sites and in breeding areas. In Delaware Bay, for instance, Red Knots, Ruddy Turnstones, and Sanderlings arrive in May coincident with the energy-rich spawn of horseshoe crabs. Myers and Lester (1992) were the first to address the possible impact of global warming on shorebirds. At northerly latitudes, the principal impacts include (1) altering the timing of emergence of insect prey such that young hatch and food is less available, and (2) reduction in total breeding habitat owing to both rising sea level inundating coastal habitats and a reduction in tundra due to the northward shift in vegetation zones. At southerly latitudes, and especially during the nonbreeding season, additional impacts may stem from (1) rising sea level, resulting in a reduction of intertidal foraging habitat, especially where seawalls preclude upslope shifts in estuarine zones; and (2) drought-induced reductions in the availability of freshwater wetlands in continental areas.

Galbraith et al. (2002) estimated potential habitat losses owing to varying scenarios of sea-level rise. They concluded that under worst-case scenarios some major coastal estuaries (such as San Francisco Bay) may sustain reductions of greater than 80% of intertidal habitats by 2100. Where seawalls exist to contain tidal effects on surrounding landscapes (such as in the Netherlands), there is bound to be substantial loss of important intertidal habitats used by shorebirds and other wildlife. The impact of these climate-induced changes on populations will be strong.

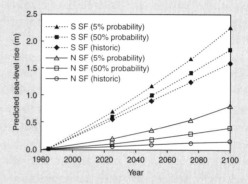

Predicted sea-level changes at San Francisco Bay, California, under conditions of no change (historic), 5% and 50% probability. By 2100, these scenarios result in approximately 80% reductions in the area of intertidal habitats. From Galbraith et al. (2002).

LITERATURE CITED

Alerstam, T., and Å. Lindström. 1990. Optimal bird migration: The relative importance of time, energy, and safety. In *Bird migration: Physiology and ecophysiology,* ed. E. Gwinner, 331–351. Berlin: Springer-Verlag.

Austin, G. E., and M. M. Rehfisch. 2005. Shifting nonbreeding distributions of migratory fauna in relation to climate change. *Global Change Biology* 11: 31–38.

Barter, M. 2002. *Shorebirds of the Yellow Sea: Importance, threats and conservation status.* Wetlands International Global Series 9. International Wader Studies 12. Canberra, Australia: A.C.T.

Battley, P. F. 2006. Consistent annual schedules in a migratory shorebird. *Biology Letters* 22: 517–520.

Battley, P. F., and T. Piersma. 2005. Body composition and flight ranges of Bar-tailed Godwits (*Limosa lapponica baueri*) from New Zealand. *The Auk* 122: 922–937.

Bishop, M. A., P. M. Meyers, and P. F. McNeley. 2000. A method to estimate migrant shorebird numbers on the Copper River Delta. *Journal of Field Ornithology* 71: 627–637.

Bishop, M. A., N. Warnock, and J. Y. Takekawa. 2006. Spring migration patterns in Western Sandpipers *Calidris mauri.* In *Waterbirds around the world,* ed. G. C. Boere, C. A. Galbraith, and D. A. Stroud, 545–550. Edinburgh: The Stationary Office.

Bock, W. J. 1958. A generic review of the plovers (Charadriinae: Aves). *Bulletin of the Museum of Comparative Zoology* 118: 27–97.

Boere, G. C., C. A. Galbraith, and D. A. Stroud. 2007. *Waterbirds around the world.* Edinburgh: TSO Scotland.

Boere, G. C., and D. A. Stroud. 2006. The flyway concept: What it is and what it isn't. In *Waterbirds around the world*, ed. G. C. Boere, C. A. Galbraith, and D. A. Stroud, 40–47. Edinburgh: The Stationary Office.

Boland, J. M. 1988. The ecology of North America shorebirds: Latitudinal distributions, community structure, foraging behaviors, and interspecific competition. Ph.D. dissertation, University of California, Los Angeles.

———. 1990. Leapfrog migration in North American shorebirds: Intra- and interspecific examples. *The Condor* 92: 284–290.

Brown, S., C. Hickey, B. Harrington, and R. Gill. 2001. *United States Shorebird Conservation Plan*. 2nd ed. Manomet, MA: Manomet Center for Conservation Sciences.

Buehler, D. M., A. J. Baker, and T. Piersma. 2006. Reconstructing palaeoflyways of the late Pleistocene and early Holocene Red Knot *Calidris canutus*. *Ardea* 94: 485–498.

Burns, J. G., and R. C. Ydenberg. 2002. The effects of wing-loading and gender on escape flights of Least (*Calidris minutilla*) and Western (*Calidris mauri*) sandpipers. *Behavioral Ecology Sociobiology* 52: 128–136.

Butler, R. W., N. C. Davidson, and R. I. G. Morrison. 2001. Global-scale shorebird distribution in relation to productivity of near-shore ocean waters. *Waterbirds* 24: 224–232.

Butler, R. W., T. D. Williams, N. Warnock, and M. A. Bishop. 1997. Wind assistance: A requirement for migration of shorebirds? *The Auk* 114: 456–466.

Castro, G., and J. P. Myers. 1989. Flight range estimates for shorebirds. *The Auk* 106: 474–476.

Colwell, M. A., and J. R. Jehl, Jr. 1994. Wilson's Phalarope (*Phalaropus tricolor*). In *The birds of North America*, ed. A. Poole and F. Gill, no. 83. Philadelphia/Washington, DC: The Academy of Natural Sciences/The American Ornithologists' Union.

Colwell, M. A., S. E. McAllister, C. B. Millett, A. N. Transou, S. M. Mullin, Z. J. Nelson, C. A. Wilson, and R. R. LeValley. 2007. Philopatry and natal dispersal of the Western Snowy Plover. *Wilson Journal of Ornithology* 119: 378–385.

Colwell, M. A., and L. W. Oring. 1988. Sex ratios and intrasexual competition for mates in a sex-role reversed shorebird, Wilson's phalarope (*Phalaropus tricolor*). *Behavioral Ecology and Sociobiology* 22: 165–173.

Davidson, N. C. 1984. How valid are flight range estimates for waders? *Ringing and Migration* 5: 49–64.

Delany, S., D. Scott, T. Dodman, and D. Stroud, eds. 2009. *An atlas of wader populations in Africa and western Eurasia*. Wageningen, the Netherlands: Wetlands International.

Demers, S., M. A. Colwell, J. Y. Takekawa, and J. T. Ackerman. 2008. Breeding stage influences space use of female American Avocets in San Francisco Bay, California. *Waterbirds* 31: 365–371.

Dunn, P. O., T. A. May, M. A. McCollough, and M. A. Howe. 1988. Length of stay and fat content of migrant Semipalmated Sandpipers in eastern Maine. *The Condor* 90: 824–835.

Evans, P. R., J. D. Goss-Custard, and W. G. Hale, eds. 1984. *Coastal waders and wildfowl in winter*. Cambridge: Cambridge University Press.

Farmer, A. H., and A. H. Parent. 1997. Effects of the landscape on shorebird movements at spring migration stopovers. *The Condor* 99: 698–707.

Farmer, A. H., and J. A. Wiens. 1999. Models and reality: Time-energy trade-offs in Pectoral Sandpiper (*Calidris melanotos*) migration. *Ecology* 80: 2566–2580.

Galbraith, H., R. Jones, R. Parks, J. Clough, S. Herrod-Julius, B. Harrington, and G. Page. 2002. Global climate change and sea level rise: Potential losses of intertidal habitat for shorebirds. *Waterbirds* 25: 173–183.

García-Peña, G. E., G. H. Thomas, J. D. Reynolds, and T. Székely. 2009. Breeding systems, climate, and the evolution of migration in shorebirds. *Behavioral Ecology* 20: 1026–1033.

Gill. F. B. 2007. *Ornithology*. 3rd ed. New York: W. H. Freeman.

Gill, R. E., Jr., T. Piersma, G. Hufford, R. Servranckx, and A. Riegen. 2005. Crossing the ultimate ecological barrier: Evidence for a 11000-km-long nonstop flight from Alaska to New Zealand and eastern Australia by Bar-tailed Godwits. *The Condor* 107: 1–20.

Gill, R. E., Jr., T. Tibbitts. D. Douglas, C. Handel, D. Mulcahy, J. Gottschalck, N. Warnock, B. McCaffery, P. Battley, and T. Piersma. 2009. Extreme endurance flights by landbirds crossing the Pacific Ocean: Ecological corridor rather than barrier? *Proceedings of the Royal Society of London, Series B* 276: 447–457

Guglielmo, C. G., and T. D. Williams. 2003. Phenotypic flexibility of body composition in relation to migratory state, age, and sex in the Western Sandpiper (*Calidris mauri*). *Physiological and Biochemical Zoology* 76: 84–98.

Handel, C. M., and R. E. Gill, Jr. 2000. Mate fidelity and breeding site tenacity in a monogamous sandpiper, Black Turnstone. *Animal Behaviour* 60: 471–481.

Hedenström, A., and T. Alerstam. 1992. Climbing performance of migrating birds as a basis for estimating limits for fuel-carrying capacity and muscle work. *Journal of Experimental Biology* 164: 19–38.

Henningsson, S. S., and T. Alerstam. 2005a. Barriers and distances as determinants for the evolution of bird migration links: The Arctic shorebird system. *Proceedings of the Royal Society of London, Series B* 272: 2251–2258.

———. 2005b. Patterns and determinants of shorebird species richness in the circumpolar Arctic. *Journal of Biogeography* 32: 383–396.

Hickey, C., N. Warnock, J. Y. Takekawa, and N. Ahearn. 2007. Space use by Black-necked Stilts in the San Francisco Bay Estuary. *Ardea* 95: 275–288.

Hicklin, P. W. 1993. The feeding ecology of Semipalmated Sandpipers in the Bay of Fundy. *Wader Study Group Bulletin* 39: 48.

Hochbaum, H. A. 1956. *Travels and traditions of waterfowl*. Minneapolis: University of Minnesota Press.

Hockey, P. A. R., R. A. Navarro, B. Kalejta, and C. R. Velasquez. 1992. The riddle of the sands: Why are shorebird densities so high in southern estuaries? *American Naturalist* 140: 961–979.

Holmes, R. T. 1966. Breeding ecology and annual cycle adaptations of the Red-backed Sandpiper (*Calidris alpina*) in northern Alaska. *The Condor* 68: 3–46.

Holmgren, N., H. Ellegren, and J. Pettersson. 1993. Stopover length, body mass and fuel deposition rate in autumn migrating adult Dunlins *Calidris alpina*: Evaluating the effects of moulting status and age. *Ardea* 81: 9–20.

Isleib, M. E. 1979. Migratory shorebird populations on the Copper River Delta and eastern Prince William Sound, Alaska. *Studies in Avian Biology* 2: 125–129.

Iverson, G. C., S. E. Warnock, R. W. Butler, M. Bishop, and N. Warnock. 1996. Spring migration of Western Sandpipers along the Pacific Coast of North America: A telemetry study. *The Condor* 98: 10–21.

Jehl, J. R., Jr. 1988. Biology of the Eared Grebe and Wilson's Phalarope in the nonbreeding season: A study of adaptations to saline lakes. *Studies in Avian Biology* 12: 1–74.

Jenni, L., and S. Jenni-Eiermann. 1998. Fuel supply and metabolic constraints in migrating birds. *Journal of Avian Biology* 29: 521–528.

———. 1999. Fat and protein utilisation during migratory flight. In *Proceedings of the 22nd International Ornithological Congress, Durban, South Africa*, ed. N. J. Adams and R. H. Slotow, 1437–1449. Johannesburg: BirdLife South Africa.

Jenni-Eiermann, S., L. Jenni, A. Kvist, A. Lindström, T. Piersma, and G. H. Visser. 2002. Fuel use and metabolic response to endurance exercise: A wind tunnel study of a long-distance migrant shorebird. *Journal of Experimental Biology* 205: 2453–2460.

Joseph, L., E. P. Lessa, and L. Christidis. 1999. Phylogeny and biogeography in the evolution of migration: Shorebirds of the *Charadrius* complex. *Journal of Biogeography* 26: 329–342.

Klaassen, M. 1996. Metabolic constraints on long-distance migration in birds. *Journal of Experimental Biology* 199: 57–64.

Kvist, A., and Å. Lindström. 2003. Gluttony in migratory waders—unprecedented energy assimilation rates in vertebrates. *Oikos* 103: 397–402.

Landys, M. M., T. Piersma, G. H. Visser, J. Jukema, and A. Wijker. 2000. Water balance during real and simulated long-distance migratory flight in the Bar-tailed Godwit. *The Condor* 102: 645–652.

Landys-Ciannelli, M. M., T. Piersma, and J. Jukema. 2003. Strategic size changes in internal organs and muscle tissue in the Bar-tailed Godwits during fat storage on a spring stopover site. *Functional Ecology* 17: 151–159.

Lank, D. B. 1983. Migratory behavior of Semipalmated Sandpipers at inland and coastal staging areas. Ph.D. dissertation, Cornell University.

Lank, D. B., R. W. Butler, J. Ireland, and R. C. Ydenberg. 2003. Effects of predation danger on migration strategies of sandpipers. *Oikos* 103: 303–319.

Lehnen, S. E., and D. G. Krementz. 2005. Turnover rates of fall-migrating Pectoral Sandpipers in the lower Mississippi alluvial valley. *Journal of Wildlife Management* 69: 671–680.

Lincoln, F. C. 1935. *The waterfowl flyways of North America*. Circular 342. Washington, DC: U.S. Department of Agriculture.

Lindström, Å., and T. Piersma. 1993. Mass changes in migrating birds: The evidence for fat and protein storage re-examined. *The Ibis* 135: 70–78.

Lyons, J. E., and S. M. Haig. 1995. Fat content and stopover ecology of spring migrant Semipalmated Sandpipers in South Carolina. *The Condor* 97: 427–437.

McNeil, R., and F. Cadieux. 1972. Numerical formulae to estimate flight range of some North American shorebirds from fresh weight and wing length. *Bird-Banding* 43: 107–113.

Meltofte, H., T. Piersma, H. Boyd, B. McCaffery, B. Ganter, V. V. Golovnyuk, K. Graham, et al. 2007. Effects of climate variation on the breeding ecology of Arctic shorebirds. *Meddelelser om Grønland (Bioscience)* 59: 1–48.

Morrison, R. I. G. 1975. Migration and morphometrics of European Knot and Turnstone on

Ellesmere Island, Canada. *Bird-Banding* 46: 290–301.

———. 1984. Migration systems of some new world shorebirds. In *Shorebirds: Migration and foraging behavior,* ed. J. Burger and B. L. Olla, 125–202. New York: Plenum Press.

Moser, M. E. 1988. Limits to the numbers of grey plovers *Pluvialis squatarola* wintering on British estuaries: An analysis of long-term population trends. *Journal Applied Ecology* 25: 473–485.

Mundkur, T. 2006. Flyway conservation in the Central Asian flyway: Workshop introduction. In *Waterbirds around the world,* ed. G. C. Boere, C. A. Galbraith, and D. A. Stroud, 263. Edinburgh: The Stationary Office.

Myers, J. P. 1981. Cross-seasonal interactions in the evolution of sandpiper social systems. *Behavioral Ecology and Sociobiology* 8: 195–202.

Myers, J. P., and R. T. Lester. 1992. Double jeopardy for migrating animals: Multiple hits and resource asynchrony. In *Global warming and biodiversity,* ed. R. L. Peters and T. E. Lovejoy, 193–200. New Haven, CT: Yale University Press.

Myers, J. P., R. I. G. Morrison, P. Z. Anatas, B. A. Harrington, T. E. Lovejoy, M. Sallaberry, S. E. Senner, and A. Tarak. 1987. Conservation strategy for migratory species. *American Scientist* 75: 19–26.

Nebel, S., D. B. Lank, P. D. O'Hara, G. Fernandez, B. Haas, F. Delgado, F. A. Estela, et al. 2002. Western Sandpipers during the nonbreeding season: Spatial segregation on a hemispheric scale. *The Auk* 119: 922–928.

Niles, L. J., H. P. Sitters, A. D. Dey, P. W. Atkinson, A. J. Baker, K. A. Bennett, R. Carmona, et al. 2008. *Status of the Red Knot* (Calidris canutus rufa) *in the Western Hemisphere,* ed. C. D. Marti. Studies in Avian Biology No. 36. Camarillo, CA: Cooper Ornithological Society.

O'Hara, P. D., G. Fernández, F. Becerril, H. de la Cueva, and D. B. Lank. 2005. Life history varies with migratory distance in Western Sandpipers *Calidris mauri. Journal of Avian Biology* 36: 191–202.

Oring, L. W., and D. B. Lank. 1982. Sexual selection, arrival times, philopatry and site fidelity in the polyandrous Spotted Sandpiper. *Behavioral Ecology and Sociobiology* 10: 185–191.

Page, G., and A. L. A. Middleton. 1972. Fat deposition during autumn migration in the Semipalmated Sandpiper. *Bird-Banding* 43: 85–96.

Pienkowski, M. W., and P. R. Evans. 1984. Migratory behavior of shorebirds in the western Palearctic. In *Shorebirds: Migration and foraging behavior,* ed. J. Burger and B. L. Olla, 73–123. New York: Plenum Press.

Pierce, R. J. 1989. Breeding and social patterns of Banded Dotterels (*Charadrius bicincuts*) at Cass River. *Notornis* 36: 13–23.

Piersma, T. 1987. Hop, skip or jump? Constraints on migration of arctic waders by feeding, fattening, and flight speed. *Limosa* 60: 185–194.

———. 2007. Using the power of comparison to explain habitat use and migration strategies of shorebirds worldwide. *Journal of Ornithology* 148 (Suppl.): S45–S59.

Piersma, T., and R. E. Gill, Jr. 1998. Guts don't fly: Small digestive organs in obese Bar-tailed Godwits. *The Auk* 115: 196–203.

Piersma, T., G. A. Gudmundsson, and K. Lilliendahl. 1999. Rapid changes in the size of different functional organ and muscle groups during refueling in a long-distance migrating shorebird. *Physiological and Biochemical Zoology* 72: 405–415.

Piersma, T., and Å. Lindström. 2004. Migrating shorebirds as integrative sentinels of global environmental change. *The Ibis* 146 (Suppl. 1): S61–S69.

Piersma, T., D. I. Rogers, P. M. Gonzalez, L. Swarts, L. J. Niles, I. de Lima, S. do Nascimento, C. D. T. Minton, and A. J. Baker. 2005. Fuel storage rates before northward flights in Red Knots worldwide. In *Birds of two worlds,* ed. R. Greenberg and P. P. Marra, 262–273. Baltimore: Johns Hopkins University Press.

Reynolds, J. D., M. A. Colwell, and F. Cooke. 1987. Sexual selection and spring arrival times of Red-necked and Wilson's phalaropes. *Behavioral Ecology and Sociobiology* 18: 303–310.

Sandercock, B. K., D. B. Lank, and F. Cooke. 1999. Seasonal declines in the fecundity of Arctic-breeding sandpipers: Different tactics in two species with an invariant clutch size. *Journal of Avian Biology* 30: 460–468.

Sanzenbacher, P. M., and S. M. Haig. 2002. Regional fidelity and movement patterns of wintering Killdeer in an agricultural landscape. *Waterbirds* 25: 16–25.

Schamel, D. 2000. Female and male reproductive strategies in the Red-necked Phalarope, a polyandrous shorebird. Ph.D. thesis, Simon Fraser University.

Schamel, D., D. M. Tracy, and D. B. Lank. 2004. Male mate choice, male availability and egg production as limitations on polyandry in the red-necked phalarope. *Animal Behavior* 67: 847–853.

Senner, S. E., and M. A. Howe. 1984. Conservation of Nearctic shorebirds. In *Shorebirds: Breeding behavior and populations,* ed. J. Burger and B. L. Olla, 379–421. New York: Plenum Press.

Senner, S. E., G. C. West, and D. W. Norton. 1981. The spring migration of Western Sandpipers and Dunlins in southcentral Alaska: Numbers, timing, and sex ratios. *Journal of Field Ornithology* 52: 271–284.

Skagen, S. K., D. A. Granfors, and C. P. Melcher. 2008. On determining the significance of ephemeral continental wetlands to North American shorebirds. *The Auk* 125: 20–29.

Skagen, S. K., and F. L. Knopf. 1993. Toward conservation of mid-continental shorebird migrations. *Conservation Biology* 7: 553–541.

———. 1994. Residency patterns of migrating sandpipers at a mid-continental stopover. *The Condor* 96: 949–958.

Stein, R. W., A. R. Place, T. Lacourse, C. G. Guglielmo, and T. D. Williams. 2005. Digestive organ sizes and enzyme activities of refueling Western Sandpipers (*Calidris mauri*): Contrasting effects of season and age. *Physiological and Biochemical Zoology* 78: 434–446.

Stenzel, L. E., J. C. Warriner, J. S. Warriner, K. S. Wilson, F. C. Bidstrup, and G. W. Page. 1994. Long-distance breeding dispersal of Snowy Plovers in western North America. *Journal of Animal Ecology* 63: 887–902.

Straw, P. J., K. B. Gosbell, and C. D. T. Minton. 2006. Shorebird research in the East Asian-Australasian Flyway: Looking to the future. In *Waterbirds around the world*, ed. G. C. Boere, C. A. Galbraith, and D. A. Stroud, 328–331. Edinburgh: The Stationary Office.

Stroud, D. A., D. Chambers, S. Cook, N. Buxton, B. Fraser, P. Clement, P. Lewis, I. McLean, H. Baker, and S. Whitehead, eds. 2001. *The UK SPA network: Its scope and content.* Peterborough, United Kingdom: JNCC.

Summers, R. W., and M. Waltner. 1978. Seasonal variations in the mass of waders in southern Africa, with special reference to migration. *The Ostrich* 50: 21–37.

van Tuinen, M., D. Waterhouse, and G. J. Dyke. 2004. Avian molecular systematics on the rebound: A fresh look at modern shorebird phylogenetic relationships. *Journal of Avian Biology* 35: 191–194.

Vézina, F., K. M. Jalvingh, A. Dekinga, and T. Piersma. 2007. Thermogenic side effects to migratory predisposition in shorebirds. *American Journal of Physiology—Regulatory, Integrative, and Comparative Physiology* 292: 1287–1297.

Warnock, N., and M. A. Bishop. 1998. Spring stopover ecology of migrant Western Sandpipers. *The Condor* 100: 456–467.

Warnock, N., C. Elphick, and M. Rubega. 2001. Shorebirds in the marine environment. In *Biology of marine birds,* ed. J. Burger and B. A. Schreiber, 581–615. Boca Raton, FL: CRC Press.

Warnock, N., and R. E. Gill, Jr. 1996. Dunlin (*Calidris alpina*). In *The Birds of North America*, ed. A. Poole and F. Gill, no. 203. Philadelphia/Washington, DC: Academy of Natural Sciences/American Ornithologists' Union.

Warnock, N., J. Y. Takekawa, and M. A. Bishop. 2004. Migration and stopover strategies of individual Dunlin along the Pacific Coast of North America. *Canadian Journal of Zoology* 82: 1687–1697.

Whelam, C. V. J. 1994. Flight speeds of migrating birds: A test of maximum range speed predictions from three aerodynamic equations. *Behavioral Ecology* 5: 1–8.

Williams, T. D., N. Warnock, J. Takekawa, and M. A. Bishop. 2007. Flyway-scale variation in plasma triglyceride levels as an index of refueling rate in spring-migrating Western Sandpipers (*Calidris mauri*). *The Auk* 124: 886–897.

Ydenberg, R. C., R. W. Butler, D. B. Lank, C. G. Guglielmo, M. Lemon, and N. Wolf. 2002. Trade-offs, condition dependence and stopover site selection by migrating sandpipers. *Journal of Avian Biology* 33: 47–55.

Zwarts, L., B. J. Ens, M. Kersten, and T. Piersma. 1990. Moult, mass and flight range of waders ready to take off for long-distance migrations. *Ardea* 78: 339–364.

Foraging Ecology and Habitat Use

CONTENTS

LIKE MOST BIRDS, SHOREBIRDS spend a larger percentage of their time foraging so as to ingest sufficient energy to maintain the exceptionally high metabolic rates associated with flight, homeothermism, and the demands of reproduction. Nonbreeding shorebirds experience especially rigorous energetic demands associated with their long migrations, periodic molts, and simply maintaining themselves in sometimes cold, wet weather. Because of these demands, shorebirds feed much of the daylight hours and occasionally at night. They routinely forage in open, unvegetated habitats on soft-bodied invertebrates. In coastal regions, activity patterns are strongly dictated by daily tidal rhythms. Consequently, individuals often must maximize the time they spend feeding and their energy intake rates when foraging habitats are available to them.

Several lines of evidence highlight the evolutionary significance of food to shorebird ecology. First, shorebirds spend a large proportion of their daily time-activity budget feeding, especially during migration and in the depths of winter at northerly latitudes. The rate at which they capture and acquire energy from their prey is strongly influenced by a suite of abiotic and biotic factors that constrain access to food. Second, observational and experimental studies indicate that shorebirds (and other species such as fish and invertebrates that consume the same invertebrate prey) sometimes reduce the abundance and availability of prey over short and long periods. Consequently, food may occasionally be in short supply such that competition for limiting resources increases. In some instances, this reduction in food has been shown to have consequences for individual survival. Third, the spatial distributions and social organization of foraging shorebirds are strongly influenced by the density of their prey. Shorebirds often concentrate in areas where prey are most abundant or available. This positive correlation exists at a variety of spatial scales, including global perspectives. Fourth, over ecological and evolutionary time, competition for food is the most plausible explanation for interspecific differences in habitat use and behavior exhibited in mixed-species flocks during the nonbreeding season. Closely related species often appear to partition their niches by feeding in subtly different ways, presumably to reduce competition. These interspecific differences occasionally manifest themselves in clear community patterns. For instance, the morphology of bills and hindlimbs belies a long evolutionary history with food. Collectively, these observations illustrate that many facets of the nonbreeding ecology of shorebirds are closely intertwined with the spatial and temporal availability of food resources, and that competition for food has shaped the biology of shorebirds over evolutionary time. In this chapter, I review the literature on shorebird diets and foraging ecology, emphasizing those features that are most affected by anthropogenic habitat loss and degradation, and hence are subject to management actions.

DIETS

With few exceptions, the diets of shorebirds are dominated by invertebrates, especially soft-bodied prey. Skagen and Oman (1996) reviewed 75 papers that characterized the diets of 45 species, mostly from the Western Hemisphere. Shorebirds ate mostly freshwater and marine invertebrates from the classes Insecta, Malacostraca, Gastropoda, Polychaeta, and Bivalvia. There were, however, obvious differences in diet composition associated with habitat when these invertebrate taxa were summarized for the 10

most-studied shorebirds: a group of five small calidridines, two plovers, two larger tringines, and a dowitcher. At inland sites, shorebirds ate mostly dipterans and coleopterans, whereas in coastal areas they consumed mainly bivalves, amphipods, and polychaetes. Larger shorebirds, especially thick-knees, curlews, and some tringine sandpipers, occasionally ate small vertebrates such as fishes, amphibians, reptiles, and even small mammals (Cramp 1983; del Hoyo et al. 1992). By contrast, the diets of some sandpipers, especially smaller calidridines, occasionally include seeds or tubers (Alexander et al. 1996). The seedsnipes of South America feed extensively on vegetation (Fjeldså and Krabbe 1990). Fruits are eaten occasionally by some Arctic-breeding species such as the Bristle-thighed Curlew. Skagen and Oman (1996) concluded that, as a group, shorebirds are euryphagic with broad dietary flexibility, which suggests opportunism. This conclusion was based on species-level assessment of diet breadth and the observation that shorebirds alter their diets in association with the patchy distribution and abundance of their prey in the habitats they occupy as they wend their way between breeding and nonbreeding areas.

Although shorebirds may be opportunistic, there is ample evidence that individuals of a variety of species often specialize on particular prey types for foraging bouts varying in length from hours to the life of individuals. Over intervals spanning a few hours to months, for instance, Long-billed Curlews feeding in intertidal areas specialized on just a few prey types. Some curlews occasionally ingested hundreds of small bivalves over a 2-hour period, whereas neighboring curlews specialized on burrowing shrimp, shore crabs, or polychaetes (Leeman et al. 2001; Colwell et al. 2002). Probably the best studied case of diet specialization, however, involves the Eurasian Oystercatcher. Individually marked birds that differ in bill morphology use quite different foraging strategies to feed on different prey species (Sutherland et al. 1996). Short-term evidence for diet specialization comes from analyses of consecutive captures of different

OYSTERCATCHER
BILL MORPHOLOGY
AND DIET SPECIALIZATION

Oystercatcher bills are among the strongest of shorebirds owing to the hard-shelled prey that they consume. During the nonbreeding season, the bill morphology of the Eurasian Oyster-catcher takes three distinct forms: pointed, chisel-like, and blunt (Swennen et al. 1983; Sutherland et al. 1996). Each of these shapes is associated with different foraging techniques. Individuals with pointed bills forage in soft intertidal sediments for worms and small bivalves; those with chisel-like bills stab at mussels to open the shell; those with blunt tips hammer open the shells of bivalves. There are also sex differences in foraging behavior and bill morphology. Most (70%) females tend to have pointier bills than males, the vast majority (90%) of which have blunt bill tips; the sex ratio of chisel-like bills is 50:50 (Le V. dit Durrell et al. 1993). Oystercatcher bills grow at the rate of 0.44 mm per day, which is equal to one bill-length every 6 months (Sutherland et al. 1996). Consequently, the oystercatcher bill-tip shape can change dramatically, principally with changes in diet associated with hard and soft prey. These differences in prey affect abrasion on the bill tip and cause changes in the shape of bills. In a sample of 56 oystercatchers that were captured during the winter and again in the breeding season, the bill shape became shorter during the breeding season. The individuals that showed the most dramatic seasonal change in bill shape tended to be females, especially those with more pointed tips. Sutherland et al. (1996) suggested that this stemmed from the greater effects of abrasion on the distal bill tip of females. Abrasion also shapes the chisel-like and blunt tips of male oystercatcher bills, but less so. Perhaps surprisingly, Le V. dit Durell (2007) reported no differences in survival of oyster-catchers that specialized on different prey types. However, females survived less well than males, perhaps owing to differences in social status.

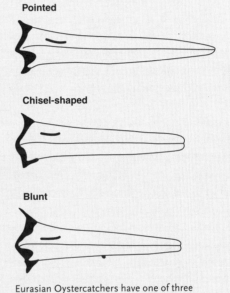

Pointed

Chisel-shaped

Blunt

Eurasian Oystercatchers have one of three types of bill specialized for feeding in different ways on different prey. From Hulscher et al. (1996).

prey types. Observations of breeding oyster-catchers feeding continuously during a low-tide interval revealed that the chance that two consecutive prey captures were the same species was highest (95%) for the bivalve *Macoma* and lowest (47%) for the burrow-dwelling lug-worm, *Arenicola*. Long-term studies of color-marked individuals indicate that oystercatchers foraged quite selectively on a single prey type, including cockles, mussels, shore crabs, and polychaetes. Moreover, the composition of diets of individual oystercatchers was consistent between breeding seasons and among years, indicating that individual specialization persisted for years.

Two principal factors appear to influence the breadth of a species' diet, as gauged by the taxonomic diversity of prey consumed (and not necessarily by prey size). First, dietary breadth may be associated positively with the diversity of habitats frequented by a species (Skagen and Oman 1996). The relationship between dietary breadth and habitat diversity is probably related to whether a species is a permanent resident in one habitat as opposed to a globe-trotting migrant. For example, diet specialization is typical of many of the 12 species of oystercatcher, most of which are nonmigratory (Hockey 1996). Although the diets of oystercatchers include many species of invertebrate, their dietary diversity is rather limited when viewed in terms of prey morphology. This is especially true of black morphs that are largely restricted to rocky intertidal habitats, where they consume mostly mollusks such as mussels, limpets, snails, and chitons (Hockey 1996). Over shorter intervals, oystercatchers typically feed on a limited number of prey within their home ranges in intertidal habitats. By contrast, greater diet breadth characterizes many of the globe-trotting sandpipers that move among varying wetland types during their annual migrations between hemispheres (Skagen and Oman 1996).

A second factor that may influence interspecific differences in diet breadth is body size and associated variation in bill morphology. In reality, bill length and shape are probably the chief morphological characters that affect diet breadth because of the close relationship with the depth to which an individual can probe. Bill size also influences the ability to handle large-bodied prey; a large gape permits a bird to swallow large prey. Bill gape is larger in large-bodied species such as thick-knees, oystercatchers, and curlews. By contrast, small calidridine sandpipers and plovers are constrained by their shorter bills and smaller gapes to feed on smaller prey. Small species may not be able to probe deeply into substrates to acquire prey, and individuals may be unable to effectively handle large prey. For example, the diverse diet of the Semipalmated Sandpiper, as characterized by 22 studies (Skagen and Oman 1996), is dominated by small prey from 37 invertebrate families. By comparison, the diet of Lesser Yellowlegs includes 65 invertebrate families, despite half the number of studies (Skagen and Oman 1996). One final example illustrates diet breadth in a large shorebird occupying a single location during the nonbreeding season. The largest North American shorebird, the Long-billed Curlew, occasionally

SEX DIFFERENCES IN FORAGING

In scolopacid sandpipers, strong sexual size dimorphism exists, with females occasionally 1.3 times the mass of males. Females also have longer bills and legs (Jönsson and Alerstam 1990). These morphological differences sometimes produce habitat segregation of the sexes in a pattern that rivals interspecific differences: females and males feed as if they were two different species. Sex differences in habitat use are well known from wintering studies of the Eurasian Curlew. In coastal areas of England, females occasionally defend intertidal feeding territories over dependable food resources that lie deep in the intertidal substrates. Females are better able to access this food because they have a longer bill than males. By contrast, males are more likely to be encountered feeding in coastal pastures and arable lands, where their smaller bill allows them to feed on earthworms and other invertebrates (Townshend 1981). Male and female Bar-tailed Godwits also differ in size, with 10 cm bills for females versus 8 cm for males, which achieve higher capture rates of the lugworm (*Arenicola*; Smith 1975).

picks small amphipods (*Megalorchestia*) from the surface of sandy substrates. Curlews also probe for both small bivalves and polychaetes and into burrows to capture large shrimp (Leeman et al. 2001; Colwell et al. 2002).

FORAGING MANEUVERS AND HABITAT USE

The foraging maneuvers and habitat use of shorebirds are quite varied, especially among the various families and tribes (Table 7.1). Shorebirds use habitats that range from highly marine and aquatic to increasingly terrestrial and even aerial. Many taxa, however, feed to a great extent in edge habitats of wetlands, including rocky intertidal shores, sandy beaches, tidal flats, and sparsely vegetated wetlands. Shorebirds use a limited array of foraging maneuvers to acquire prey. Many feed on prey detected using visual cues and either glean or pick prey from the water's surface or just below it. A large number of species detect prey using tactile sensation and grasp prey from below the surface of wet substrates. Terrestrial-feeding shorebirds search for prey visually, which they grasp from the surface of substrates or vegetation. Lastly, the aerial-feeding pratincoles hawk insects, like swallows. The following is a more detailed account of the summary provided in Table 7.1, organized by habitats.

Only a few shorebirds, notably the phalaropes, are truly aquatic, feeding almost exclusively on prey gleaned at or near the water's surface. In most instances, phalaropes use visual cues to detect prey in the water column. In areas of high prey density, Red-necked Phalaropes often insert the open bill into the water column and rely on the physical properties of water adhesion to the maxilla and mandible to draw water droplets encapsulating small prey (such as cladocerans) into their gullet (Rubega and Obst 1993). Avocets occasionally feed from a floating position in deep water, often plunging their head below the surface to grasp prey.

Jacanas forage from the surface of the water and vegetation supported by long toes that enable them to distribute their weight evenly on floating vegetation in tropical wetlands. Many shorebirds wade in shallow water in unvegetated habitats and probe deeply for prey using tactile cues. This group includes a diverse array of sandpipers, including godwits, dowitchers, painted snipes, many calidridines, and occasionally curlews. Godwits and dowitchers detect prey using tactile cues. They insert their bills to varying depths of soft substrates and either probe singly or in a rapid stitching motion until they detect prey. Movements are often localized in areas of high prey density, as evidenced by the remnant patterns of stitching dowitchers which were occasionally observed in a recently desiccated wetland (Fig. 7.1). Curlews use both visual

and tactile cues when foraging. During the nonbreeding season, Long-billed Curlews range over their feeding territories picking crabs and small amphipods from the surface or grasping small fish from pools of water. Curlews also move methodically across exit holes of large burrowing shrimp and insert their bill into the opening to detect prey tactilely. Curlews occasionally concentrate over areas of high bivalve density and probe deeply, using tactile cues to extract large numbers of prey in short intervals (Colwell et al. 2002). Calidridine sandpipers principally use tactile cues to locate prey, and their foraging maneuvers are typified by probing of the bill to varying depths to extract prey. Sanderlings probe to varying depths in the sandy substrates of the wash zone of ocean-fronting beaches (Myers et al. 1980). Recently, images from a scanning electron microscope have revealed that the tip of a Western Sandpiper tongue has a brushlike structure, which suggests that individuals sometimes lap nutritious biofilm from the surface of fine substrates (Elner et al. 2005). Many species forage in wet substrates that enhance foraging. In intertidal habitats, avocets scythe their bill through water and fine sediments, often swallowing prey suspended in a slurry of mud and then allowing the gut to process the meal (Hamilton 1975; Quammen 1982).

Most terrestrial-feeding shorebirds detect food using visual cues and peck at surface substrates or vegetation to capture prey. Plovers, large and small, exhibit a characteristic walk-stop-peck sequence of foraging behaviors. The short, stout bills of plovers are used to peck at prey at the surface of substrates, even when they feed at night illuminated by a full moon. Plovers occasionally remain stationary and "foot-stir" to liquefy fine sediments and render prey more vulnerable to pecking and shallow probing. In coastal estuaries, Black-bellied Plovers occasionally feed on large burrowing shrimp, which they acquire by probing into burrows. The Wrybill's asymmetrical bill (the tip always curves to the right) facilitates a left to right sweeping motion as the bird forages for invertebrates and fish

TABLE 7.1

Summary of foraging maneuvers in shorebird tribes, families, or subfamilies arranged in relation to principal feeding habitats covering a continuum from aquatic to terrestrial feeding. Shading indicates degree to which maneuver has been observed in a particular foraging habitat: not observed □, rare □, occasional ▨, and common ■.

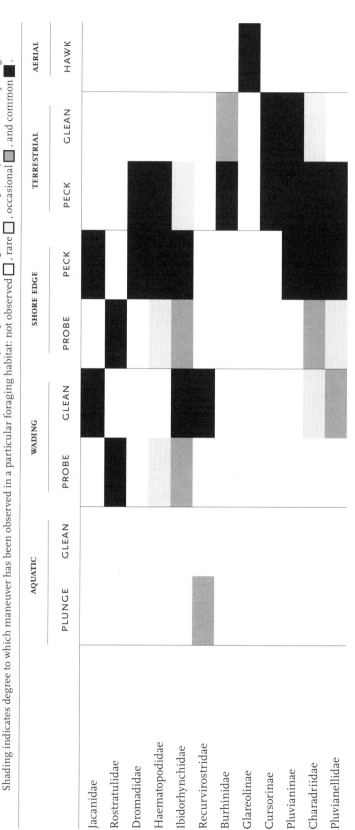

Habitats: Aquatic = supported by the water's surface. Wading = standing in shallow water. Shore edge = walking on wet substrates in close proximity to water's edge. Terrestrial = foraging on dry substrates or upland vegetation far from the water's edge. Aerial = flying.
Foraging maneuvers: Plunge = dipping bill into water column or moving it side to side. Glean = picking prey at or near the surface of water or vegetation. Probe = inserting bill into substrate. Peck = striking prey at substrate's surface. Hawk = gathering prey with open gape.

Scolopacini
Gallinagonini
Limnodromini
Numeniini
Tringini
Arenariini
Calidridini
Phalaropodini
Pedionomidae
Thinocoridae
Chionididae

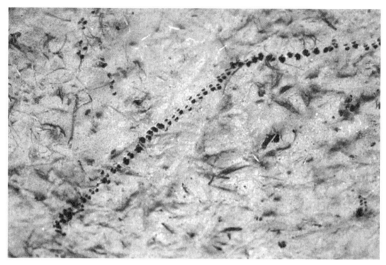

FIGURE 7.1. Recently dewatered freshwater wetland showing the track of a dowitcher bill. Dowitchers probe in fine, wet substrates by rapidly inserting their bill to varying depths in search of benthic prey. The track shows the slight gape of maxilla and mandible as the bird foraged for chironomid larvae in soft, muddy sediments. Photo credit: M. Colwell.

eggs, in a clockwise fashion around and under rocks in braided streambeds (Pierce 1979). The shore-feeding Magellanic Plover often concentrates its foraging in a limited area and uses its feet to repeatedly jump forward, scraping the substrate to uncover food. The Magellanic Plover also flips over small stones in aquatic habitats, much like a turnstone. Turnstones appear to flip over small rocks by inserting the bill beneath the stone and opening it, rather than using a lifting motion of the head. The Ibisbill is often observed wading in shallow water and gleaning prey from the undersurface of large stones in high elevation riverine habitats where it breeds. Tringine sandpipers also forage visually in a variety of wetland habitats. They capture prey by pecking or probing to shallow depths. Some species, such as the Greater Yellowlegs, occasionally sweep their bill side-to-side in shallow puddles or fine substrates. Yellowlegs often chase small fishes in shallow pools.

Several groups of shorebirds are highly terrestrial in their habits, including thick-knees and stone curlews, the Plains-wanderer, seedsnipes, and sheathbills. Virtually all of these groups detect food visually and peck at prey. The large-bodied thick-knees and stone curlews feed on large prey, even taking small mammals and lizards. Little is known of the foraging behavior and diet of the Plains-wanderer, perhaps owing to its nocturnal habits and rarity. Seedsnipes feed in grasslands on insects and seeds. Lastly, the omnivorous sheathbills forage in coastal areas, where they feed on invertebrates and opportunistically steal food in penguin colonies.

The pratincoles are truly special among shorebirds in feeding on the wing, like swallows. In northwestern Australia, an enormous concentration of Oriental Pratincoles was observed during a recent nonbreeding season. These birds were feeding on insects (principally locusts) amassed by weather fronts during a period of especially wet weather elsewhere in much of their Australian range (Sitters et al. 2004).

ACQUIRING ENERGY

Shorebirds spend considerable time foraging to maintain the energy reserves essential for survival and reproduction as well as preparing for migration. A multitude of factors interact to affect an individual's ability to acquire energy, including aspects of the individual and the environment. Several excellent summaries of time-activity budgets (Puttick 1984) and intake rates (Goss-Custard 1984; Pienkowski et al. 1984) of nonbreeding shorebirds have been published. In the following section, I provide an overview of factors contributing to variation in activity budgets and intake rates, and summarize literature addressing how the predictability of some environmental features and the vagaries of others influence energy acquisition. Later, I address evidence concerning age-related variation in foraging proficiency, especially in species that feed on prey that are difficult to capture and, hence, require some learning.

TIME-ACTIVITY BUDGETS

Depending on the time of year, shorebirds allocate varying amounts of time and energy to various essential behaviors. These patterns of daily behavior are summarized in a time-activity budget. The budget characterizes the percentage of daylight hours or, preferably, 24-hour interval, apportioned among various behavioral categories such as feeding and resting. Obviously, feeding is a critical element of the daily budget. Feeding activity, the percentage of time an individual spends searching for, handling, and digesting prey, has been shown to vary over the annual cycle. During the breeding season, time spent foraging is inevitably reduced somewhat owing to the competing demands of breeding activities such as courtship, incubation, and brood-rearing. Nevertheless, breeding birds may make up for lost foraging time by feeding more actively and being less vigilant.

During the nonbreeding season, shorebirds spend more time feeding in order to acquire the energy to fuel migration and molt as well as to maintain the energy reserves sufficient to get through taxing weather conditions. Still, there is considerable variation in the percentage of time that shorebirds feed, and this variation is clearly associated with three factors that are inextricably linked to where they spend the nonbreeding season. These three factors are (1) day length, (2) weather patterns, specifically temperature, precipitation, and wind velocity, and (3) tidal regime in coastal areas. For species that

occupy nontidal habitats for much of the year (such as the Mountain Plover), the latter effect is obviously irrelevant. These factors interact to affect directly the amount of feeding time and intake rate or to alter an individual's energy requirements. For instance, Rock Sandpipers wintering in Cook Inlet, Alaska, may feed only during daylight hours of winter and when tidal action exposes the soft sediments where they feed on small bivalves. Moreover, the amount of time sandpipers need to feed may increase during inclement weather. On top of this, feeding activity patterns may be influenced strongly by how a species detects and acquires prey. Species that feed using tactile cues tend to be less affected by environmental conditions compared with visual feeders because the latter are reliant on ambient light and unfettered surface conditions of water or substrate to be effective foragers.

The geographical location where a bird winters strongly affects its time-activity budget via a relationship between latitude and day length (Puttick 1984). Early research indicated that many shorebirds wintering at northern latitudes foraged for 80% to 100% of daylight hours. For example, Ruddy Turnstones, Common Redshanks, and Red Knots wintering at various locations in Great Britain foraged nearly 100% of daylight hours (Goss-Custard 1969, 1979; Baker 1981). Goss-Custard et al. (1977) reported that the time-activity budgets of several species wintering on The Wash in England exceeded 95%. This stands in contrast to shorebirds wintering in tropical and southerly latitudes where the percentage of time feeding is often much lower. For example, during the austral summer (October through March), Curlew Sandpipers wintering in South Africa fed for 55% to 65% of daylight hours (Puttick 1984). Similar patterns have been reported for nonbreeding shorebirds elsewhere in Africa. A diverse assemblage of 11 shorebird species winters on the Kenyan coast. Fasola and Biddau (1997) reported that the percentage of time foraging ranged from 25% to 83%. Most species fed less than 60% of the 24-hour period. Zwarts et al. (1990) studied 14 species on the Banc d' Arguin, Mauritania, during

AGE DIFFERENCES IN FORAGING

The food of shorebirds varies greatly in abundance, prey size, and the ease with which individuals may acquire it. Many small sandpipers, for instance, feed by pecking or probing for small, soft-bodied macroinvertebrates, whereas some large species, such as oystercatchers and curlews, regularly capture large crustaceans or mollusks, which may be difficult and dangerous to handle (Tsipoura and Burger 1999). In species that use complex foraging maneuvers to capture and handle prey, individuals may require considerable time to learn how to forage. Consequently, juveniles and adults may differ in feeding behavior and success. The complexity of foraging maneuvers thus requires a period of learning, during which individuals become increasingly proficient at prey capture. Several observational studies have demonstrated this in species that are easily aged by plumage differences between adults and juveniles in their first months of life. Juvenile Ruddy Turnstones are less successful foraging than adults in rocky intertidal and tidal flat habitat (Groves 1978). Avocets use a distinctive scything behavior to feed by touch on nektonic or benthic prey (Hamilton 1975). Scything is a complex foraging behavior. In American Avocets, scything becomes an increasingly important part of the foraging repertoire of prefledging young until it dominates the feeding methods (Becker et al. 2002). Juvenile avocets are less proficient foragers than adults (Burger and Gochfeld 1986); similar findings have been reported for Black-necked Stilts (Burger 1980).

Age-related changes in foraging behavior and efficiency are best documented for the Eurasian Oystercatcher (Goss-Custard and Le V. dit Durrell 1987; Goss-Custard et al. 1998). In their first autumn, juvenile (first winter) oystercatchers occasionally steal food from conspecifics, but this behavior diminishes with time until individuals feed almost exclusively in the conventional manner of stabbing, hammering, or probing for prey. The foraging proficiency of juvenile oystercatchers was 44% that of adults early in the nonbreeding season, whereas that of the immatures (2 to 4 years old) was comparable to adults. By midwinter, juvenile foraging success was comparable to adults. This suggests that foraging proficiency of these young birds had increased.

The Crab Plover is probably the most intriguing example of a shorebird with complex foraging coupled with large, challenging prey. The Crab Plover feeds on an array of invertebrates taken from the surface and benthos of intertidal habitats. The single-egg clutch hatches early in spring, and adults quickly move juveniles to wintering areas. On the wintering grounds, juveniles have the same rate of prey detection but are less proficient foragers compared with adults, especially when attempting to capture prey hidden in mud substrates (Fasola et al. 1996). Juvenile Crab Plovers also beg for food from conspecifics (Sanctis et al. 2005). These relationships between prey type, foraging maneuvers, and age-related feeding success are undoubtedly related to the tendency for some species of shorebird to exhibit delayed sexual maturity.

Sanderlings feed almost exclusively in the wash-zone of high energy ocean beaches. This bird has captured a burrowing mole crab (*Emerita analoga*). Photo credit: Noah Burrell.

late winter and early spring as individuals prepared for migration. The average percentage time that shorebirds fed during daylight hours in February ranged from 66% in the Bar-tailed Godwit to 98% in the Common Ringed Plover. By April, 10 of 14 species fed for greater than 90% of daylight hours. In the southern Great Plains of North America, feeding dominated the activity budgets of migrant shorebirds (Davis and Smith 1998). The percentage of time feeding ranged from 40% in the American Avocet to 60% to 90% in the Long-billed Dowitcher, Least Sandpiper, and Western Sandpiper. In each case, feeding constituted a greater percentage of the activity budget during spring than during late summer and fall.

NOCTURNAL FORAGING

Early research on shorebird foraging ecology characterized activity budgets and impacts on prey populations using observations of birds foraging during daylight hours. It has long been known, however, that some species of shorebird often and regularly feed at night (Pienkowski et al. 1984). Night feeding occurs across a variety of latitudes during the breeding and nonbreeding season. More recently, a growing body of literature has shown that shorebirds of a variety of species occasionally or regularly feed at night; this information has been enhanced by improved technologies of radio telemetry and thermal imaging. A thorough review of the patterns of nocturnal feeding in shorebirds, waterfowl, and wading birds is presented by McNeil et al. (1992).

Several factors, some characteristic of species and others relating to the environment, may influence the extent to which shorebirds forage during the day and night. First, feeding success at night may be related to a species' principal means of detecting prey. Plovers and stilts have high densities of retinal receptors that enhance foraging at night, even under low light conditions (Rojas de Azuaje et al. 1999). But foraging success of tactile-feeding sandpipers is not constrained by light levels. Consequently, species may differ in nocturnal foraging based on the extent to which they detect prey visually versus tactilely and how environmental conditions (for instance, ambient light affected by moon phase or winds altering surface cues) influence prey availability.

Interspecific variation in body size and energetic demands in general may affect the extent to which species forage at night. Small-bodied

NOCTURNAL FORAGING AND SEX DIFFERENCES IN INCUBATION BEHAVIOR

During the breeding season, in species that exhibit shared incubation, one sex routinely incubates during the day and the other at night. The sex with diurnal incubation must forage at night to meet the energy demands of diurnal incubation. In the Snowy Plover, females incubate during the day and males at night. Talitrid amphipods (such as *Megalorchestia*) are important prey of plovers on coastal beaches where they breed in western North America; these amphipods are most active at night (Geppetti and Tongiorgi 1967). It is possible that incubation patterns (females by day; males at night) stem from the advantages to females of feeding nocturnally on abundant prey and incubating during warmer daylight hours. This argument assumes that females are taxed energetically by egg production, and that nocturnal foraging by females has evolved to capitalize on abundant prey. The fact that females also require more energy to replace failed clutches lends further support to the idea that incubation patterns may be intertwined with the energy benefits of nocturnal foraging.

Talitrid amphipods are a common prey item of Snowy Plovers that breed on ocean beaches. Amphipods are more abundant at night, when females are on incubation recess. Photo credit: Mark Colwell.

species with high surface area to volume ratios will deplete energy reserves and lose mass faster during prolonged periods of inclement weather compared with larger species. This is especially true in winter at northern latitudes where short days interact with tidal regimes to restrict daytime feeding opportunities. At other times of year, the energy demands associated with migration and molt increase the need for food. Consequently, individuals may feed across a greater proportion of daylight hours and also into the night. For example, migration and a complete prebasic molt occur over the autumn months in many species that breed at northern latitudes; these energy demands may cause birds to feed both day and night (Mouritsen 1994). Additionally, the demographic makeup of a population, especially the presence of large numbers of juveniles, may affect foraging patterns. The prevalence of juveniles that are generally inexperienced at foraging may result in increased numbers feeding at night if they are unable to meet their energy demands during the day.

Seasonal variation in environmental conditions, especially tides, may constrain individuals to feed at night at certain times of year. At Humboldt Bay, California, an estuary with mixed semidiurnal tides, nocturnal foraging by a suite of species was most prevalent during autumn. This coincided with an interval in which low tides exposed tidal flats for a relatively short interval during the day. By contrast, the lower of the two daily low tides occurred at night, coinciding with greater shorebird use (Dodd and Colwell 1996, 1998). Alternative (terrestrial or freshwater wetland) habitats were generally unavailable as foraging areas for shorebirds during autumn.

A rich literature has examined nocturnal feeding during the nonbreeding season. The prevalence (frequency of occurrence) with which individuals feed at night varies greatly, within and among seasons, and across wintering areas. At northerly latitudes, a variety of wintering shorebirds have been shown to feed at night during all seasons. For example, Black-bellied Plovers wintering in northeast England commonly feed at night (Wood 1983). But shorebirds wintering in tropical habitats also feed extensively at night (Robert et al. 1989; Zwarts et al. 1990; Fasola and Biddau 1997).

Two nonmutually exclusive hypotheses have been proposed to explain nocturnal foraging in shorebirds. An assumption of these explanations is that the baseline condition is one of diurnal foraging behavior, and that nocturnal behavior is a response of individuals to differences in the environment that either force them to feed at night or make it more profitable to do so. The supplemental feeding hypothesis posits that individuals unable to meet their daily energy demands are forced to supplement their diet by feeding at night. Evidence for this hypothesis comes from several sources, including the seasonal observation that nocturnal feeding occurs more at night in coastal areas during autumn when tides do not afford sufficient feeding time during the day (Dodd and Colwell 1996). Additionally, detailed energy studies indicate that nocturnal feeding occurs when individuals cannot meet their daily energy requirements by feeding during daylight hours (Dugan 1981; Zwarts et al. 1990).

The preferential feeding hypothesis argues that nocturnal foraging occurs owing to advantages accrued by individuals feeding at night. These advantages may include a lower risk of predation by diurnal predators such as falcons. If this explanation were true then it would predict that individuals would forego daytime feeding owing to the high risk of predation and only feed at night. Alternatively, birds may prefer to feed nocturnally because of increased intake rates accompanying greater availability of prey at night. For example, the nocturnal feeding rates of Black-bellied Plovers foraging mostly on polychaetes were lower than the diurnal values; however, prey were more available at night (Dugan 1981). Kuwae (2007) used thermal imaging to show that, in winter, night-feeding Kentish Plovers at Tokyo Bay, Japan, had higher defecation rates (and hence higher intake rates) than those feeding during the day. Plovers achieved 79% of their prey captures at night in the winter at this northern site. It is possible that visual-feeding plovers took advantage of greater prey availability (principally nereid polychaetes) owing to activity near the substrate surface at night to achieve higher intake rates. Regardless, the observations indicate that night feeding contributed more to the daily activity budget of the plover.

EFFECTS OF BODY SIZE, WEATHER, AND TIDES

In addition to the observation that individuals wintering at northern latitudes feed more than those at southerly latitudes, details hint that the energy costs of maintenance for individuals vary with body size. Several studies have shown an inverse relationship between feeding activity and body size when a diverse community has been studied at a single location. Zwarts et al. (1990) reported this inverse relationship for an assemblage of 14 species wintering on the Banc d'Arguin, situated at 20° north latitude. The climate on the west coast of sub-Saharan Africa is arid, with low temperatures averaging 19°C in February under the influence of northerly winds off the Atlantic Ocean (Engelmoer et al. 1984). This is clearly a rather benign climate. Still, species fed in a pattern suggesting that energy conservation was important. In February, Eurasian Oystercatchers fed for 4 hours of the 24-hour period, whereas Little Stints fed for 17 hours. The strength of this relationship diminished in the spring as individuals of all species dramatically increased their feeding in preparation for migration. Even in equatorial Kenya (4° south latitude), where climate is even more benign, large-bodied species fed for less time than smaller species (Fasola and Canova 1993; Fasola and Biddau 1997). In general, one can predict the amount of time a species spends feeding during the nonbreeding season with knowledge of body size, latitude, and season: smaller species feed for longer intervals than larger shorebirds, especially at northerly latitudes and when preparing for migration.

A second environmental factor that varies with latitude and influences time-activity budgets is weather. The changing seasons of the long nonbreeding period add additional variation in weather conditions. This variation acts on an individual's ability to thermoregulate, and it may compromise the availability of food. Cold, windy, and wet weather, for example, increases the energy budget of individuals in northerly latitudes during the boreal winter. By contrast, shorebirds wintering at southerly latitudes (during the austral summer) generally experience relatively benign conditions. An exception to this may be some species that stage in hot coastal salt pans, where individuals must rid themselves of excessive heat loads at some times of year (Battley et al. 2003). Accordingly, time-activity budgets vary seasonally in association with weather and latitude. Additionally, the food available to shorebirds wintering at northerly latitudes diminishes in winter (Evans and Dugan 1984). Coupled with the greater energy demands, this makes for increased foraging activity. Additionally, prey populations may not replenish themselves during the winter such that the effect of foraging is to decrease food availability.

In coastal environs, tide is a third factor that dictates daily activity budgets. This is readily apparent to anyone observing the predictable local movements of shorebirds between intertidal feeding areas exposed at low tide and high-tide roosts. Flooding and ebbing tides alternately inundate and expose intertidal habitats, which directly affects the availability of food. Consequently, in virtually all studies conducted in coastal habitats, individuals move to roosts where they are comparatively inactive at high tide but feed to varying degrees throughout the subsequent low-tide interval. Within the low-tide sequence, however, there may be substantial variation in the extent to which individuals feed. Zwarts et al. (1990) showed that the feeding activity was highest during the few hours bracketing low tide.

The predictability of tidal regimes often supersedes the effect of day length: birds cannot feed during the day if their main feeding areas are under water; conversely, they must feed at night when foraging habitats are exposed. Consequently, activity patterns of shorebirds in coastal habitats take on an ultradian rhythm, dictated by tidal periodicity. There is, however, considerable variation in the relationship between tide and shorebird activity patterns. In some cases, feeding activity is strongly dictated by tidal rhythm and independent of daylight (dark:light) conditions (Fasola and Biddau 1997). In other cases, tidal regime exerts a strong influence, but diurnal time-activity budgets change seasonally as individual energy demands increase during the premigratory period. For instance, during much of the nonbreeding season, activity patterns of shorebirds wintering in the Berg River estuary, South Africa, were influenced most strongly by tide. But several species fed increasingly at night as they prepared for migration (Kalejta 1992). This was also true of the 14 species studied by Zwarts et al. (1990) on the Banc d'Arguin. In other studies of shorebirds in coastal areas, tides have less of an impact on the daily activity patterns. For instance, Robert et al. (1989) studied a mixed assemblage of shorebirds wintering in a mangrove estuary in coastal Venezuela, where tidal amplitude was relatively small (30 cm). The abundance of some species was clearly associated with tidal rhythms, whereas others were associated with the dark: light regime. Not surprisingly, the diurnal activity patterns of shorebirds in habitats unaffected by tides are largely driven by circadian rhythms of photoperiod (Davis and Smith 1998).

Tides vary greatly around the world in daily timing and amplitude. However, three types of tidal pattern occur worldwide. Most coastal areas experience two high and two low tides in a little more than 24 hours. In the mixed semidiurnal pattern, the highs and lows are of unequal height; semidiurnal tides are of equal amplitude. In a few regions, diurnal tides occur with a single high and low in approximately 24 hours. Tidal amplitude also varies across the globe, with very large differences between consecutive high and low tides occurring at high latitudes, especially at the far reaches of long fjords, inlets, and bays. The world's most extreme tides occur in the Minas Basin of the Bay of Fundy, Nova Scotia, where tidal amplitude averages 13 m and occasionally reaches 16 m. In Cook Inlet, Alaska, the maximum tidal range is 11 m, with a tidal bore that travels at a speed of 4 m s^{-1}. The gravitational pull of the moon exerts the strongest influence on tides; hence, it is not surprising that substantial variation in tidal amplitude is associated with predictable patterns in the lunar cycle that follow a monthly pattern. Spring tides occur twice a month when the gravitational pull of the sun and the (full or new) moon are aligned to create the highest and lowest tides. By contrast, neap tides of moderate amplitude correspond to periods when the sun and quarter moons act in opposition to one another. Each of these conditions, spring and neap tides, predictably occur twice monthly.

The monthly variation in amplitude of spring and neap tides subtly affects shorebird activity patterns. During neap tides, the tide edge moves slowly as it ebbs and floods tidal flats; at high tide, water may not fully inundate the upper reaches of tidal flats. This allows shorebirds to feed longer at the upper margins of tidal flats. By contrast, the fast-moving edge of a spring tide creates a more dynamic situation, with the upper reaches of intertidal habitat inundated early on a rising tide and remaining flooded for longer. The contrast in feeding activity between spring and neap tides was documented by Goss-Custard et al. (1977), who studied feeding conditions of five nonbreeding shorebirds on The Wash in England. High spring tides fully inundated feeding areas and birds roosted for extended periods during autumn, winter, and spring. By contrast, a narrow ribbon of foraging habitat bordering salt marsh was exposed during neap tides. Consequently, some shorebirds regularly fed throughout the high tide period of neap tides, especially during winter.

INTAKE RATE

Diet composition, foraging behavior, including habitat use, and activity budgets are integral aspects of shorebird foraging ecology. However, intake rate is arguably the most critical information of any foraging study. Intake rate is typically quantified as the amount of energy (in kilojoules) ingested per unit time, typically over a 24-hour interval. To achieve a given intake rate, individuals may alter the amount of time they forage, increase the rate at which they feed, or change their diet.

From a theoretical perspective, intake rate is intimately related to a predator's functional response (Holling 1959), which describes the rate of prey eaten by a shorebird in relation to variation in prey density. We will return to this topic in Chapter 8. There is great applied value in knowing intake rate because it allows an assessment of habitat quality from the perspective of the individual. For example, foraging areas vary in food availability, and this can be judged by measuring the intake rate of a species sampled across a variety of habitats. High-quality habitats are those where individuals achieve a high intake rate compared with areas where foraging is less profitable. Intake rate may also be useful in evaluating anthropogenic habitat degradation. For example, under conditions of no human activity, intake rate provides a baseline measure of habitat quality against which a researcher can judge the effects of human disturbance. Specifically, the intake rate of a continuously foraging shorebird can be characterized and then compared with conditions when varying numbers of humans are present at different distances from foraging shorebirds. The reduction in intake rate from baseline measures stands as a direct estimate of anthropogenic disturbance.

Intake rate is assessed by observing foraging birds and recording their attempts at prey capture, their successes, and the size of the prey consumed. This can prove challenging for species that feed on small prey probed from the benthos and then swallowed quickly. In these circumstances, it may be difficult to judge whether an individual was successful or not in capturing prey. Success is sometimes judged by observing swallowing behavior. For example, Goss-Custard (1969, 1970) studied wintering Common Redshanks feeding on a large polychaete and two smaller prey species, the gastropod mollusk *Hydrobia* and the burrowing amphipod *Corophium*. He used observational data of foraging birds, gizzard contents, and prey sampling to estimate intake rate. Goss-Custard could read-

ily observe the rate at which redshanks ate polychaetes because of the large size of worms, but the small size of the other prey made this nearly impossible, especially at long distances. So he used the proportion of prey in the gizzard and esophagus of a sample of collected birds to establish the relative percentage of the diet consisting of these two prey species. Then he apportioned observations of small prey (both *Hydrobia* and *Corophium*) accordingly. In some instances, especially with larger prey, the size of prey can be estimated by comparing a food item with the bill length of a focal individual. This requires practice by the observer and that birds handle their prey for sufficiently long to effectively estimate size. When prey are small, estimation of size may be difficult or impossible. Under these conditions, knowledge of average prey size available to foraging birds is used as a surrogate. Feeding rate and prey size are then converted to a rate of consumption of biomass (g h^{-1}) or energy (kJ h^{-1}).

FOOD AVAILABILITY

Food is not always readily accessible to foraging shorebirds because of myriad environmental factors, including physical features of the habitat and interactions with conspecifics. Factors that affect intake rate can be divided into two general categories: those *directly* affecting intake rate by compromising an individual's ability to detect and reach prey, as opposed to factors acting *indirectly* to alter the behavior of prey, which make food less available (Goss-Custard 1984). Tide is a good example of an environmental factor that directly influences prey availability. Ebbing and flooding tides predictably expose and inundate foraging habitats, thus directly influencing prey availability. Not surprisingly, the daily activity patterns of coastal shorebirds are driven by tidal rhythms (Burger et al. 1977; Hötker 1999). But tides affect the availability of prey in more subtle ways, too. Prey tend to burrow deeply into substrates as the interval lengthens since a receding tide first exposed and desiccated tidal flats. Consequently, invertebrates

ESTIMATING INTAKE RATE

The eight species of curlew that occur worldwide all are ideal subjects for studies of intake rate. During the nonbreeding season, curlews often defend feeding territories or occupy predictable home ranges where they consume a rather limited array of large-bodied prey species. Because these large prey items require substantial handling time, it is relatively easy to estimate the size of prey in comparison to bill length, especially when curlews feed in close proximity to observers who are using high-quality optical equipment. A first step in quantifying the energy budget of a curlew is to conduct focal observations (1 to 2 hours is probably sufficient to ensure that they forage across the full range of intertidal habitats within their territory or home range). Observers tally prey consumed in the

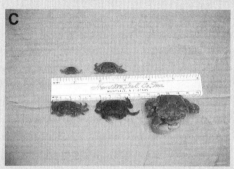

In coastal northern California, Long-billed Curlews feed in intertidal habitats on just a few types of prey. These images show the six main prey of curlews. Curlews (A) sometimes peck at the surface or under algal mats for fish (B: *Clevelandia*) and shore crabs (C: *Hemigrapsus*, *Pachygrapsus*); they also probe into burrows of shrimp (D: *Callianassa*), and into soft sediments to feed by touch on polychaete worms (E: *Nereis*) and bivalves (F: *Macoma*). Photo credits: Linda and Thomas Leeman.

order they are taken, and estimate the size of each item consumed with knowledge of both size variation of different prey species and the sexual dimorphism that exists in curlews. Females are larger than males, hence their bills are longer. The next step is to convert the observations of prey consumed to a rate of energy intake, typically expressed as Kcal h^{-1}. This requires that the observed lengths (a surrogate for size) of prey be converted to weights and then to energy content. This requires published estimates of the energy content of prey, which are relatively easy to find in the litera-ture. Regression models are constructed to estimate dry weights of prey from observed lengths. The published energy values of prey are used to estimate the caloric value of each prey item eaten. These data can then be summed across the interval of time to yield an estimate of the daily energy budget of a sample of individuals. During high tides and after winter rains soften pastures, curlews feed extensively on earthworms. Individual curlews had similar energy intake rates despite feeding on very different prey. See Leeman et al. (2001) for details.

become less available to foraging shorebirds, and intake rates decrease (Evans 1979, 1988). The following is a summary of the direct and indirect factors that affect shorebird foraging, including a brief summary of the biological features of invertebrate prey, the physical characteristics of shorebird foraging habitats, and the interactions with other species, including competitors and predators.

PREY POPULATION BIOLOGY

The availability of food to foraging shorebirds varies seasonally owing to the demographic characteristics of invertebrates. Evans and Dugan (1984) diagrammed a generalized overwinter decline in food availability in northern estuaries stemming from two main causes. First, invertebrates tend to breed in seasonal pulses, which results in high recruitment during spring and summer followed by an overwinter interval during which invertebrate densities decrease. Second, this decrease in food availability arises because shorebirds (and other predators) remove some of the standing crop through their feeding activity. Additionally, invertebrates tend to burrow deeper into substrates during winter, which further reduces their availability to shorebirds. For shorebirds wintering at southern latitudes, this pattern is reversed in the austral summer, which may explain the comparatively

higher densities of shorebirds at these southern estuaries (Hockey et al. 1992).

ABIOTIC HABITAT FEATURES

Of the numerous physical aspects of the environment that affect shorebird foraging behavior and intake rates (Piersma 1987), two physical characteristics of habitats play a prominent role in affecting where and how effectively shorebirds feed. On a fine spatial scale, substrate particle size and wetness influence foraging success. On a coarse spatial scale, the underwater topography or bathymetry of a wetland interacts with water levels to constrain where a bird feeds. These two habitat features act slightly differently to alter the attractiveness of foraging habitats by influencing the profitability of feeding in different habitats via effects on intake rates.

Particle size and water-holding capacity of substrates interact to affect shorebird foraging. Much of the foraging behavior of shorebirds involves individuals picking, probing, or scything for prey across substrates ranging from clayey mud to sand and cobble. Aside from the obvious relationships between these physical features of the sediment and the abundance of various invertebrate prey species, particle size may affect the attractiveness of habitats to different species of shorebird and alter their foraging behaviors. Probing shorebirds favor fine, soft sediments compared with species that routinely peck at surface prey on coarse, firm sediments. For example, Gerritsen and van Heezik (1985) examined substrate preferences of three calidridine sandpipers (the Dunlin, Sanderling, and Purple Sandpiper) that foraged preferentially on mudflats, sandy shores, and rocky shores, respectively. The intake rates of these species varied with habitat and foraging behavior. On a fine spatial scale, foraging Dunlin favored soft, muddy sediments where prey were more accessible to probing (Kelsey and Hassall 1989). The relationship between particle size and foraging behavior is supported by experimental work. Quammen (1982) added sand to muddy substrates and showed that several species altered the percentage time and rate of feeding. American Avocets spent less time in sand-enhanced plots, where they increased the frequency of pecking but decreased scything maneuvers. Similar changes occurred in Western Sandpipers and dowitchers.

Particles of different size vary in their water-holding capacity. Fine sediments hold water better than coarse ones, which makes it easier for a shorebird to probe in fine substrates. As a result, substrate penetrability increases, and individuals have an easier time acquiring prey in fine, wet substrates. Foraging rates may also be influenced by the penetrability of substrates varying in compaction. Myers et al. (1980) held Sanderlings in captivity and measured foraging rates under conditions of varying substrate compaction as well as the size, density, and spatial distribution of prey. Sanderlings were less successful capturing prey in more compacted sand, but their foraging rates were also affected by prey characteristics. In the wild, Sanderlings characteristically feed in the wash zone of ocean-fronting beaches. It is here that they achieve highest foraging success, in wetted substrates where prey exist at high density near the surface. By comparison, foraging success decreased upslope in drier substrates where prey burrowed deeper into the sand and beyond the reach of the average Sanderling bill.

At a much coarser scale, tidal flats are not two-dimensional habitats. Rather, they often vary subtly in topography. Even broad expanses of tidal flat vary in microtopography, with films of water interspersed amid shallow puddles and ridges of drier sand. Add channels to the mix, and the diversity of foraging habitats transforms dramatically into a complex three-dimensional array. This fine- and coarse-scale variation in bathymetry interacts with sediment particle size and water-holding capacity to influence the ability of shorebirds to probe effectively for prey, with consequences for intake rate. Short-billed Dowitchers tend to probe more on the crests of ripples than in troughs of intertidal sand flats, and these probes required 53% to 70% less force to penetrate compared with adjacent troughs (Grant 1984). It is unknown whether the difference in energy required to probe on crests produced a higher intake rate.

Two prominent features of wetland topography, slope and channelization, influence the amount of edge along which shorebirds can feed most profitably. The amount of foraging habitat suitable for any species (with a particular bill and hindlimb morphology) increases as wetland slope decreases because a broader band of shallow water and moist substrates exists. In other words, the change in water depth that strongly influences where species forage is more gradual on a gently sloping plane compared with a site with steep bathymetry and abrupt edges (Fig. 7.2). In intertidal habitats, this is readily apparent as a neap tide slowly ebbs and exposes a broad band of shallow water and moist sediments. Pooling of water on intertidal flats can have similar effects on habitat availability. Dense, mixed-species flocks of shorebirds often assemble under these conditions. By contrast, steeply sloped tidal flats afford much less area over which individuals can forage.

Channels further increase the total area of habitat available to foraging shorebirds. A greater amount of habitat exists along a complex of channels compared with a similar-sized area with a simple tidal edge with few or no channels. This increase should accommodate more individuals. In support of this, Lourenço et al. (2005) showed that the distributions of seven

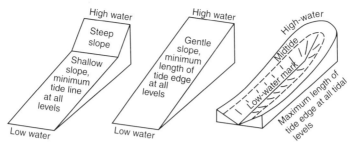

FIGURE 7.2. Tidal flat topography can strongly affect the length of the tide edge and hence the foraging of shorebirds. Steep slopes afford less edge than shallow ones, and intertidal channels incised into tidal flats increase the amount of edge. In addition, channels create sheltered areas where individuals may feed more effectively when windy conditions may affect their ability to detect prey. After Evans and Dugan (1984).

species of shorebird foraging in the Tagus estuary, Portugal, were influenced by the presence of tidal channels. Highest densities of all species occurred within 1 to 2 m of channels, with 44% of birds feeding on just 12% of area within 5 m of channels. Additionally, prey were more abundant near channels, which caused some species to feed more successfully near channels. Deeply incised channels also have been shown to benefit foraging individuals, especially during inclement weather. Some Black-bellied Plovers wintering on the Tees Estuary, England, defended intertidal feeding territories where deep channels cut into tidal flats to varying extents. Plovers occupying territories with channels were able to feed during cold, windy periods when buffeting winds made foraging difficult for individuals on exposed flats (Townshend 1981).

WEATHER

Rain, wind, and temperature can affect shorebird foraging at any time of year, but especially during winter at northern latitudes. Large spatial shifts in winter distributions of shorebirds in the Severn estuary of England were prompted by occasional storms (Ferns 1983). Strong winds associated with these storms eroded and mobilized intertidal sediments. At the same time, shorebirds abandoned their main foraging areas and moved to more sheltered reaches of the estuary with greater food availability. Weather-related shifts in the distribution of foraging

shorebirds are almost certainly prompted by changes in the profitability of feeding in different habitats. In support of this, numerous studies have documented the effects of wind, rain, and temperature on shorebird foraging success, either via compromised ability to detect prey or by changes in prey behavior directly. For instance, Pienkowski (1983) showed that the prey of visual-feeding Black-bellied Plovers and Common Ringed Plovers were more active at the surface when temperatures were warmer.

In coastal areas where winter weather reduces food resources in intertidal habitats, precipitation may actually increase food availability in terrestrial habitats. Along the Pacific Coast of North America, from California north to British Columbia, a variety of shorebirds forage in coastal pastures and agricultural lands adjacent to estuaries. The most abundant species, the Dunlin, moves on a daily basis to feed in terrestrial habitats at locations such as Bolinas Lagoon (Warnock et al. 1995), Humboldt Bay (Conklin and Colwell 2007), and the Fraser River Delta (Shepherd and Lank 2004; Evans Ogden et al. 2006). Dunlin that winter on the Fraser River Delta acquire a significant percentage of their diet from terrestrial habitats, as evidenced by stable isotope analysis.

BIOTIC INTERACTIONS

The flocks of shorebirds feeding on invertebrates in any wetland are a dynamic, mixed-species assemblage of individuals seeking to acquire prey and avoid predators. Individuals in flocks of varying density are constrained in their ability to feed by interactions with conspecifics and other shorebirds. There are, however, potential benefits from flocking with others, such as increased vigilance for predators and safety in numbers. Here, I highlight the costs of feeding in flocks as measured by reduced intake rates associated with competitive interactions.

When shorebirds aggregate in high-quality foraging areas characterized by high prey availability, individuals may experience a reduction in intake rate stemming from competition. These competitive interactions may take one of two forms, exploitation of a common prey and interference between foraging individuals. Both forms are conveniently measured as a reduction in intake rate, although the strength of competition is often assessed in other ways, including individual condition (such as reduced mass), lowered survival or reproduction, or decreased population size (Wiens 1989). Exploitative competition is the reduction in prey availability, and hence intake rate, that arises from combined effects of individuals feeding for varying durations in an area. Over short intervals of intense shorebird predation, invertebrates may be incapable of reproducing at rates sufficient to replace those lost to predators. This is especially the case in northern winters; consequently, prey populations decline (Evans and Dugan 1984).

Interference competition, by contrast, is the short-term, reversible decline in intake rate due to the presence of competitors (Goss-Custard 1980; Goss-Custard 1984; Sutherland 1983). It is measured as the proportional reduction in intake rate associated with a proportional change in density of conspecifics (Triplet et al. 1999). Interference can be the result of both direct and indirect effects of conspecific density (Goss-Custard 1984). It is important to note that the negative effects of interference do not occur because shorebirds have consumed and reduced the standing stock of invertebrates in an area. Rather, birds alter the availability of prey. In either case, the effect of interference is almost im-

PARASITES INFLUENCE AVAILABILITY OF PREY

The burrowing amphipod *Corophium* is an important prey item of many shorebirds, especially some tringines and many small calidridine sandpipers. The Semipalmated Sandpiper forages extensively on *Corophium* during its southbound migration when it stages in large numbers on the Bay of Fundy, Nova Scotia, Canada. *Corophium* construct burrows from which they filter-feed detritus from the water column using their antennae. Breeding male amphipods emerge from their burrows in search of females, which makes them more vulnerable to predation by sandpipers, especially during the day. Semipalmated Sandpipers achieve high feeding success across a gradient of soft-sediment intertidal flats (Wilson 1991). A parasitic nematode (*Skrjabinoclava morrisoni*) infects *Corophium* and alters the daily activity pattern of both males and females, making them more prone to crawl during daylight hours. This increased diurnal activity facilitates the transmission of the nematode to its final host, the Semipalmated Sandpiper, which feeds using visual cues during the day rather than at night.

mediately reversible in that when densities of feeding birds decline, intake rate increases (Goss-Custard 1984).

Interference can act in two main ways to reduce intake rates. It can act indirectly via the reduction in prey availability owing to other individuals feeding on a common resource. In this case, several mechanisms operate to reduce prey availability and hence intake rates. First, the presence of predators may elicit escape behavior in prey, which reduces food availability. Intake rates of the Common Redshank declined as conspecific density increased because their main prey (*Corophium*) left the substrate surface and burrowed deeper to evade shorebird predators. On the other hand, shorebirds may occasionally reduce prey availability via their forag-

FIGURE 7.3. Exit burrows of lugworms (*Arenicola*), which occur in high densities on intertidal flats around Lindisfarne, England. The burrows are visible as a dimpled texture to the sandy substrate covered with a shallow film of water. Black-bellied Plovers detect lugworms as they back up to the exit burrows and defecate. Photo credit: Mark Colwell.

ing activity and thus reduce the intake rate of conspecifics. This is more likely to occur at low prey densities (Goss-Custard 1984). For example, Goss-Custard (1980, 1984) showed that only a small percentage (1%) of *Corophium* was actually available to foraging Common Redshanks. Moreover, individuals may take a large proportion of available prey, thus reducing the amount of food for conspecifics. The effect of increased shorebird density on behaviorally induced reductions in prey availability probably acts strongest in shorebirds that detect prey visually. The reason for this is that visual predators must detect prey at or near the substrate surface for effective capture. Hence, if burrowing behavior increases with shorebird density, fewer prey will be available. Good examples of this come from Common Redshanks searching for *Corophium* on the mud surface and Bar-tailed Godwits and Black-bellied Plovers searching for lugworms (*Arenicola*) defecating near the surface of their burrows (Fig. 7.3). Indirect interference is less likely for tactile-feeding shorebirds. This is because prey may not be able to burrow and escape beyond the extent to which a shorebird can probe (Goss-Custard 1984).

Shorebirds may also directly reduce the foraging success of conspecifics via two behavioral interactions, kleptoparasitism and dominance. In these cases, interference is more frequent at high densities because birds are more likely to encounter one another. In the case of kleptoparasitism, interference is more likely to occur when food items are large and require long handling time, which increases the opportunity of another bird to approach and successfully steal food. Empirical evidence suggests that large prey such as decapods, burrowing shrimp, fishes, and bivalves are stolen frequently from foraging oystercatchers and Long-billed Curlews (Ens and Cayford 1996; Leeman et al. 2001). Interestingly, Dunlins rarely feed visually on large prey, but when they do they may lose a substantial proportion of the food to gulls and plovers (Warnock 1989). By contrast, smaller prey that require short handling times are less likely to be stolen. Behavior-based models confirm that attack distance is a strong predictor of interference via kleptoparasitism: large food items that require long handling times, and hence allow longer approach distances, enhance the abilities of kleptoparasites to steal food (Stillman et al. 2002).

At high densities, shorebirds are more likely to interact aggressively with one another, which results in reduced intake rates. This observation has been noted by numerous individuals studying aggression in dynamic, high-density flocks of migrant shorebirds (e.g., Recher and Recher 1969). In other species, long-term, stable dominance interactions among long-lived individuals occur with greater frequency as densities increase and individuals forage near one another. This latter form of interference has been especially well studied in the Eurasian Oystercatcher (Ens and Goss-Custard 1984; Goss-Custard and Le V. dit Durell 1988; Ens and Cayford 1996; Triplet et al. 1999; Stillman et al. 2002). A unique feature of the foraging ecology of oystercatchers is that they capture large prey that require long handling times. This affords other oystercatchers greater opportunity to steal food, which occurs more often at low prey densities (Triplet et al. 1999). When feeding on bivalves, both the mussel (*Mytilus*) and occasionally the cockle (*Cerastoderma*), dominance interactions increase with oystercatcher density, resulting in lower intake rates (Ens and Goss-Custard 1984; Goss-Custard and Le V. dit Durell 1988; Triplet et al. 1999). When foraging densities were high, oystercatcher intake rates decreased for several reasons. First, individuals were more aware of conspecifics and the possibility that another oystercatcher may steal food. This increased vigilance has a cost in lowered intake rate. Additionally, there were clear dominance relationships among individuals feeding on particular mussel or oyster bed, such that individuals were simply more wary of others when more conspecifics were present. Ironically, kleptoparasitism was more likely to occur when cockles were rare, whereas dominance-based interference increased dramatically when densities of oystercatchers were high (Triplet et al. 1999). In summary, interactions among conspecifics, and occasionally with other shorebirds, are more likely in areas of high prey density. These interactions, coupled with indirect effects that further compromise prey availability, regularly depress the intake rate of individual shorebirds below that predicted by simple relationships with prey density. These observations have strong implications for the carrying capacity of estuaries and other wetlands that are essential habitats for the shorebird populations that rely on them for much of the annual cycle.

Considering all the biotic and abiotic factors that act to constrain foraging, an important question is whether there is evidence that shorebirds maximize their intake rate? Goss-Custard (1985) provided the following evidence in support of an affirmative answer to this question. First, most studies indicate that individuals choose the most profitable prey when given a choice as to where to feed, and from among food items of varying size, availability, and digestibility. Second, optimal choice of food does not necessarily indicate that an individual is foraging at a maximum rate. Also, there is considerable evidence that intake rates vary with tide cycle, as shorebirds prepare for migration during the breeding season. When individuals do not feed at maximum rates, alternative explanations for their "maladaptive" choice are plausible (Goss-Custard 1985).

In summary, many abiotic and biotic factors interact to dictate the instantaneous intake rate of a foraging shorebird. A few simple observations are pertinent to understanding the complexity of these relationships (Evans and Dugan 1984). First, the invertebrates that comprise the diet of shorebirds are characterized by seasonal pulses of reproduction, which vary with latitude. At southern latitudes during the austral summer (boreal winter), invertebrates may breed prolifically and offset losses to shorebirds and other predators. This is less likely at northern sites, where invertebrate populations do not grow or reproduce during winter. Second, for a variety of reasons, the standing crop of invertebrate biomass available to shorebirds may not be the same as the density measured in the field. This results from numerous environmental factors that cause some prey to burrow deeper than others. This decreased food availability may be especially pronounced for shorebirds with bills too short to reach buried prey. Even if all prey are available to shorebirds, size may influence their profitability to a predator. From the perspective of optimal foraging, abundant, small prey may not be of sufficient energy value to elicit capture by shorebirds; very large prey may be easily stolen and not easily digested. Coupled to these environmental effects are features of shorebirds that may limit intake rate. Intake rate may vary owing to the selectivity of foraging individuals confronted by prey populations that vary in density, size, and accessibility, which all affect the profitability of foraging in a particular habitat. Lastly, shorebird density may alter the profitability of foraging in a particular habitat because intake rate varies with interference from conspecifics, which may include kleptoparasitism by conspecifics and birds of other species.

INDIVIDUAL VARIATION

Some studies have demonstrated considerable intraspecific variation in foraging behavior, especially as it relates to the repertoire of individuals of different age (experience) feeding on prey that are challenging to acquire. The circumstances are especially apparent in species that forage using maneuvers that require practice and learning to gain proficiency. For example, Black Oystercatchers use their chisel-like bill to pry limpets (*Collisella, Patella*) from rocky intertidal substrates, slit adductor muscles so that they may extract mussels (*Mytilus*) from their shells, and probe in coarse sand of the wash zone for large sand crabs (*Emerita*). These foraging maneuvers require varying degrees of practice. Moreover, prey have evolved responses to thwart oystercatcher predation. In rocky intertidal habitats, Eurasian Oystercatchers feed on limpets by prying individuals from the hard surfaces they occupy, but oystercatchers have been observed to attempt to capture only a single limpet amid an aggregation of many. Clearly, oystercatchers would do better to remain in an aggregation of limpets and continue to feed. However, recent evidence shows that limpets sense the vibrations of an oystercatcher feeding on a neighboring limpet and respond by clamping down

tightly to the substrate. As a result, the force required to remove limpets adjacent to a depredated one is significantly increased. Oystercatchers appear to know this because they typically leave a patch after feeding on a single limpet and move to another, distant aggregation to remove another limpet that does not sense what is about to happen (Coleman 2008).

Factors intrinsic to individuals (such as age, gender, and dominance) may contribute to intraspecific variation in intake rate. Age differences in foraging behavior and success have been documented for a few species of shorebird. Juveniles often are less successful than adults because they have not had sufficient time to perfect the foraging behaviors necessary to feed effectively. This is especially true in species (such as the Crab Plover and oystercatchers) that forage on difficult-to-capture prey (Goss-Custard and Le V. dit Durell 1987; Goss-Custard et al. 1998; Fasola et al. 1996), but it has also been shown for small sandpipers feeding on small macroinvertebrates (Dann 2000). Gender is another factor that may contribute to intraspecific variation in foraging behavior. For species that exhibit strong sexual dimorphism in body size and associated morphological characters that influence feeding (bill and hindlimb length), females and males may differ in habitat use, diet composition, foraging behavior, and intake rate. Larger females may be more capable of acquiring food that is otherwise unavailable to males who cannot probe as deeply for prey. This appears to be the case for several species of curlew (Townshend 1981). Finally, it is important to note that because the energetic demands of molt, migration, and breeding vary seasonally, foraging behavior will vary accordingly.

CONSERVATION IMPLICATIONS

Estimates of energy intake rates are of great utility to management and conservation, as they are intimately related to assessment of habitat quality (Goss-Custard et al. 2006). For example, by knowing whether or not shorebirds achieve the maximum intake rate, we can examine the effects of degradation of foraging habitat owing to human activity. In this case, the consequences of human disturbance may be gauged as the lost energy owing to human disruption of feeding. This may be particularly important for large-bodied shorebirds, such as oystercatchers or curlews, which respond at greater distances to human disturbance (Davidson and Rothwell 1993). However, if small-bodied species have less margin of error in their energy budgets, then they are more likely to exhibit the effects (in terms of reduced mass or increased mortality) of disturbance.

LITERATURE CITED

Alexander, S. A., K. A. Hobson, C. L. Gratto-Trevor, and A. W. Diamond. 1996. Conventional and isotopic determination of shorebird diets at an inland stopover: The importance of invertebrates and *Potamogeton pectinatus* tubers. *Canadian Journal of Zoology* 74: 1057–1068.

Baker, J. M. 1981. Winter feeding rates of redshank *Tringa totanus* and turnstone *Arenaria interpres* on a rocky shore. *The Ibis* 123: 85–87.

Battley, P. F., D. I. Rogers, T. Piersma, and A. Koolhaas. 2003. Behavioural evidence for heat-load problems in Great Knots in tropical Australia fuelling for long-distance flight. *The Emu* 103: 97–103.

Becker, M., M. A. Rubega, and L. W. Oring. 2002. Development of feeding mechanics in growing birds: Scything behavior in juvenile American Avocets (*Recurvirostra americana*). *Bird Behavior* 15: 1–10.

Burger, J. 1980. Age differences of foraging Black-necked Stilts in Texas. *The Auk* 97: 633–636.

Burger, J., and M. Gochfeld. 1986. Age differences in foraging efficiency of American Avocets *Recurvirostra americana*. *Bird Behavior* 6: 66–71.

Burger, J., M. A. Howe, D. C. Hahn, and J. Chase. 1977. Effects of tide cycles on habitat selection and habitat partitioning in migrating shorebirds. *The Auk* 94: 743–758.

Coleman, R. A. 2008. Overestimations of food abundance: Predator responses to prey aggregations. *Ecology* 89: 1777–1783.

Colwell, M. A., R. L. Mathis, L. W. Leeman, and T. S. Leeman. 2002. Space use and diet of territorial Long-billed Curlews (*Numenius americanus*) during the nonbreeding season. *Northwestern Naturalist* 83: 47–56.

Conklin, J. R., and M. A. Colwell. 2007. Diurnal and nocturnal roost site fidelity of Dunlin (*Calidris alpina pacifica*) at Humboldt Bay, California. *The Auk* 124: 677–689.

Cramp, S., and S. E. L. Simmons, eds. 1983. *Birds of the Western Palearctic*. Vol. 3. Oxford: Oxford University Press.

Dann, P. 2000. Foraging behavior and diets of red-necked stints and curlew sandpipers in southeastern Australia. *Wildlife Research* 27: 61–68.

Davidson, N. C., and P. I. Rothwell. 1993. Disturbance to waterfowl on estuaries: Conservation and management implications of current knowledge. *Wader Study Group Bulletin* 68: 97–105.

Davis, C. A., and L. M. Smith. 1998. Behavior of migrant shorebirds in playas of the southern High Plains. *The Condor* 100: 266–276.

del Hoyo, J., A. Elliot, and J. Sargatal, eds. 1996. *Handbook of the birds of the world*. Vol. 3, *Hoatzin to Auks*. Barcelona: Lynx Edicions.

Dodd, S. L., and M. A. Colwell. 1996. Seasonal variation in diurnal and nocturnal distributions of nonbreeding shorebirds at north Humboldt Bay, California. *The Condor* 98: 196–207.

———. 1998. Environmental correlates of diurnal and nocturnal foraging patterns of nonbreeding shorebirds. *Wilson Bulletin* 110: 182–189.

Dugan, P. J. 1981. The importance of nocturnal foraging in shorebirds: A consequence of increased invertebrate prey activity. In *Feeding and survival strategies of estuarine organisms,* ed. N. V. Jones, and W. J. Wolff, 251–260. New York: Plenum Press.

Elner, R. W., P. G. Beninger, D. L. Jackson, and T. M. Potter. 2005. Evidence of a new feeding mode in Western Sandpiper (*Calidris mauri*) and Dunlin (*Calidris alpina*) based on bill and tongue morphology and ultrastructure. *Marine Biology* 146: 1223–1234.

Engelmoer, M., T. Piersma, W. Altenburg, and R. Mes. 1984. The Banc d'Arguin (Mauritania). In *Coastal waders and wildfowl in winter*, ed. P. R. Evans, J. D. Goss-Custard, and W. G. Hale, 293–310. Cambridge: Cambridge University Press.

Ens, B. J., and J. T. Cayford. 1996. Feeding with other oystercatchers. In *The oystercatcher,* ed. J. D. Goss-Custard, 77–104. Oxford: Oxford University Press.

Ens, B., and J. D. Goss-Custard. 1984. Interference among oystercatchers, *Haematopus ostralegus,* feeding on mussels, *Mytilus edulis,* on the Exe estuary. *Journal of Animal Ecology* 53: 217–231.

Evans, P. R. 1979. Adaptations shown by foraging shorebirds to cyclic variations in the activity and availability of their intertidal invertebrate prey. In *Cyclical phenomena in marine plants and animals*, ed. E. Naylor and R. G. Hartnoll, 357–366. Oxford/New York: Pergamon Press.

———. 1988. Predation of intertidal fauna by shorebirds in relation to time of the day, tide and year. In *Behavioural adaptations to intertidal life*, ed. G. Chelazzi and M. Vannini, 65–78. New York: Plenum Press.

Evans, P. R., and P. J. Dugan. 1984. Coastal birds: Numbers in relation to food resources. In *Coastal waders and wildfowl in winter*, ed. P. R. Evans, J. D. Goss-Custard, and W. G. Hale, 8–28. Cambridge: Cambridge University Press.

Evans Ogden, L. J., K. A. Hobson, D. B. Lank, and S. Bittman. 2006. Stable isotope analysis reveals that agricultural habitat provides an important dietary component for nonbreeding Dunlin. *Avian Conservation and Ecology* 1: 3.

Fasola, M., and L. Biddau. 1997. An assemblage of wintering waders in Kenya: Activity budget and habitat use. *African Journal of Ecology* 35: 339–350.

Fasola, M., and L. Canova. 1993. Diel activity of resident and immigrant waterbirds at Lake Turkana, Kenya. *The Ibis* 135: 442–450.

Fasola, M., L. Canova, and L. Biddau. 1996. Foraging habits of Crab Plovers *Dromas ardeola* overwintering on the Kenyan coast. *Colonial Waterbirds* 19: 207–213.

Ferns, P. N. 1983. Sediment mobility in the Severn Estuary and its influence on the distribution of shorebirds. *Canadian Journal of Fisheries and Aquatic Sciences* 40 (Suppl. 1): 331–340.

Fjeldså, J., and N. Krabbe. 1990. *Birds of the high Andes*. Copenhagen: University of Denmark.

Geppetti, L., and P. Tongiorgi. 1967. Nocturnal migrations of *Talitrus saltator* (Crustacea Amphipoda). *Monitore Zoologica Italiano* 1: 37–40.

Gerritsen, A. F. C., and Y. M. van Heezik. 1985. Substrate preferences and substrate related foraging behaviour in three *Calidris* species. *Netherlands Journal of Zoology* 35: 671–692.

Goss-Custard, J. D. 1969. The wintering feeding ecology of the Redshank *Tringa totanus*. *The Ibis* 111: 338–356.

———. 1970. The responses of Redshank (*Tringa totanus* (L.)) to spatial variations in the density of their prey. *Journal of Animal Ecology* 39: 91–113.

———. 1979. The energetics of foraging by Redshank, *Tringa totanus*. *Studies in Avian Biology* 2: 247–257.

———. 1980. Competition for food and interference among waders. *Ardea* 68: 31–52.

———. 1984. Intake rates and food supply in migrating and wintering shorebirds. In *Shorebirds: Migration and foraging behavior*, ed. J. Burger and B. L. Olla, 233–270. New York: Plenum Press.

———. 1985. Foraging behaviour of wading birds and the carrying capacity of estuaries. In *Behavioural ecology*, ed. R. M. Sibly and R. H. Smith, 169–188. Oxford: Blackwell.

Goss-Custard, J. D., J. T. Cayford, and S. E. G. Lea. 1998. The changing trade-off between food finding and stealing in juvenile oystercatchers. *Animal Behavior* 55: 745–760.

Goss-Custard, J. D., R. A. Jenyon, R. E. Jones, P. E. Newberry, and R. L. Williams. 1977. The ecology of The Wash. II. Seasonal variation in the feeding conditions of wading birds (Charadrii). *Journal of Applied Ecology* 14: 701–719.

Goss-Custard, J. D., R. E. Jones, and P. E. Newberry. 1977. The ecology of The Wash. I. Distribution and diet of wading birds (Charadrii). *Journal of Applied Ecology* 14: 681–700.

Goss-Custard, J. D., and S. E. A. Le V. dit Durell. 1987. Age-related effects of oystercatchers, *Haematopus ostralegus*, feeding on mussels, *Mytilus edulis*. I. Foraging efficiency and interference. *Journal of Animal Ecology* 56: 521–536.

———. 1988. The effect of dominance and feeding method on the intake rates of oystercatchers, *Haematopus ostralegus*, feeding on mussels. *Journal of Animal Ecology* 57: 827–844.

Goss-Custard, J. D., A. D. West, M. G. Yates, R. W. G. Caldow, R. A. Stillman, L. Bardsley, J. Castilla, et al. 2006. Intake rates and the functional response in shorebirds (Charadriiformes) eating macroinvertebrates. *Biological Review* 81: 501–529.

Grant, J. 1984. Sediment microtopography and shorebird foraging. *Marine Ecology Progress Series* 19: 293–296.

Groves, S. 1978. Age-related differences in Ruddy Turnstone foraging and aggressive behavior. *The Auk* 95: 95–103.

Hamilton, R. B. 1975. Comparative behavior of the American Avocet and the Black-necked Stilt (recurvirostridae). *Ornithological Monographs* 17: 1–98.

Hockey, P. A. R. 1996. *Haematopus ostralegus* in perspective: Comparisons with other oystercatchers. In *The oystercatcher*, ed. J. D. Goss-Custard, 251–285. Oxford: Oxford University Press.

Hockey, P. A. R., R. A. Navarro, B. Kalejta, and C. R. Velasquez. 1992. The riddle of the sands: Why are shorebird densities so high in southern estuaries? *American Naturalist* 140: 961–979.

Holling, C. S. 1959. Some characteristics of simple types of predation and parasitism. *Canadian Entomologist* 91: 385–398.

Hötker, H. 1999. What determines the time-activity budgets of avocets (*Recurvirostra avosetta*)? *Journal of Ornithology* 140: 57–71.

Hulscher, J. B. 1996. Food and feeding behaviour. In *The Oystercatcher,* ed. J. D. Goss-Custard, 7–29. Oxford: Oxford University Press.

Jönsson, P. E., and T. Alerstam. 1990. The adaptive significance of parental care role division and sexual size dimorphism in breeding shorebirds. *Biological Journal of the Linnean Society* 41: 301–314.

Kalejta, B. 1992. Time budgets and predatory impacts of waders at the Berg River estuary, South Africa. *Ardea* 80: 327–342.

Kelsey, M. G., and M. Hassall. 1989. Patch selection by Dunlin on a heterogeneous mudflat. *Ornis Scandinavica* 20: 250–254.

Kuwae, T. 2007. Diurnal and nocturnal feeding rate of Kentish plovers *Charadrius alexandrinus* on an intertidal flat as recorded by telescopic video system. *Marine Biology* 151: 663–673.

Le V. dit Durell, S. E. A. 2007. Differential survival in adult Eurasian Oystercatchers *Haematopus ostralegus. Journal of Avian Biology* 38: 530–535.

Le V. dit Durell, S. E. A., J. D. Goss-Custard, and R. W. G. Caldow. 1993. Sex-related differences in diet and foraging behavior in the oystercatcher *Haematopus ostralegus. Journal of Animal Ecology* 62: 205–215.

Leeman, L. W., M. A. Colwell, Leeman, T. S., and R. L. Mathis. 2001. Diets, energy intake, and kleptoparasitism of nonbreeding Long-billed Curlews in a northern California estuary. *Wilson Bulletin* 113: 194–201.

Lourenço, P. M., J. P. Granadeiro, and J. M. Palmeirim. 2005. Importance of drainage channels for waders foraging on tidal flats: Relevance for the management of estuarine wetlands. *Journal of Applied Ecology* 42: 477–486.

McNeil, R., P. Drapeau, and J. D. Goss-Custard. 1992. The occurrence and adaptive significance of nocturnal habits in waterfowl. *Biological Reviews* 67: 381–419.

Mouritsen, K. N. 1994. Day and night feeding Dunlins *Calidris alpina*: Choice of habitat, foraging technique, and prey. *Journal of Avian Biology* 25: 55–62.

Myers, J. P., S. L. Williams, and F. A. Pitelka. 1980. An experimental analysis of prey availability for Sanderlings (Aves: Scolopacidae) feeding on sandy beach crustaceans. *Canadian Journal of Zoology* 58: 1564–1574.

Pienkowski, M. W. 1983. Surface activity of some intertidal invertebrates in relation to temperature and the foraging behavior of their shorebirds. *Marine Ecology Progress Series* 11: 141–150.

Pienkowski, M. W., P. N. Ferns, N. C. Davidson, and P. R. Evans. 1984. Balancing the budget: Measuring the energy intake and requirements of shorebirds in the field. In *Coastal waders and wildfowl in winter,* ed. P. R. Evans, J. D. Goss-Custard, and W. G. Hale, 29–56. Cambridge: Cambridge University Press.

Pierce, R. J. 1979. Foods and feeding of the Wrybill (*Anarhynchus frontalis*) on its riverbed breeding grounds. *Notornis* 26: 1–21.

Piersma, T. 1987. Production of intertidal benthic animals and limits to their reduction by shorebirds: A heuristic model. *Marine Ecology Progress Series* 38: 187–196.

Puttick, G. M. 1984. Foraging and activity patterns in wintering shorebirds. In *Shorebirds: Migration and foraging behavior,* ed. J. Burger and B. L. Olla, 203–231. New York: Plenum Press.

Quammen, M. L. 1982. Influence of subtle substrate differences on feeding by shorebirds on intertidal mudflats. *Marine Biology* 71: 339–343.

Recher, H. F., and J. Recher. 1969. Some aspects of the ecology of migrant shorebirds. II. Aggression. *Wilson Bulletin* 81: 140–154.

Robert, M., R. McNeil, and A. Leduc. 1989. Conditions and significance of night feeding in shorebirds and other water birds in a tropical lagoon. *The Auk* 106: 94–101.

Rojas de Azuaje, L. M., R. McNeil, T. Cabana, P. Lachapelle. 1999. Diurnal and nocturnal visual capabilities in shorebirds as a function of their feeding strategies. *Brain, Behavior and Evolution* 53: 29–43.

Rubega, M. A., and B. S. Obst. 1993. Surface-tension feeding in phalaropes: Discovery of a novel feeding mechanism. *The Auk* 110: 169–178.

Sanctis, De. A., L. Biddau, and M. Fasola. 2005. Post-migratory care of young by Crab Plovers *Dromas ardeola. The Ibis* 147: 490–497.

Shepherd, P. C. F., and D. B. Lank. 2004. Marine and agricultural habitat preferences of Dunlin wintering in British Columbia. *Journal of Wildlife Management* 68: 61–73.

Sitters, H., C. Minton, P. Collins, B. Etheridge, C. Hassell, and F. O'Connor. 2004. Extraordinary numbers of Oriental Pratincoles in NW Australia. *Wader Study Group Bulletin* 103: 26–31.

Skagen, S. K., and H. D. Oman. 1996. Dietary flexibility of shorebirds in the Western Hemisphere. *Canadian Field-Naturalist* 110: 419–444.

Smith, P. C. 1975. A study of the winter feeding ecology and behaviour of the Bar-tailed Godwit *Limosa lapponica*. Ph.D. thesis, University of Durham.

Stillman, R. A., A. E. Poole, J. D. Goss-Custard, R. W. G. Caldow, M. G. Yates, and P. Triplet. 2002. Predicting the strength of interference more quickly using behaviour-based models. *Journal of Animal Ecology* 71: 532–541.

Sutherland, W. J. 1983. Aggregation and the ideal free distribution. *Journal of Animal Ecology* 52: 821–828.

Sutherland, W. J., B. J. Ens, J. D. Goss-Custard, and J. B. Hulscher. 1996. Specialization. In *The oystercatcher*, ed. J. D. Goss-Custard, 56–76. Oxford: Oxford University Press.

Swennen, C., L. L. N. deBruijn, P. Duivan, M. F. Leopold, and E. C. L. Marteijn. 1983. Differences in the bill form of the Oystercatcher *Haematopus ostralegus*; a dynamic adaptation to different foraging techniques. *Netherlands Journal of Sea Research* 17: 57–83.

Townshend, D. J. 1981. The importance of field feeding to the survival of wintering male and female curlews *Numenius arquata* on the Tees Estuary. In *Feeding and survival strategies of estuarine organisms*, ed. N. V. Jones and W. J. Wolff, 261–273. New York: Plenum Press.

Triplet, P., R. A. Stillman, and J. D. Goss-Custard. 1999. Prey abundance and the strength of interference in a foraging shorebird. *Journal of Animal Ecology* 68: 254–265.

Tsipoura, N., and J. Burger. 1999. Shorebird diet during spring migration stopover on Delaware Bay. *The Condor* 101: 635–644.

Warnock, N. 1989. Piracy by Ring-billed Gulls on Dunlins. *Wilson Bulletin* 101: 96–97.

Warnock, N., G. W. Page, and L. E. Stenzel. 1995. Non-migratory movements of Dunlins on their California wintering grounds. *Wilson Bulletin* 107: 131–139.

Wiens, J. A. 1989. *The ecology of bird communities.* Vol. 1. Cambridge: Cambridge University Press.

Wilson, W. H., Jr. 1991. The foraging ecology of migratory shorebirds in marine soft-sediment communities: The effects of episodic predation on prey populations. *American Zoologist* 31: 840–848.

Wood, A. G. 1983. Grey plover time budgets—a hard day's night. *Wader Study Group Bulletin* 39: 51.

Zwarts, L., A. Blomert, and R. Hupkes. 1990. Increase of feeding time for waders preparing for spring migration from the Banc d'Arguin, Mauritania. *Ardea* 78: 237–256.

8

Shorebirds as Predators

CONTENTS

THE RATHER SIMPLE DIETS OF shorebirds coupled with the comparative ease with which most species can be observed and counted while feeding in open habitats makes for ideal conditions for studying both theoretical and applied aspects of predation. From a theoretical perspective, shorebirds often feed in dense flocks where prey are most available; conversely, shorebirds tend to avoid areas where food is sparse. Moreover, intake rates often vary greatly across habitats of varying prey density. Individuals are most successful in habitats where prey are densest, although interference may compromise intake rates at high conspecific densities. These observations are the essence of the functional and numerical responses of a predator to variation in the density of a prey species. Consequently, considerable interest has been directed at studying foraging shorebirds across a range of habitats of varying prey density. These relationships have strong applied value as well. The notion of carrying capacity, or the amount of food resource necessary to sustain a local population occupying an estuary during the nonbreeding season, is the foundation of wetland conservation efforts and strategies to manipulate habitats to increase food availability. Consequently, over several decades, considerable research has addressed evidence that shorebirds are food-limited during the nonbreeding season. The extent to which this is true has consequences for the survival of individuals and population size. In this chapter, I examine the functional and numerical responses of shorebird predators to variation in the density of their prey. I summarize what is known about the extent to which shorebirds reduce prey populations to low levels with negative consequences for individuals and populations. Finally, I explore various ecological patterns that suggest that food limitation has played an important role in shaping ecomorphological patterns of bills and assemblages of nonbreeding species at wetlands during the winter and on migration.

SHOREBIRD PREDATORS AND THEIR PREY

As evidenced by ecomorphology, the feeding ecology of shorebirds is strongly influenced by their prey. As predators of macroinvertebrates, shorebirds respond in two ways to variation in prey abundance. First, individual shorebirds exhibit a functional response, in which their feeding or energy intake rates vary with prey density. Feeding rates of individuals have been shown to increase predictably to some upper limit in association with increases in prey abundance. These behavioral responses of individuals manifest themselves in another ecological relationship, that of a numerical response of a predator to its prey: the individual-based habitat choices translate into population-level responses. Traditionally, the numerical response of a predator refers to the changing size of a population as resources become increasingly abundant. Individuals that feed more effectively convert resources into offspring, which translates into population growth. Predators often aggregate in areas of high prey density, and this facet of the numerical response is readily apparent in the observation that shorebird distributions are often positively correlated with the abundance of prey populations. These two facets of predator–prey relationships have received considerable treatment in the literature on shorebirds, perhaps owing to the value of understanding these responses as measures of habitat quality and the consequences to populations of human-caused loss and degradation of habitat.

FUNCTIONAL RESPONSE

A fundamental ecological question concerns variation in the feeding rate of a predator in response to variation in the density of its prey (Holling 1959). The functional response of a shorebird is quantified by observing foraging individuals and estimating the rate at which prey are eaten (h^{-1}) or energy is consumed ($kJ\ h^{-1}$). The functional response is the outcome of the sum of behaviors related to searching for, acquiring, and handling individual prey items as well as the rate at which food is digested. Ecologists recognize three types of functional response (Fig. 8.1; Krebs 2009). A type 1 functional response is characterized by a linear increase in the number of prey eaten as prey density increases. This response is rarely or never reported in studies of vertebrates because predators eventually reach an upper limit in their abilities to handle and digest prey. In other words, predators eventually become satiated when feeding in areas of high prey density. As a result, the feeding rate plateaus, which is evident in the other two types of functional response. These two other types of functional response are usually reported for vertebrates. A type 2 functional response is characterized by a gradual increase in feeding rate that decelerates at intermediate prey densities. This is the typical functional response exhibited by most shorebirds feeding on invertebrates. Alternatively, a type 3 functional response is characterized by a more abrupt change in feeding behavior at intermediate prey densities such that the predator switches its searching behavior to forage almost exclusively on a single, abundant prey. Behavioral ecologists refer to predators developing a "search image," which corresponds to this abrupt change in searching behavior at intermediate prey density. No shorebird has been shown to have a type 3 response.

Virtually all studies of shorebirds that eat approximately the same size of prey (that is, the prey do not vary in energy value, only density) have reported that the general shape of the functional response is characterized by a decelerating rise to a plateau (Goss-Custard et al. 2006). The type 2 response is typical of predators feed-

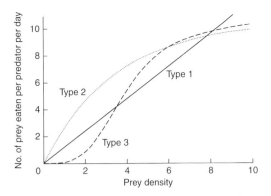

FIGURE 8.1. Three types of functional response that characterize the feeding rates of predators in response to variation in the density of prey. A Type 1 response is a linear increase in the number of prey consumed per unit time. A Type 2 response is commonly observed in shorebirds feeding on a variety of invertebrates, with their rates of prey consumption reaching a plateau set by the time required to handle prey or digestive rate. A Type 3 response is typical of a predator switching to a particular prey type and developing a search image at intermediate prey densities.

ing on a single species of prey, where the asymptote of the curve is reached when an individual's intake rate is constrained by the time required to handle a prey item, rather than locate food. Several well-chosen examples illustrate the methods, results, and application of quantifying the functional response of shorebirds.

The Eurasian Oystercatcher is probably the best-studied shorebird owing to its large size and its diet of large prey. Goss-Custard et al. (2006) summarized results of many studies of oystercatchers feeding principally on hard-shelled bivalves, either cockles (*Cerastoderma*) or mussels (*Mytilus*). Oystercatchers exhibited a classic type 2 functional response; however, intake rate varied with attributes of prey. At a given prey density, oystercatchers had higher intake rates when feeding on large prey, especially during the breeding season; conversely, intake rates were lower when oystercatchers fed on mussels.

In some cases, a single, uniform-sized food item dominates the diet for short intervals, providing the opportunity to quantify in remarkable detail the functional response of shorebirds using different foraging maneuvers. Such was the case with a suite of species that stage each spring at

Delaware Bay, where shorebirds feed on the small (2 to 3 mm diameter) but energy-rich resource of horseshoe crab eggs. Gillings et al. (2007) examined the intake rates of three calidridine sandpipers (the Semipalmated Sandpiper, Sanderling, and Red Knot) that numerically dominated foraging flocks. The researchers experimentally manipulated density and availability of food by placing eggs on the surface of a sand-filled tray or with a majority buried in the sand at depths up to 5 cm; they placed trays in the path of foraging sandpipers. These two conditions mimic natural factors that influence the availability of crab eggs. Spawning crabs bury their eggs at high tide, but the action of waves and other crabs as well as feeding by shorebirds cause eggs to be brought to the surface. All three species of shorebird exhibited a classic type 2 functional response whether eggs were placed on the substrate surface or buried. But intake rates were higher when shorebirds pecked eggs from the surface. Pecking was twice as successful as probing, which indicates that intake rates of visual versus tactile feeding shorebirds differed substantially. This was evidenced by the shallower functional response curves of probing versus pecking sandpipers. Gillings et al. (2007) applied their findings to estimate the amount of time required for Red Knots to fulfill their daily requirement of 24,000 eggs. They concluded that few beaches where crabs spawned provided buried eggs at densities that were profitable to foraging shorebirds.

A final example of a shorebird functional response was presented by Gill et al. (2001) who examined the intake rate of Black-tailed Godwits wintering at six estuaries in southeast England. During focal observations, individual godwits exhibited quite consistent patterns of prey selection, with the vast majority of their foraging attempts focused on a single prey type, principally small bivalves (*Scrobicularia*, *Macoma*, and *Mya*). Godwits exhibited a classic type 2 functional response to variation in bivalve density (Fig. 8.2). At low prey densities, intake rate declined rapidly as birds searched for prey. At high densities, however, intake rate leveled off because godwits were constrained by the

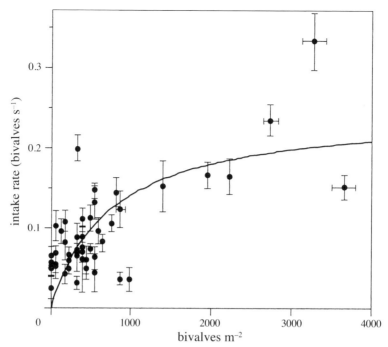

FIGURE 8.2. A typical Type 2 functional response of a shorebird predator to variation in the density of its invertebrate prey. In this case, the shorebird predator is the Black-tailed Godwit feeding almost exclusively on small (5 to 20 mm shell width) bivalves (*Scrobicularia*, *Macoma*, and *Mya*) at estuaries on the southeast coast of England. From Gill et al. (2001).

time required to handle or swallow prey rather than by time required to search for prey.

Goss-Custard et al. (2006) reviewed a wealth of studies with the aim of characterizing the functional response of shorebirds in hopes of developing a clear predictive relationship so that future researchers could avoid the costly efforts of collecting field data for every species. Several important findings come from their work. First, most studies demonstrated that shorebirds exhibit a classic type 2 functional response. At low prey density, individuals have low intake rates owing to infrequent encounters with prey. As prey density increases, these encounters rise and so does intake rate. With increasing prey density, however, intake rate decelerates and eventually levels off. This plateau is often reached at quite low densities of prey. The low asymptote in intake rate may occur for several reasons. First, individuals may be limited by their handling time, especially when feeding on large prey. Second, intake rates may be low because of

physiological constraints on the time required to digest large prey. In other words, the digestive machinery of the shorebird gut may not be able to keep pace with feeding rate. This has been shown for the Eurasian Oystercatcher feeding on cockles (Kersten and Visser 1996), and it is likely that other large shorebirds (such as curlews) feeding on large prey (such as burrowing shrimp and polychaetes) exhibit similar digestive constraints. For example, during low-tide intervals, Long-billed Curlews occasionally stop feeding and either rest or preen for up to 10 minutes after ingesting large prey (Leeman et al. 2001). Finally, Goss-Custard et al. (2006) suggested that the low intake rates of shorebirds may also be related to the perceptual constraints of foraging using visual cues. Specifically, some shorebirds have to perceive prey before they can handle and digest it, and this perceptual constraint may limit encounter rates.

Another important finding reported by Goss-Custard et al. (2006) was that intake rates often varied greatly over a range of prey densities; moreover, there was considerable variation even at a particular prey density. In most examples, intake rates leveled off at low densities of prey, considerably lower than expected. This was accompanied by the observation that shorebirds in 11 separate studies spent only 51% of their time actually feeding (pecking at or handling prey). The remainder of the time, individuals searched for prey. This suggested that individuals did not achieve the maximum intake rate predicted based on prey density. In a comparative analysis across species that varied greatly in body size, much variation in intake rate was explained by the mass of a shorebird and its prey mass. Larger species had higher intake rates because they could use their larger gape to consume larger prey.

These observations illustrate that a variety of factors affect the functional response of shorebirds and hence alter the foraging proficiency of individuals. Clearly, characteristics of prey (such as size), behaviors of shorebirds (visual versus tactile feeding), and environment (such as the depth of prey or the weather) can directly influence the form of a functional response by a shorebird predator. As a consequence, individuals may not achieve maximum energy intake rate at a particular density of prey. The problems of energy intake associated with the functional response are exacerbated when humans disturb foraging shorebirds and further compromise their ability to feed effectively by increasing the amount of time the shorebirds spend being vigilant.

NUMERICAL RESPONSE

The density of prey may also influence the abundance of a predator. In its original formulation, the numerical response described the change in the per capita reproductive rate of a predator population in relation to food density (Soloman 1949). However, researchers commonly examine the aggregative (behavioral) response of predators to variation in their prey as a surrogate for reproductive rate. In this case, a shorebird population may increase over short intervals owing to the concentration of individuals in high-quality habitats where food is rich and readily available. Over longer intervals, populations may increase owing to the positive effect of increased food on survival and reproduction. As illustrated earlier, there is a direct relationship between the functional and numerical response of shorebirds: birds aggregate in areas of high prey density where intake rate is high.

For shorebirds, prey density is an important factor that influences the tendency for foraging individuals to use certain habitats and avoid others. At a variety of spatial scales, shorebirds "map onto" populations of their prey, as evidenced by a positive correlation between density of predator and prey populations (Evans and Dugan 1984; Colwell and Landrum 1993); this aggregative response is evident at a variety of spatial scales. At a global scale, Hockey et al. (1992) addressed the "riddle of the sands," or why densities of wintering shorebirds were higher at southern estuaries in the East Atlantic Flyway. They found that shorebird densities correlated positively with food availability. In the New World, Schott (1931a, 1931b) and Butler et al. (2001) observed that the main migratory routes and wintering areas of Nearctic shorebirds occurred in areas of high productivity in coastal areas; Butler et al. (2001) speculated that this was true of other areas of the world. At the scale of estuaries, many studies have shown that shorebirds aggregate in habitats where prey density is high. For example, over the many kilometers' reach of the Firth of Forth, Scotland, the densities of six shorebird species correlated positively with their principal prey species (Bryant 1979). Similar results were shown for shorebirds feeding on diverse prey at other estuaries worldwide.

The generalization that shorebirds aggregate in areas of high prey abundance applies to habitats other than estuaries and at finer spatial scales. Dugan et al. (2003) reported that densities of the Black-bellied and Snowy Plovers correlated positively with the density of their principal prey, burrowing amphipods (*Megalorchestia*), across a sample of sandy, ocean-fronting beaches in southern California. At spatial scales spanning meters and including diverse substrates, densities of small calidridine sandpipers correlated positively with their prey (Colwell and Landrum 1993), although not always (Wilson 1990). Even when significant positive correlations existed, however, the densities of shorebirds varied greatly and were often low in areas of high prey density (Colwell and Landrum 1993), perhaps because high densities represented small, unprofitable prey. In conclusion, shorebird densities correlate positively with the densities of their prey at all spatial scales of analysis, but the correlations are strongest at coarse scales.

The relatively straightforward numerical relationship between shorebird predators and densities of their prey is made more complex by the various features of the habitat and the prey species themselves that affect how successfully shorebirds feed. In fact, most studies reporting significant numerical relationships highlight that there is considerable unexplained variation about the regression line relating predator and prey densities. This probably results from simple measures of prey density that fail to account for differences in the size and hence the energy value of different prey. These attributes of prey are readily apparent to shorebirds selecting from among foraging sites that differ in profitability. It may also be that subtleties of substrate particle size, compactness, and wetness compromise the availability of prey in some microhabitats.

PREDICTING WETLAND USE

The conservation value of understanding functional and numerical responses is realized when this knowledge is applied to predicting the population sizes of shorebirds that can be supported by wetlands that differ in size and amount of food resources. Simply stated, estuaries with rich and replenishing food supplies should support more birds than wetlands of lower quality and diminishing food resources. Moreover, the negative consequences of anthropogenic habitat loss and degradation, and hence reduction in food availability, should be apparent in the reduced population size of overwintering shorebirds. These observations also highlight the value of quantifying the extent to which shorebirds reduce the abundance of their food supplies to levels that have consequences for individuals and populations. As such, they are the foundation of the concept of carrying capacity and the notion that competition for a limiting food resource limits local population size. For decades, a major conservation objective directed at shorebirds has entailed understanding the relationships between food supply and the number of birds wintering in estuaries and other wetlands worldwide (see Evans and Dugan 1984; Senner and Howe 1984; Goss-Custard 1985; van Gils et al. 2004). In principle, knowledge of prey density, predator intake rate, and shorebird density allows one to predict the population of shorebirds that can be sustained by food resources at estuaries of varying size.

CARRYING CAPACITY

The concept of carrying capacity is the foundation of conservation and management directed at habitats occupied by shorebird populations during the nonbreeding season (Hale 1980; Evans 1984; Evans and Dugan 1984; Goss-Custard 1985). So it is important to be clear about the meaning of carrying capacity, as well as the limitations of its use in this context. Newton (1998) articulated that carrying capacity encapsulates the idea that in any habitat (estuary) of given size, resources (invertebrate food supply) must limit the number of individuals that can overwinter there. Here, limitation means specifically that periods of food shortage act to set an upper bound to the number of individuals that can occupy an estuary. When food is in short supply, individuals either leave a site in search of other suitable wintering areas or they die because they cannot find sufficient food. In this case, competition is the ecological process that limits population size. The mechanism by which competition acts is via density dependence. That is, as individuals increase in number, the effect of the population on food is to decrease its availability, either directly via exploitation or indirectly via interference. Food shortage acts to increase the per capita rate of mortality: a greater proportion of the population dies as competition for food intensifies. As pointed out earlier, evidence is strong that these mechanisms operate in some populations of wintering shorebirds, although the strength of evidence varies greatly depending on species, location, season, and year. When food is sufficiently short in supply, especially for long durations, the effect of limiting food supplies may be particularly noticeable in the condition of individuals and in population size.

Over the years, several lines of evidence have been used to evaluate the idea that shorebirds are limited by their food supplies and hence that populations occupying particular wintering sites are at carrying capacity. Virtually all of these arguments have been directed at coastal estuaries, owing to their importance to populations of shorebirds and because of the dire threats to the existence and quality of these foraging habitats posed by anthropogenic habitat loss and degradation. To date, evidence varies from cursory examination of variation in population sizes to increasingly detailed arguments involving linkages between food supply, individual intake rate, and population consequences.

POPULATION COUNTS

One approach commonly used to address whether a population residing at an estuary is near carrying capacity is to analyze population trends. Specifically, researchers have examined long-term data sets documenting annual variation in shorebird abundance at individual estuaries and reasoned that if habitats are full, then a population should be relatively stable from year to year. By contrast, substantial year-to-year variation in numbers suggests that, at least in years of low numbers, estuaries must not be at carrying capacity. This approach is strengthened when combined with regional trends showing an overall increase in population size.

The British maintain arguably the best long-term database on shorebird populations wintering at individual estuaries (Birds of Estuaries Enquiry, Wetland Birds Survey; see Chapter 10). Several authors have capitalized on this information to address questions of local carrying capacity. Hale (1980) presented selected examples of

nationwide and estuary-specific population estimates. For example, a 3-year (1971 to 1973) comparison of the Black-bellied Plover population in the British Isles showed considerable annual variation. Observers tallied approximately 16,000 plovers in October 1973, which was nearly four times the tally 2 years earlier. The annual differences persisted throughout the subsequent 6 months of each year. Hale (1980) argued that the fourfold increase in numbers in 1973 was evidence that coastal estuaries could not have been at carrying capacity in the previous 2 years. Similar arguments have been applied to local populations of shorebirds at individual estuaries. Hale (1980) presented several years of data (from Prater 1977) for the total number of shorebirds wintering at 33 estuaries around Great Britain. The relative constancy of year-to-year totals for these sites, as well as a trend for larger estuaries to support greater numbers of wintering shorebirds, are evidence, albeit weak, that these estuaries had reached carrying capacity. Although this approach seems plausible, the absence of direct measures of food supply and intake rate as well as details on weather conditions made it difficult to draw firm conclusions about carrying capacity. From the perspective of wintering populations of shorebirds, these observations emphasize that carrying capacity of a habitat varies over time in association with the productivity of invertebrates and all the weather-related phenomena that may affect their availability to shorebirds. Weather also varies in its effect on the energy requirements of overwintering birds. During severe winters, shorebirds require more food to survive whereas in mild years energy requirements are less. In other words, carrying capacity is a dynamic, moving target, which makes the utility of the concept even more challenging.

An excellent use of estuary and national counts was provided by Moser (1988), who analyzed the winter abundance of Black-bellied Plovers at 45 estuaries around Great Britain. Data for some estuaries go back decades. Plover populations appeared to have stabilized at many estuaries along the south and west of England. This conclusion was based on regressions of maximum counts over time; a slope that was not significantly different from zero was evidence of stable numbers. At other estuaries, mostly in the north and east, plover populations continued to add birds. Moser (1988) argued that, because the national population index for the plover continued to increase at the same time that numbers were steady at some estuaries, these local populations had reached carrying capacity. The Black-bellied Plover example is illustrative because it combines long-term estuary counts, regional population estimates, and knowledge of the species' foraging behavior and social system to make a case that local estuaries are at carrying capacity. Specifically, because plovers use visual cues to detect prey, they are more likely to suffer negative consequences of interference as bird density increases (Goss-Custard 1985). Moreover, the nonbreeding social organization of the plover involves some individuals defending long-term feeding territories (Townshend 1981). As a result, territorial individuals are likely to limit the settlement of newly arriving birds each winter. However, this argument does not apply well to most other shorebirds, which tend to feed in flocks (Goss-Custard 1985).

There are challenges of applying the previous population approach to evaluating carrying capacity because a local population does not feed with equal intensity across all habitats of an estuary. Rather, individuals distribute themselves in patches of habitat characterized by food resources of varying quality. When this happens, local population trends may not accurately represent an estuary's carrying capacity. This appears to be true of the Eurasian Oystercatcher feeding on mussel beds of varying quality in the Exe estuary in England (Goss-Custard et al. 1998). Long-term population indices showed a regionwide increase in oystercatcher populations during 1976 to 1994. The number of oystercatchers wintering on the Exe estuary, however, remained stable over this same period. This would suggest that the local population had reached a carrying capacity dictated by bivalve food supply. However, detailed sampling of bird use of different mussel beds of varying quality

showed that densities of oystercatcher on prime beds increased over this period. This increase occurred despite an unchanged food supply and increased disturbance by humans and corvids. Goss-Custard et al. (1998) argued that because bird densities on high-quality mussel beds continued to increase, the local population could not have been food-limited, and the estuary habitats had not reached carrying capacity (despite stable numbers).

Researchers have used population counts in a second method of evaluating the carrying capacity of an estuary. In this case, however, the analysis compares shorebird population size at a site before and after habitat loss. This approach makes several important assumptions. First, it assumes that a local population was at carrying capacity before habitat conversion. Second, it assumes that the loss of habitat equates to a comparable reduction in food supply. The reasoning is that loss of habitat translates directly to a reduced number of birds that can be supported by a smaller total area of habitat because food supplies are insufficient to support the same number of birds as before the habitat was compromised. Finally, this quasi experiment assumes that nothing else has changed between time intervals (before and after) that would affect population size, including annual variation in the productivity of breeding populations.

Several studies have used this before–after reasoning to assess carrying capacity and the consequences of habitat loss for local populations. Evans (1979b) reported that, after port development and loss of 61% (360 to 140 hectares) of mudflat in the Tees estuary in northeast England, local populations of shorebird declined by 32% to 88%. This reduction in local population size occurred at the same time that nationwide population estimates were increasing for the same shorebirds. Laursen et al. (1983) reported similar findings for the Danish Wadden Sea. Beginning in 1980, 1,100 hectares of intertidal foraging habitats and important prey populations of shorebirds were removed. Populations of 8 of 12 species declined by up to 85% subsequent to this loss of habitat. Finally, introduced plant species sometimes alter estuarine feeding habitats in significant ways such that foraging areas are effectively lost. The spread of an introduced cordgrass (*Spartina angelica*) in some British estuaries caused a significant reduction in intertidal foraging habitats because this plant colonized tidal flats and converted them to saltmarsh. Dunlin populations declined significantly at estuaries that experienced significant losses of foraging habitat (Goss-Custard and Moser 1988).

MATHEMATCAL MODELS

Recently, attempts to understand the carrying capacity of estuaries have been extended to include resource depletion models (Sutherland and Anderson 1993). This approach uses detailed data sampling of invertebrate prey and observations of intake rate to predict shorebird density supported by food resources. Although resource depletion models have been applied to fine-scale habitat questions, in principle they can be applied to differences among estuaries that differ in food resources. Sutherland and Anderson's (1993) model makes use of the functional response of shorebirds, especially two parameters—prey-handling time and searching efficiency—to predict bird use of habitats. The classic type 2 functional response typical of shorebirds indicates that intake rate is constrained by prey-handling time rather than the time required to search for prey. Accordingly, at intermediate to high prey densities, shorebirds are not capable of depleting prey because they cannot feed fast enough. Consequently, the carrying capacity of an estuary is predicted to be high where prey-handling time does not permit the predator to deplete a food resource over the winter. By contrast, at lower prey densities, shorebirds are more capable of depleting food resources, which results in lower shorebird populations. The value of this approach in predicting shorebird density is clear. If, as pointed out by Gill et al. (2001), one can predict predator density using this model, then it will be possible to predict the population consequences of habitat loss or degradation.

Gill et al. (2001) provided an excellent example of this approach in their study of Black-tailed Godwits wintering at six estuaries across the southeast coast of England. Godwits wintering in this region are the *islandica* subspecies. Godwits feed nearly exclusively in the nonbreeding season on either three species of bivalve (*Scrobicularia, Macoma*, and *Mya*) or polychaetes (*Neries*) (Moreira 1994). To predict godwit use of habitats, researchers made use of the following data. First, they sampled godwit prey from multiple locations within and among estuaries to quantify variation in prey density. This sampling was done in midautumn and late winter to characterize the extent to which prey populations had declined, presumably owing to predation by godwits. Second, they quantified intake rate by observing godwits feeding in habitats of varying prey density. These observations were facilitated by the comparatively large size of prey eaten by godwits. These data enabled researchers to characterize the functional response of godwits, including estimates of prey-handling time and search time, which were necessary for the resource depletion model. Additionally, these data provided an estimate of the lower (threshold) density of bivalves necessary to maintain a population of godwits. Third, observers, including participants in the British Trust for Ornithology's Wetland Bird Survey, conducted repeated counts of godwits within the estuaries to characterize bird use. Finally, they combined these data in the resource depletion model (Sutherland and Anderson 1993) to predict godwit use of different habitats within and among estuaries. The predicted densities were compared with the observed densities to evaluate how well the resource depletion model worked. The results were encouraging.

Godwits varied greatly in density within and across estuary habitats, and this variation correlated positively with the density of bivalves. This, in itself, is evidence that the six estuaries varied in their ability to support godwit populations. In other words, the carrying capacity of the estuaries differed. The functional response of godwits fit a type 2 curve, with intake rate leveling off at intermediate prey densities (Fig. 8.2).

At high prey density, intake rate was constrained by the time required to consume prey rather than locate it. Moreover, foraging godwits did not use habitats where bivalve densities were less than 150 per m^2. Using this threshold, Gill et al. (2001) examined the correlation between observed and predicted godwit use of habitats within and among estuaries and found remarkably strong correlations. This indicated that the resource depletion model accurately predicted godwit use of different habitats. Godwits occurred in higher densities at sites with richer food resources. Gill et al. (2001) extended their results by discussing the consequences of anthropogenic effects on godwit use of habitats. This treatment highlights the value of the model in addressing carrying capacity because it makes additional predictions about the consequences of habitat loss and deterioration.

COMPETITION AND FOOD LIMITATION

As predators, large numbers of shorebirds routinely forage where prey are abundant. When high densities of shorebirds persist for long periods in food-rich areas, their consumption undoubtedly reduces prey populations. This is especially true in northern winters when invertebrates exhibit seasonal breeding, and hence populations do not replace individuals lost to predators (Evans 1988). The availability of invertebrates may also be lower than the "standing crop" because invertebrates tend to burrow deeper into substrates in winter, beyond the depth to which some species can probe (Reading and McGrorty 1978; Pienkowski 1983; Evans 1979a, 1988; Evans and Dugan 1984). This variation in food availability coupled with predation by shorebirds (and other organisms) raises several questions that are central to the ecology of shorebirds. First, to what extent does predation reduce populations of invertebrate prey? And, more important from a demographic perspective, are food supplies depleted such that limiting resources effectively reduce shorebird populations via an increase in mortality or reduction in fecundity? The answers to these questions probably vary geographically and across the annual cycle. For example, it is unlikely that shorebirds have a significant effect on populations of invertebrates on their breeding grounds owing to the tremendous flush of productivity that occurs each spring and summer in northern latitudes. This may also be generally true for shorebirds wintering in southern estuaries where invertebrate prey populations reproduce and grow during the austral summer. By contrast, shorebirds may occasionally reduce food availability in the winter at northern latitudes.

In essence, the issue of food limitation is about competition for a limiting resource (Wiens 1989). A critical issue in discussions of food limitation, and hence whether competition acts on individuals and populations, is to distinguish between prey reduction and depletion. In many cases, shorebirds clearly reduce prey populations over short and long intervals (Goss-Custard 1984), but they may not deplete these food resources to the extent that there are consequences for individuals and populations. In other cases, food may be reduced to the point where some individuals die. If conditions of food shortage are extreme and persist for long enough, this may result in a detectable decrease in population size.

The effects of limiting food resources can be gauged by observing shorebird behavior, measuring the condition of individuals, and quantifying population-level responses such as local movements and mortality. Ultimately, these effects may result in decreased population sizes, depending on the severity of food limitation. Food limitation may manifest itself in the behaviors of shorebirds in several ways. First, birds may alter their patterns of habitat use to seek other, more profitable foraging areas. At an extreme, this may take the form of occasional large-scale emigrations from an area of low food availability to a region where food is more abundant. More predictable local movements between adjacent habitats are also possible. For example, Dunlin wintering on the California coast move both short and long distances coincident with the onset of winter rains. In this case, precipitation probably plays a dual role in influencing food availability. Winter precipitation may reduce the profitability of intertidal foraging areas while simultaneously increasing the quality of freshwater and pasture habitats (Warnock et al. 1995; Kelly 2001; Colwell and Dodd 1997; Conklin and Colwell 2007).

A second behavioral manifestation of food shortages may be evidenced by altered time-activity budgets and intake rates. As indicated earlier, abundant evidence shows that shorebirds overwintering at northern latitudes increase the amount of time they spend feeding during inclement weather. The percentage of time that Pied Avocets fed increased during winter periods of cold temperatures but decreased when winds exceeded 10 km h^{-1} (Hötker 1999). Additionally, individuals may increase their intake rate. The behavioral adjustments made by individuals may be effective at dealing with food shortages, but the success of these strategies is evidenced by whether or not the condition of individuals worsens. Condition is normally gauged by the change in mass of an individual. For example, during a prolonged winter cold snap in northeast England, Common Redshanks increased the amount of time they spent foraging but simultaneously lost mass. As a result, individuals died at rate far exceeding normal winters (Burton et al. 2006). Finally, from a population perspective, the consequences of food shortage may occasionally be seen in decreased population size.

PREY REDUCTION

Competition requires that resources are reduced to a level such that it results in a measurable negative effect on individuals and populations. A first step in evaluating food limitation is demonstrating that predation by shorebirds actually reduces prey populations. Numerous studies have been conducted over several decades to address the extent to which shorebirds reduce populations of their prey. Summaries of these studies are provided by Goss-Custard (1984) and Székely and Bamberger (1992).

Researchers have employed two principal methods to gauge the amount of prey consumed by shorebirds as a means for estimating prey reduction. One method involves observers sampling the foraging behavior of shorebirds to determine the number of prey consumed during an interval. Another method involves sampling food before and after shorebird predators have had their effect on prey populations. In the latter case, researchers often couple invertebrate sampling with the use of caged exclosures to directly compare predation in areas where shorebirds have fed with adjacent sites where they have been kept from feeding. Both these approaches entail methodological challenges, and the ideal study combines the two and provides a comparative perspective on prey reduction. Several authors (e.g., Baird et al. 1985; Székely and Bamberger 1992) offer a thorough treatment of these approaches in their study of shorebird predation on prey populations.

Observational evidence requires several key pieces of information. First, researchers must estimate the density of a species feeding in an area, as well as the percentage of time that an individual feeds. This is typically done by scan sampling a study plot at regular intervals (for instance, 15 to 30 minutes) and recording the number of feeding birds. Using focal sampling, observers also record the intake (or success) rate of individuals, ideally noting the identity of prey consumed. An estimate of prey removed from a habitat is obtained from the product of species' densities, foraging rates, and time spent feeding. This is done separately for each species of shorebird and their prey. Various assumptions are incorporated into these components that may affect the validity of prey reduction estimates. For instance, foraging rates often assume that a high percentage of attempts are successful, especially if food items are small and not easily observed. Estimates of density typically use average abundances and ignore substantial temporal and spatial variation in shorebird distribution. Finally, studies often assume that birds feed throughout the 24-hour period (both day and night). Estimates based on observational data are compromised by the extent to which these assumptions are invalid.

A second approach to estimating prey reduction involves sampling invertebrates before and after predation by shorebirds. This method uses the difference in density (and biomass) of invertebrates to gauge the effect of predation by shorebirds. This approach suffers from several drawbacks. First, depending on the length of the sampling interval (the days, weeks, or months transpiring between before and after), invertebrates may reproduce such that their recruitment masks the predation effect. This methodological problem is important to recognize in distinguishing between reduction versus depletion of prey. This may not be important from the perspective of estimating food limitation, but it clouds estimates of the amount of prey eaten by shorebirds. A second problem with simple before–after sampling is that it does not account for the effects of other predators (such as fishes and crabs), which may be especially important in tidal habitats. Again, this is a problem when a researcher wishes to estimate the proportion of prey reduction attributable to shorebirds and not when attempting to estimate whether prey are reduced or depleted. Another problem with sampling invertebrates concerns the natural variation in invertebrate density that can occur over spatial scales less than a meter. Substantial variation often requires large replicate samples to characterize prey density accurately. Finally, prey density is often sampled using a coring device, which assumes that prey at varying depths of the substrate are equally available to shorebird predators.

Many studies have incorporated a more rigorous before–after/control–impact (BACI) design using caged exclosures to keep shorebirds and other predators from feeding in plots. Especially elaborate exclosure designs allow separation of predation by shorebirds from the effects of fishes and crabs in tidal habitats (Quammen 1981). Exclosures are paired with control plots where predators are free to forage. The paired comparison controls for spatial variation in invertebrate abundance. This approach to estimat-

ing prey reduction also has its shortcomings. For instance, the structure of exclosures may have effects on the prey by creating refuges or altering habitat, especially on the perimeter of the cage. Overcoming these difficulties in study design is critical to estimating the effects of shorebird predators on the populations of their prey (Baird et al. 1985; Sewell 1996).

Virtually all studies that have used observational or prey sampling to examine the effects of shorebirds on their prey have been conducted during the nonbreeding season. This stems from the fact that high productivity, and hence low likelihood of food limitation, characterizes northern breeding areas during the boreal spring and summer. During the nonbreeding season, studies have varied in length. Some studies encompassed rather short intervals (days to weeks) during migration windows when large numbers of shorebirds aggregate at staging areas. Studies conducted during the boreal winter, regardless of latitude, often have lasted for much longer.

Evidence from studies conducted during spring migration indicates that shorebirds sometimes, but not always, dramatically reduce prey populations. For example, over a 22-day interval in Iceland, Eurasian Golden-Plovers reduced the density of earthworms in pastures by approximately 50% (Bengston et al. 1976). In coastal wetlands of South Carolina, shorebirds reduced the density and biomass of invertebrates (mostly polychaetes and oligochaetes) by 48% and 50%, respectively (Weber and Haig 1997). By contrast, Sewell (1996) was unable to demonstrate that predation by Western Sandpipers significantly reduced numbers of their invertebrate prey at several sites in the Fraser River Delta, British Columbia. This finding was especially remarkable because the single-day estimate for the abundance of sandpipers during the month-long study was 1.5 million birds.

During late summer and early fall, migrants moving south from breeding grounds have been shown to have strong effects on prey populations. Schneider and Harrington (1981) used caged exclosures to estimate the reduction in prey density on tidal flats in Massachusetts during summer migration. They determined that prey abundance occasionally declined by up to 90% during the 2 months of their study. Székely and Bamberger (1992) conducted what may be the most interesting study of shorebird predation on invertebrates. Their work occurred over a 13-day, midsummer interval in a desiccating seasonal wetland in Hungary. The researchers used both observational and invertebrate sampling to study prey reduction in an assemblage of 10 migrant shorebirds. Prey were largely chironomid larvae, which existed at very high densities at the start of the study. Observational data suggested that shorebirds, principally the Black-tailed Godwit, Spotted Redshank, and Ruff, consumed roughly 7,000 prey/m^2 over the short interval. Interestingly, before–after sampling of invertebrates coupled with exclosures indicated that shorebirds consumed roughly 10,300 prey, which was an 87% reduction over 13 days. The differences in these estimates probably are associated with inaccuracies in observational and prey sampling associated with assumptions discussed earlier.

In the boreal winter, evidence suggests that shorebirds can sometimes dramatically reduce populations of their prey, but this effect may be strongest at northern sites. Evans et al. (1979) estimated that shorebirds consumed 90% of prey in the Tees estuary, a site where port development had significantly reduced the amount of intertidal foraging area. Elsewhere, Eurasian Oystercatchers reduced winter populations of bivalves by 28% to 40% (Zwarts and Drent 1981; Sutherland 1982). In one study, Black Oystercatchers had only a negligible effect on the abundance of their prey, the mussel (*Mytilus*) (Hartwick and Blaylock 1979), whereas in another they reduced populations of several species of limpet (*Notoacmea*) by upward of 90% in a 7-month period (Frank 1982).

At the same time, but during the austral summer, there is mixed evidence that shorebirds reduce their prey populations. In a classic paper, Duffy et al. (1981) reported that there was little evidence that a community of shorebirds wintering on the Paracas Peninsula in coastal Peru had

a measurable effect on the density of their prey. In particular, densities of several prey groups actually tended to increase over a short interval (January and February) coincident with highest shorebird densities. Their conclusions, however, were challenged by Myers and McCaffery (1984). Elsewhere in the Southern Hemisphere there have been mixed results regarding the impact of shorebirds on the populations of their invertebrate prey. Curlew Sandpipers and Black-bellied Plovers winter in large numbers in the Berg River estuary, South Africa, where they feed extensively on various polychaetes worms. Kalejta (1993) used observational data to estimate that Curlew Sandpipers removed nearly 21% of the standing crop of worms each month from January into April. By the end of the 3 months, sandpipers had reduced the density and biomass of nereids by 58% and 77%, respectively. The effect of plovers on worms was much less owing to their smaller population size. Remarkably, however, there was no difference in the worm populations between exclosures and control plots because worms reproduced prolifically at a time of intense predation by shorebirds. By contrast, other studies at southern estuaries indicate a different story. In eastern Australia, Bar-tailed Godwits greatly impacted their prey populations, principally a *Mictyris* crab. Observational data showed that godwits reduced crab numbers by 88% whereas invertebrate sampling showed a 90% reduction, a remarkably similar estimate (Zharikov and Skilleter 2003).

In conclusion, results of many studies suggest that shorebirds have varying effects on the populations of their prey. The strength of the predation effect depends on season, latitude, and habitat. It appears that shorebirds wintering at northerly latitudes can sometimes have a significant impact on prey populations, which may elicit reciprocal effects on shorebird populations. Significant reductions (depletion) of prey populations are less likely at southerly latitudes because invertebrates may be able to overcome the strong predation effects via reproduction. In this case, it is unlikely that shorebirds will be food-limited to the extent that there are consequences for individuals and populations. These observations are supported by findings that winter densities of shorebirds are higher in southern estuaries compared with more northern sites (Hockey et al. 1992). Finally, the effects of shorebird predators may be strong in some habitats and much less noticeable in nearby areas. Goss-Custard (1984) recognized this when he categorized studies into those examining principal versus secondary feeding areas. He argued that the effects of prey reduction would be expected to be strongest in habitats preferred by shorebirds, as evidenced by their high densities and intake rates.

COMMUNITY ECOLOGY

By now it should be clear that food plays an important role in the ecology of shorebirds. Most species have rather flexible diets that readily change as individuals move among habitats (Skagen and Oman 1996). Despite this opportunism, the bill morphologies and foraging behaviors are adapted for feeding on a subset of potential prey, especially during the nonbreeding season (Baker and Baker 1979; Lifjeld 1984). As a result, some individuals may feed for varying durations on just a few prey types. The functional relationships between bill morphology and diet are probably stronger in species that are specialized on a limited array of prey species. Daily time-activity budgets give evidence that shorebirds expend considerable energy foraging and they engage in frequent agonistic interactions over food or space necessary to acquire it (Recher and Recher 1969). Intake rates are strongly influenced by prey density, as mediated by numerous abiotic and biotic features of the environment (Puttick 1984; Goss-Custard 1984). The biotic interactions include interference by conspecifics over a shared resource. Collectively, shorebirds sometimes exert strong effects on the populations of their prey, which occasionally result in resource limitation. Consequently, shorebird populations occasionally decline during periods of inclement weather, especially at northerly latitudes when food is clearly in short supply and energy demands are high (Evans and Pienkowski 1984). A logical extension of these ecological patterns is that competition has played an important role in structuring shorebird communities. The search for causation among the ecological patterns mentioned above is not, however, without difficulties (Wiens 1989). Nevertheless, competition for limiting food resources is a reasonable cause of these patterns. Below, I examine several ecological patterns that are consistent with the view that competition structures shorebird communities.

ECOMORPHOLOGICAL PATTERNS

A cursory examination of a mixed-species shorebird flock feeding on intertidal flat reveals substantial interspecific variation in body size and associated morphology of bills and hindlimbs. On high-elevation flats exposed earliest by the ebbing tide, a female Long-billed Curlew walks purposefully about drier substrates pocked by exit burrows of mud shrimps and probes her long decurved bill to tactilely detect prey. In a loose flock, Black-bellied Plovers search visually for prey over the same high-elevation flats in a distinct walk-stop-peck manner. At the upper margins of the ebbing tide edge, Least Sandpipers forage by picking at small invertebrates on wetted substrate. In the last vestiges of watery film, an American Avocet scythes its bill through fine sediments and shallow water. Western Sandpipers lap biofilm from the surface with their brushlike tongue, while Dunlins wade on longer legs and probe with their decurved bills. A cluster of dowitchers wade belly-deep a little farther from the shore and rapidly stitch for prey buried deeper in the substrates. Marbled Godwits perform similar, albeit often single, purposeful probes in even deeper water.

Embedded in this descriptive image of a community of foraging shorebirds are several ecomorphological patterns that have been well known to ornithologists for decades. These patterns include differences in bill and hindlimb structure, variation in habitat use and foraging behavior, and nonrandom associations in mixed-species flocks. Co-occurring species appear to sort themselves out along a habitat gradient in a predictable pattern that relates directly to the lengths of a species' bill and hindlimb. Shorebirds commonly aggregate in dense, mixed-species flocks during the nonbreeding season. These assemblages occur in seemingly simple (two-dimensional) wetland habitats, which has prompted considerable interest in the ways that shorebirds coexist in conditions that appear to favor intense competition. In San Francisco Bay, for example, Recher and Recher (1966, 1969) examined mixed-species flocks and concluded that competition for food was reduced by inter-specific differences in habitat use, timing of migrations, and morphology between closely related species, which were presumed to compete for food.

Evidence for the importance of competition in the morphology of shorebirds comes from examination of bill and hindlimb structures. Burton (1974) examined in detail the skeletal features and musculature of the skull in five species with very different bill morphologies and feeding habits. He commented generally on the advantages of the various bills, ranging from the long, decurved bill of the Eurasian Curlew to the short, straight bill of the Eurasian Golden-Plover. However, it was Baker (1979) who explicitly addressed the functional relationships between bill and hindlimb morphology and habitat use in a community of six species. Working in breeding areas of northern Canada and wintering areas in Florida, he reported strong positive relationships between the lengths of the bill and tarsometatarsus and the foraging water depth, especially in winter. He reasoned that these differences stemmed from competitive interactions over food. Baker and Baker (1973) expanded the comparison between breeding and winter habitats by examining niche relationships in the same group of six species. This assemblage (Least Sandpiper, Semipalmated Sandpiper, Dunlin, Short-billed Dowitcher, Lesser Yellow-legs, and Semipalmated Plover) included species from different families and one set of four congeners. Consequently, the morphological differences among species in bill and hindlimb

morphology were large, as was the range of habitats used. Species exhibited diverse foraging methods and high niche overlap on the breeding grounds when food was abundant. By contrast, species overlapped less in resource use and had narrower niches in winter. Baker and Baker (1973) concluded that food limitation more likely influenced community patterns in the winter. Lifjeld (1984) extended this reasoning and addressed relationships between body size, bill length, and handling time of prey of different sizes for a group of five Palearctic waders feeding in coastal wrack during autumn migration in Norway. Species differed in selection of prey (larvae of a single dipteran, *Scatophaga*) of different sizes, with larger-bodied species (such as the Ruff, Curlew Sandpiper, and Dunlin) feeding on larger prey of a wider size range. By contrast, small-bodied species (such as the Common Ringed Plover and Little Stint) fed on smaller prey and were less variable in their selection of prey. Bill length, however, did not correlate with diet composition, even when Lifjeld (1984) controlled statistically for the effects of body size.

SIZE RATIOS

Analysis of ecomorphological patterns as evidence for competition was spawned by Hutchinson's (1959) influential work proposing that there was some minimum size difference between putative competitors necessary to allow coexistence. Ornithologists responded with an avalanche of studies comparing the size ratios (such as bill length and body size) among groups of species that shared a common food resource (see Wiens 1989). One of the first studies of limiting similarity among shorebirds was conducted by Strauch and Abele (1979). They examined three small plovers (Semipalmated, Collared, and Wilson's) that co-occur on the Pacific Coast of Panama. Plovers foraged on intertidal flats consisting mostly of sand and mud with limited exposed rocky habitats. These plovers differ in body size and bill morphology, with bill size ratios of 1.2 to 1.4 for adjacent pairs. Plover species had limited niche overlap, which, coupled with abundant food and a depauperate shorebird community, led Strauch and Abele (1979) to conclude that there was little evidence of competition for food. They also pointed out that plovers frequently departed foraging habitats for roosts when tidal flats remained exposed, which indicated that some individuals had sufficient time to feed and that their energy budgets were not constrained by food availability. Whereas this observation is informative regarding current ecological conditions, it sheds little light on the extent to which food shortages in the past may have influenced contemporary ecological patterns. However, this argument, referred to as the "ghost of competition past" (Connell 1980), is an untestable hypothesis (Wiens 1989). Nevertheless, it is important to note that many shorebirds winter in tropical environs, sometimes at high densities, which may increase competitive interactions among closely related species. However, food limitation is probably not as prevalent at southerly latitudes because invertebrate populations are productive during the austral summer (boreal winter).

Only one study has examined bill morphologies and diets of closely related shorebirds in breeding areas: Holmes and Pitelka (1968) studied the feeding ecology of four sympatric calidridine sandpipers (Pectoral Sandpiper, Dunlin, Semipalmated Sandpiper, and Baird's Sandpiper) near Barrow, Alaska. They collected hundreds of adults and young to evaluate the degree of dietary overlap. Food consisted mostly of dipterans from several taxa. There was high overlap in sandpiper diets, especially in one year of particularly abundant food. Holmes and Pitelka (1968) commented on the similarities in feeding method of these sandpipers, all of which appeared to pick invertebrates from the surface of wet tundra or probe shallowly for prey. They concluded that the high overlap in diet and abundant food suggested that there was limited evidence for competition driving these Arctic patterns. However, they ended their paper by commenting on the rather small size ratios of bills (ranging from 1.23 to 1.26) between species' pairs of similar size. They suggested that relative to theory (Hutchinson 1959; Schoener 1965),

which predicted a minimum size ratio approximating 1.3 between close competitors, the group of calidridine sandpipers exhibited bill morphologies that permitted coexistence during periods of occasionally unfavorable food availability. They concluded that the group of four species was not a random assemblage, but acknowledged that competition for food was probably greater during the nonbreeding season. This is especially true given the short tenure of sandpipers in Arctic breeding habitats, where food is often superabundant (Holmes and Pitelka 1968).

MIXED-SPECIES ASSEMBLAGES

An extension of Hutchinson's (1959) ideas about morphological similarity addresses the extent to which evidence for competition can be found in nonrandom assemblages of species. This body of evidence was the focus of considerable debate centered on the issue of appropriate null hypotheses in ecology (see Gilpin and Diamond 1984; Connor and Simberloff 1984). The debate, however, has since subsided. Briefly, evidence that species with comparable bill and hindlimb morphologies, overlapping niches, and similar diets may be subject to competition, prompted researchers to search for nonrandom associations among presumed competitors. For shorebirds, the search for pattern has been undertaken only once, on mixed-species flocks of migrating sandpipers using ephemeral wetlands in the interior of North America (Eldridge and Johnson 1988).

Eldridge and Johnson (1988) collected observations during spring migration on flocks of northern-breeding sandpipers in North Dakota and asked whether the size ratios of bills of congeners gave evidence of competition. In this case, evidence for competition would be that size ratios in observed flocks of sandpipers were greater than expected by chance. They recorded 13 species that co-occurred in flocks that were defined by associations within a 10 m radius circle. Using published data on bill morphology (as a surrogate for body size), they examined patterns of species' co-occurrence and compared these observed patterns to several randomly generated source pools (species' assemblages). These pools were an ordered subset including (1) 51 Holarctic sandpipers, (2) 33 Nearctic species, and (3) 13 species observed during migration in North Dakota. Next, they evaluated differences in bill morphologies by comparing the size ratios in observed flocks to randomly generated assemblages. Their results provided some support for the role of competition in structuring migrant flocks. On the one hand, size ratios in observed flocks appeared to be a random reshuffling of the species pool using the migration corridor through the interior of North America. On the other hand, a consistent pattern of bill size ratios spanning 1.2 to 1.3 was obtained from comparisons with Nearctic and global species pools. In their own words, "the ephemeral, mixed-species foraging flocks" observed in North Dakota were "random mixes from a nonrandom pool" (Eldridge and Johnson 1988). This pattern was consistent with that predicted by competition theory.

In summary, shorebirds spend only a small part of their annual cycle in seasonally productive habitats of the far north, for instance. At these times of year, it is unlikely that competition for food influences community patterns because food is often superabundant. In contrast, competition probably has had a stronger influence on the ecology and evolution of shorebirds on their wintering grounds. Moreover, the strength and frequency of competitive interactions during the nonbreeding season probably increase with latitude (north of the equator) owing to the strong seasonality of food in the boreal winter. These conditions are exacerbated in populations wintering at northern latitudes because the prolonged cold, wet weather makes foraging difficult and increases the energy demands on individuals. This is especially true for small-bodied shorebirds.

CONSERVATION IMPLICATIONS

The relationships between characteristics of prey (the density, availability, and size) and shorebird biology (such as foraging density and intake rate)

are of immense theoretical and applied importance. From a theoretical point of view, the availability of food and its seasonal pattern of replenishment in relation to predation by shorebirds is probably the single most important factor affecting variation in shorebird distribution and abundance during the nonbreeding season. This is evident in the simple numerical relationships at spatial scales ranging from a global perspective (Hockey et al. 1992; Butler et al. 2001) down to fine-scale decisions made by individuals to feed in microhabitats on prey of varying profitability (van Gils et al. 2004). At the spatial scale of a wintering site (such as a coastal estuary), food supply interacts with species' ecologies to set an upper limit on population size via its effect on individual intake rate. Individuals that do not feed effectively have lower probabilities of survival. In this case, competition acts in a density-dependent manner to set local population size via its effects on overwinter survival. When food is abundant, individuals survive better and populations increase; as foraging conditions worsen, mortality increases disproportionately.

Evidence that food influences the hemispheric patterns of shorebird distribution and abundance comes from several sources. Hockey et al. (1992), for example, evaluated the link between wintering densities of shorebirds in relation to the seasonal productivity of their invertebrate prey for 31 estuaries spanning 90° of latitude within the East Atlantic Flyway (western Palearctic). Shorebird densities were higher at southern than northern estuaries, and these high densities were associated with the tendency for invertebrate populations to replace themselves coincident with the period of greatest occupancy by their shorebird predators. By comparison, northern estuaries supported lower densities of wintering shorebirds, probably owing to the strong seasonality of invertebrate reproduction. Shorebirds exerted their greatest impact on populations of their prey during the winter, when invertebrates are not reproducing or growing (Evans 1984).

Several examples highlight the conservation value of detailed knowledge of species' digestive anatomy and physiology, their intake rates, and spatial variation in prey density. The Icelandic population of the Red Knot winters in northwestern Europe, where the population has declined by approximately 25% (from approximately 330,000 to 250,000) during the interval 1997 to 2003 (van Gils et al. 2006). Approximately one-third to one-half of the population winters in the Dutch Wadden Sea, where birds feed selectively on bivalves from soft sediments. The size of the gizzard of Red Knot is finely tuned to digest hard-shelled molluscs (*Mytilus, Mya, Macoma*) of different sizes. The digestive efficiency of processing large versus small bivalves varies greatly. Despite protection by the Dutch government (as a Marine Protected Area) and recognition under the Ramsar Convention, mechanical dredging by a small number of shell-fishers exploiting cockles (*Cerastoderma*) takes place within the principal foraging areas of Red Knots. This dredging has altered the prey base directly by reducing the size of bivalves 11.3% annually; also, dredging indirectly affects the sediment particle size, which influences the settlement of spat. As a result, the amount of poor-quality foraging habitat available to Red Knots has increased dramatically from 66% to 77%. Consequently, the diet of birds has declined in quality by 12% per year, with gizzard mass increasing to compensate for larger, lower quality prey. Lower resighting of color-marked Red Knots over the interval of population decline suggests that birds either emigrate or die as a result of these increasingly poor feeding conditions. Furthermore, individual Red Knots with undersized gizzards had lower survival (or higher emigration) rates. Van Gils et al. (2006) used this information to criticize the notion that Marine Protected Areas as operated by the Dutch government could effectively balance the interests of a relatively small shellfishing industry with the long-term population viability of a unique shorebird.

As pointed out by others (Evans and Dugan 1984; Goss-Custard 1985; Piersma 1987), a critical issue in the conservation and management of shorebirds is to understand whether the use of particular sites by shorebirds is limited by the benthic food supply. During the nonbreeding season, it is clear that food is critical to the decisions made by individuals regarding where to feed. Shorebirds concentrate in areas of high prey abundance and avoid areas where food is absent or nearly so. Wintering shorebirds, especially those at northern latitudes, spend much of their time foraging. The large amount of time spent feeding suggests that high-quality foraging areas are at a premium. Consequently, loss or degradation of such habitats may have consequences for individual survival and subsequent reproduction. Evidence to this effect comes from a recent study of overwinter mortality in a population of Common Redshanks, which experienced loss of its foraging habitats (Burton et al. 2006). After the construction of a tidal barrage had converted estuarine feeding areas to freshwater habitats, the body condition (mass) of redshanks declined, and overwinter mortality increased 44%. It is also possible that the subtle effects of loss and degradation of foraging habitats can result in lower reproductive success of individuals.

LITERATURE CITED

Baird, D., P. R. Evans, H. Milne, and M. W. Pienkowski. 1985. Utilization by shorebirds of benthic invertebrate production in intertidal areas. *Oceanography and Marine Biology Annual Review* 23: 573–597.

Baker, M. A. 1979. Morphological correlates of habitat selection in a community of shorebirds (Charadriiformes). *Oikos* 33: 121–126.

Baker, M. C., and A. E. M. Baker. 1973. Niche relationships among six species of shorebirds on their wintering and breeding ranges. *Ecological Monographs* 43: 193–212.

Bengston, S.-A., A. Nilsson, S. Nordström, and S. Rundgren. 1976. Effect of bird predation on lumbricid populations. *Oikos* 27: 9–12.

Bryant, D. M. 1979. Effects of prey density and site characters on estuary usage by over-wintering waders (Charadrii). *Estuarine and Coastal Marine Science* 9: 369–384.

Burton, N. H. K., M. M. Rehfisch, N. A. Clark, and S. G. Dodd. 2006. Impacts of sudden winter habitat loss on the body condition and survival of redshanks *Tringa totanus*. *Journal of Applied Ecology* 43: 464–473.

Burton, P. J. K. 1974. *Feeding and the feeding apparatus in waders: A study of anatomy and adaptations in the Charadrii*. London: Trustees of the British Museum.

Butler, R. W., N. C. Davidson, and R. I. G. Morrison. 2001. Global-scale shorebird distribution in relation to productivity of near-shore ocean waters. *Waterbirds* 24: 224–232.

Colwell, M. A., and S. L. Dodd. 1997. Environmental and habitat correlates of pasture use by non-breeding shorebirds. *The Condor* 99: 337–344.

Colwell, M. A., and S. L. Landrum. 1993. Nonrandom shorebird distribution and fine-scale variation in prey abundance. *The Condor* 95: 94–103.

Conklin, J. R., and M. A. Colwell. 2007. Diurnal and nocturnal roost site fidelity of Dunlin (*Calidris alpina pacifica*) at Humboldt Bay, California. *The Auk* 124: 677–689.

Connell, J. H. 1980. Diversity and the coevolution of competitors, or the ghost of competition past. *Oikos* 35: 131–138.

Connor, E. F., and D. Simberloff. 1984. Neutral models of species' co-occurrence patterns. In *Ecological communities: Conceptual issues and the evidence*, D. R. Strong, Jr., D. Simberloff, L. G. Abele, and A. B. Thistle, 316–331. Princeton, NJ: Princeton University Press.

Duffy, D. C., N. Atkins, and D. S. Schneider. 1981. Do shorebirds compete on their wintering grounds? *The Auk* 98: 215–229.

Dugan, J. E., D. M. Hubbard, M. D. McCrary, and M. O. Pierson. 2003. The response of macrofauna communities and shorebirds to macrophyte wrack subsidies on exposed sandy beaches of southern California. *Estuarine, Coastal and Marine Science* 58S: 25–40.

Eldridge, J. L., and D. H. Johnson. 1988. Size differences in migrant sandpiper flocks: Ghosts in ephemeral guilds. *Oecologia* 77: 433–444.

Evans, P. R. 1979a. Adaptations shown by foraging shorebirds to cyclic variations in the activity and availability of their intertidal invertebrate prey. In *Cyclical phenomena in marine plants and animals*, ed. E. Naylor and R. G. Hartnoll, 357–366. Oxford/New York: Pergamon Press.

———. 1979b. Reclamation of intertidal land: Some effects on Shelduck and wader populations in the Tees Estuary. *Verhandlungen der ornithologischen Gesellschaft in Bayern* 23: 147–168.

———. 1984. Introduction. In *Coastal waders and wildfowl in winter*, ed. P. R. Evans, J. D. Goss-Custard, and W. G. Hale, 2–7. Cambridge: Cambridge University Press.

———. 1988. Predation of intertidal fauna by shorebirds in relation to time of the day, tide and year. In *Behavioural adaptations to intertidal life*, ed. G. Chelazzi and M. Vannini, 65–78. New York: Plenum Press.

Evans, P. R., and P. J. Dugan. 1984. Coastal birds: Numbers in relation to food resources. In *Coastal waders and wildfowl in winter*, ed. P. R. Evans, J. D. Goss-Custard, and W. G. Hale, 8–28. Cambridge: Cambridge University Press.

Evans, P. R., D. M. Herdson, P. J. Knights, and M. W. Pienkowski. 1979. Short-term effects of reclamation of part of Seal Sands, Teesmouth, on wintering waders and Shelduck. I. Shorebirds' diets, invertebrate densities and the impact of predation on the invertebrates. *Oecologia* 41: 183–206.

Evans, P. R., and M. W. Pienkowski. 1984. Population dynamics of shorebirds. In *Shorebirds: Breeding behavior and populations*, ed. J. Burger and B. L. Olla, 83–123. New York: Plenum Press.

Frank, P. W. 1982. Effects of winter feeding on limpets by Black Oystercatchers, *Haematopus bachmani*. *Ecology* 63: 1352–1362.

Gill, J. A., W. J. Sutherland, and K. Norris. 2001. Depletion models can predict shorebird distribution at different spatial scales. *Proceedings of the Royal Society of London, Series B* 268: 369–376.

Gillings, S., P. W. Atkinson, S. L. Bardsley, N. A. Clark, S. E. Love, R. A. Robinson, R. A. Stillman, and R. G. Weber. 2007. Shorebird predation of horseshoe crab eggs in Delaware Bay: Species contrasts and availability constraints. *Journal of Animal Ecology* 76: 503–514.

Gilpin, M. E., and J. M. Diamond. 1984. Are species co-occurrences on islands non-random, and are null hypotheses useful in community ecology? In *Ecological communities: Conceptual issues and the evidence*, ed. D. R. Strong, Jr., D. Simberloff, L. G. Abele, and A. B. Thistle, 297–315. Princeton, NJ: Princeton University Press.

Goss-Custard, J. D. 1984. Intake rates and food supply in migrating and wintering shorebirds. In *Shorebirds: Migration and foraging behavior*, ed. J. Burger and B. L. Olla, 233–270. New York: Plenum Press.

———. 1985. Foraging behaviour of wading birds and the carrying capacity of estuaries. In *Behavioural ecology*, ed. R. M. Sibly and R. H. Smith, 169–188. Oxford: Blackwell.

Goss-Custard, J. D., J. T. Cayford, and S. E. G. Lea. 1998. The changing trade-off between food finding and stealing in juvenile oystercatchers. *Animal Behavior* 55: 745–760.

Goss-Custard, J. D., and M. E. Moser. 1988. Rates of change in the numbers of Dunlin wintering in British estuaries in relation to the spread of *Spartina angelica. Journal of Applied Ecology* 25: 95–109.

Goss-Custard, J. D., A. D. West, M. G. Yates, R. W. G. Caldow, R. A. Stillman, L. Bardsley, J. Castilla, et al. 2006. Intake rates and the functional response in shorebirds (Charadriiformes) eating macro-invertebrates. *Biological Reviews* 81: 501–529.

Hale, W. G. 1980. *Waders.* London: Collins.

Hartwick, E. B., and W. Blaylock. 1979. Winter ecology of a Black Oystercatcher population. In *Studies in Avian Biology, No. 2,* ed. F. A. Pitelka, 207–216. Lawrence, KS: Cooper Ornithological Society.

Hockey, P. A. R., R. A. Navarro, B. Kalejta, and C. R. Velasquez. 1992. The riddle of the sands: Why are shorebird densities so high in southern estuaries? *American Naturalist* 140: 961–979.

Holling, C. S. 1959. Some characteristics of simple types of predation and parasitism. *Canadian Entomologist* 91: 385–398.

Holmes, R. T., and F. A. Pitelka. 1968. Food overlap among coexisting sandpipers on north Alaskan tundra. *Systematic Zoology* 17: 305–318.

Hötker, H. 1999. What determines the time-activity budgets of Avocets (*Recurvirostra avosetta*)? *Journal of Ornithology* 140: 57–71.

Hutchinson, G. E. 1959. Homage to Santa Rosalia or why are there so many kinds of animals? *American Naturalist* 93: 145–159.

Kalejta, B. 1993. Intense predation cannot always be detected experimentally: A case study of shorebird predation on nereid polychaetes in South Africa. *Netherlands Journal of Sea Research* 31: 385–393.

Kelly, J. P. 2001. Hydrographic correlates of winter Dunlin abundance and distribution in a temperate estuary. *Waterbirds* 24: 309–322.

Kersten, M., and W. Visser. 1996. The rate of food processing in the oystercatcher: Food intake and energy expenditure constrained by a digestive bottleneck. *Functional Ecology* 10: 440–448.

Krebs, C. J. 2009. *Ecology.* 6th ed. San Francisco: Benjamin Cummings.

Laursen, K., I. Gram, and L. J. Alberto. 1983. Short-term effect of reclamation on numbers and distribution of waterfowl at Hojer, Danish Wadden Sea. In *Proceedings of the Third Nordic Congress of Ornithology, 1981,* ed. J. Fjeldså and H. Meltofte, 97–118. Copenhagen: Dansk Ornitologisk Forening Zoologisk Museum.

Leeman, L. W., M. A. Colwell, Leeman, T. S., and R. L. Mathis. 2001. Diets, energy intake, and klepto-parasitism of nonbreeding Long-billed Curlews in a northern California estuary. *Wilson Bulletin* 113: 194–201.

Lifjeld, J. T. 1984. Prey selection in relation to body size and bill length of five species of waders feeding in the same habitat. *Ornis Scandinavica* 15: 217–226.

Moreira, F. 1994. Diet, prey-size selection and intake rates of Black-tailed Godwits *Limosa limosa* feeding on mudflats. *The Ibis* 136: 349–355.

Moser, M. E. 1988. Limits to the numbers of grey plovers (*Pluvialis squatarola*) wintering on British estuaries: An analysis of long-term population trends. *Journal of Applied Ecology* 25: 473–485.

Myers, J. P., and B. J. McCaffery. 1984. Paracas revisited: Do shorebirds compete on their wintering grounds? *The Auk* 101: 197–199.

Newton, I. 1998. *Population limitation in birds.* New York: Academic Press.

Pienkowski, M. W. 1983. Surface activity of some intertidal invertebrates in relation to temperature and the foraging behavior of their shorebirds. *Marine Ecology Progress Series* 11: 141–150.

Piersma, T. 1987. Production of intertidal benthic animals and limits to their reduction by shorebirds: A heuristic model. *Marine Ecology Progress Series* 38: 187–196.

Prater, A. J. 1977. *Birds of the Estuaries Enquiry: Report on the 1974–75 counts.* Tring, United Kingdom: British Trust for Ornithology.

Puttick, G. M. 1984. Foraging and activity patterns in wintering shorebirds. In *Shorebirds: Migration and foraging behavior,* ed. J. Burger and B. L. Olla, 203–231. Plenum Press, New York, NY.

Quammen, M. L. 1981. Use of exclosures in studies of predation by shorebirds on intertidal mudflats. *The Auk* 82: 812–817.

Reading, C. J., and S. McGrorty. 1978. Seasonal variations in the burying depth of *Macoma balthica* (L.) and its accessibility to wading birds. *Estuarine and Coastal Marine Sciences* 6: 135–144.

Recher, H. F. 1966. Some aspects of the ecology of migrant shorebirds. *Ecology* 47: 393–407.

Recher, H. F., and J. Recher. 1969. Some aspects of the ecology of migrant shorebirds. II. Aggression. *Wilson Bulletin* 81: 140–154.

Schneider, D. C., and B. A. Harrington. 1981. Timing of shorebird migration in relation to prey depletion. *The Auk* 98: 801–811.

Schoener, T. W. 1965. The evolution of bill size differences among sympatric congeneric species of birds. *Evolution* 19: 189–213.

Schott, G. 1931a. Kaltes wasser vor der Kuste von Venezuela und Kolumbian. *Ann. Hydr. Mar. Met.* 59: 224–226.

———. 1931b. Der Peru-Strom und seine nordlichen Nachbargebiete in normaler und abnormaler Ausbildung. Ann. Hydr. Mar. Met. 59: 161–252.

Senner, S. E., and M. A. Howe. 1984. Conservation of Nearctic shorebirds. In *Shorebirds: Breeding behavior and populations,* ed. J. Burger and B. L. Olla, 379–421. New York: Plenum Press.

Sewell, M. 1996. Detection of the impact of predation by migratory shorebirds: An experimental test in the Fraser River Delta, British Columbia, Canada. *Marine Ecology Progress Series* 144: 23–40.

Skagen, S. K., and H. D. Oman. 1996. Dietary flexibility of shorebirds in the Western Hemisphere. *Canadian Field-Naturalist* 110: 419–444.

Soloman, M. E. 1949. The natural control of animal populations. *Journal of Animal Ecology* 18: 1–35.

Strauch, J. G., Jr., and L. G. Abele. 1979. Feeding ecology of three species of plovers wintering on the Bay of Panama, Central America. *Studies in Avian Biology* 2: 217–230.

Sutherland, W. J. 1982. Spatial variation in the predation of cockles by oystercatchers at Traeth Melynog, Anglesey. I. The pattern of mortality. *Journal of Animal Ecology* 51: 491–500.

Sutherland, W. J., and C. W. Anderson. 1993. Predicting the distribution of individuals and the consequences of habitat loss: The role of prey depletion. *Journal of Theoretical Biology* 160: 223–230.

Székely, T., and Z. Bamberger. 1992. Predation of waders (Charadrii) on prey populations: An exclosure experiment. *Journal of Animal Ecology* 61: 447–456.

Townshend, D. J. 1981. The importance of field feeding to the survival of wintering male and female curlews *Numenius arquata* on the Tees Estuary. In *Feeding and survival strategies of estuarine organisms,* ed. N. V. Jones and W. J. Wolff, 261–273. New York: Plenum Press.

van Gils, J. A., P. Edelaar, G. Escudero, and T. Piersma. 2004. Carrying capacity models should not use fixed prey density thresholds: A plea for using more tools of behavioural ecology. *Oikos* 104: 197–204.

van Gils, J. A., T. Piersma, A. Dekinga, B. Spaans, and C. Kraan. 2006. Shellfish dredging pushes a flexible avian top predator out of a marine protected area. *PLoS Biology* 4: 2399–2404.

Warnock, N., G. W. Page, and L. E. Stenzel. 1995. Non-migratory movements of Dunlins on their California wintering grounds. *Wilson Bulletin* 107: 131–139.

Weber, L. M., and S. M Haig. 1997. Shorebird-prey interactions in South Carolina coastal soft sediments. *Canadian Journal of Zoology* 75: 245–252.

Wiens, J. A. 1989. *The ecology of bird communities.* Vol. 2. Cambridge: Cambridge University Press.

Wilson, W. H., Jr. 1990. Relationships between prey abundance and foraging site selection by Semipalmated Sandpipers on a Bay of Fundy mudflat. *Journal of Field Ornithology* 61: 9–19.

Zharikov, Y., and G. A. Skilleter. 2003. Depletion of benthic invertebrates by Bar-tailed Godwits *Limosa lapponica* in a subtropical estuary. *Marine Ecology Progress Series* 254: 151–162.

Zwarts, L., and R. H. Drent. 1981. Prey depletion and the regulation of predator density: Oystercatchers (*Haematopus ostralegus*) feeding on mussels (*Mytilus edulis*). In *Feeding and survival strategies of estuarine organisms,* ed. N. V. Jones and W. J. Wolff, 193–216. New York: Plenum Press.

9

Spatial Ecology and Winter Social Organization

CONTENTS

THE CHALLENGE OF ECOLOGY IS to understand the various biotic and abiotic factors that contribute to variation in the distribution and abundance of species. During the nonbreeding season, shorebirds are ideal subjects to investigate spatial ecology, a discipline that characterizes how individuals are distributed across the landscape with the aim of understanding the selective forces that shape these patterns of dispersion. Shorebirds are ideal subjects for spatial analyses for several reasons. First, they are easily observed in open

habitats, where they feed and roost. Thus, their spatial distributions can be readily quantified and analyzed, especially given recent advancements in field methods (such as satellite telemetry, which provides real-time locations for individuals) and the improvement of analytical tools in geographic information systems and geostatistics. Second, the selective forces shaping sociality, principally food and predation, are comparatively easy to measure in shorebirds. As a result, many important studies in spatial ecology of wintering birds (e.g., Myers et al. 1979; Goss-Custard 1970; Townshend 1985) have been conducted on shorebirds. Finally, shorebirds exhibit considerable variation in space use and sociality both among and within species. This provides the opportunity for productive comparative analyses addressing the causes of this variation.

In this chapter, I review the social organization and spatial distribution of nonbreeding shorebirds and address these dispersion patterns in relation to two principal factors that are widely recognized to cause individuals to coalesce into dense flocks as opposed to feeding alone. These two factors are variation in density and quality of their prey, and the threat posed by predators, principally raptors.

QUANTIFYING SPATIAL DISTRIBUTIONS

During the nonbreeding season, a scan of any wetland, whether a coastal estuary at low tide or a large hypersaline wetland fringed by an alkali flat, reveals much about shorebird spatial ecology. At first glance, one may be struck by a large flock of calidridine sandpipers that peck and probe through shallow water and exposed substrate in search of food. The sandpipers move about quickly, almost frantically, as they forage, seemingly as one cohesive unit. But individuals occasionally jab and poke at neighbors that approach too closely. The impression of a unified social group is solidified by the highly synchronized evasive response of the wheeling and turning flock to the approach of a falcon. After a while, the flock settles to feed again, this time in a location several kilometers from where the predator disturbed them. Further observations reveal isolated shorebirds of other species that are scattered about, methodically searching for prey. A single, large-bodied Black-bellied Plover moves purposefully about a high-elevation tidal flat in its characteristic walk-stop-peck foraging pattern. Nearby, two Eurasian Curlews vocalize at one another as they approach what appears to be a territory boundary represented by the edge of a deeply incised tidal channel. A few hours later, as the flooding tide inundates this area, the countless individuals from a mix of species depart their foraging areas and fly to a nearby salt marsh island where they roost in very close proximity for the following high-tide interval. On the roost, individuals stand side-by-side, alternating between preening and resting.

These patterns of interspecific and intraspecific variation in behavior and spatial distribution prompt numerous intriguing observations and questions in shorebird spatial ecology. For example, shorebirds of different size frequent a diverse array of wetland habitats over the course of the nonbreeding season. As sandpipers depart northern breeding areas during the middle to late summer and wend their way to wintering grounds, they undoubtedly experience a dynamic suite of factors that alter the costs and benefits of foraging and antipredator behaviors. These fluid conditions involve the birds themselves (such as their varying mass associated with migration) as well as two important environmental attributes: food and danger. The various estuaries where they stage differ in the types and availability of invertebrate prey. This stems from the response of prey to a multitude of abiotic factors that shape their distributions and abundances as well as the seasonality of invertebrate recruitment and predation by a host of predators, including shorebirds. Shorebirds must alter where and how they forage to maximize their intake rate, even during a comparatively short leg of a migratory journey from sub-Arctic to temperate and tropical wetlands. These changes in feeding and antipredator behavior may be prompted by the structure of the landscape surrounding a wetland and how it affects the danger posed by predators. Foraging behavior may also be strongly influenced by the microtopography of the substrates where birds forage. As they reach their wintering grounds and undergo a costly molt that replaces the flight feathers necessary for effective escape behaviors, individuals also experience a change in the danger posed by predators. But predation danger is not uniformly distributed in space and time. The picture that emerges is one of a spatially and temporally dynamic landscape where individual behaviors give rise to varying spatial distributions of species, and these patterns are in constant flux. Moreover, the costs and benefits of these behaviors certainly vary with a species' size (mass) in combination with varying danger posed by different predators.

These behavioral observations, coupled to spatial distributions, require tools and methods for quantifying spatial patterns that are applicable across a range of spatial scales of analysis. All populations are characterized by a spatial distribution or dispersion pattern (Krebs 1999). For plants, these distributions are comparatively static and easily quantified owing to the sedentary nature of individuals. For animals, especially volant taxa such as birds, spatial distributions are much more dynamic, and therefore they present analytical challenges. Regardless of taxon, however, a species' pattern of spatial distribution may take one of three forms. A starting point for any spatial analysis is to compare a distribution to a *random* pattern, often viewed as the null case or null distribution. Some species are randomly distributed such that any location is as likely to be occupied or not by an individual. For shorebirds, another more profitable way of looking at a random distribution involves quantifying the distance between an individual and conspecifics, referred to as the nearest neighbor distance. In a random distribution, this nearest neighbor distance takes the form of the Poisson distribution, with the average for the population characterized by a variance that equals the mean. Ecologists recognize two alternatives to the random pattern. In some instances, a species may be *evenly* (the terms hyperdispersed and orchard are synonyms) distributed such that individuals occur at distances greater than one would expect by chance. For animals, an even pattern may be seen in territorial species where agonistic interactions between competing neighbors cause individuals to avoid one another with the result that they are widely distributed within a habitat. A third possibility is that individuals are *aggregated* (or clumped) within habitats such that nearest neighbors are, on average, quite close. This is a common pattern exhibited by nonbreeding shorebirds, as exemplified by the large flocks that occur, especially during migration.

The methods used to quantify a species' pattern of spatial distribution differ among sedentary and mobile organisms. Krebs (1999) provided a detailed perspective on these methods. For shorebirds, the approaches used to study

distributions have varied, probably owing to the nature of the research question posed and hence the study design and data collected. There are two general approaches used to quantify spatial patterns. Analyses of distributions of plants and sessile marine intertidal taxa commonly use sampling plots to estimate species' densities, especially when the species of interest are small. For these organisms, a sample of 1 m² randomly placed plots yields the average (and standard deviation) number of individuals per plot. In this case, the comparison to a random pattern is made with the Poisson distribution, where the expected (random) ratio of the variance to the mean number of individuals per plot equals 1. The variance:mean ratio is referred to as the INDEX OF DISPERSION (Krebs 1999). An aggregated pattern has an index that exceeds 1, whereas an even distribution is characterized by an index that is much less than unity (that is, the variance in the number of individuals per quadrat is much less than the mean). Use of a plot-based method to quantify shorebird distributions requires an appropriate-sized plot. If the plot size is too small, individuals rarely are counted. Conversely, plots that are overly large may be difficult to sample, especially if observers conduct instantaneous scans to characterize use and the distant plot edges make it difficult to ascertain whether birds are actually using a plot. See Krebs (1999) for statistical approaches to test for a deviation from a random pattern.

The utility of a plot-based approach is probably greatest in studies of species' distributions within a particular habitat type, especially when variation in species' densities is coupled with food sampling (e.g., Bryant 1979; Colwell and Landrum 1993). An example of this method (Colwell and Sundeen 2000) sampled shorebird distributions during late winter and early spring using 40 randomly located plots on sandy, ocean-fronting beach in coastal northern California. Plots were 500 m in length. A single observer visited each plot four times for an hour and recorded, from the midpoint of the plot, shorebird density at 15-minute intervals. Nine of 12 species had aggregated distributions, and these concentrations occurred on beaches that lay in close proximity to the main estuarine feeding areas of Humboldt Bay.

At the scale of the estuary, a variation on the quadrat-based approach was used to quantify the low-tide spatial distributions of American Avocets and Long-billed Curlews at Humboldt Bay, California (Fig. 9.1; Colwell et al. 2001; Mathis et al. 2006). In the field, multiple observers worked in a coordinated, simultaneous survey effort to plot the location of individuals of both species on high-resolution digital images of intertidal habitats. Later, these locations were transferred to a geographic information system, and a grid system (500 m² grid cell=quadrat) was overlaid on the image. This yielded the average number of curlews or avocets per grid cell. At the large spatial scale of the estuary, both species were aggregated in their use of intertidal habitats. Curlews were concentrated in high-elevation tidal flats. By contrast, avocets occurred in highest densities in bay habitats characterized by fine sediments, and they were absent from other areas along the main shipping channel where high tidal velocities carried away fine sediments and left coarse substrates (Danufsky and Colwell 2003). This variation in density of foraging avocets is probably associated with their avoidance of coarse sediments, which impedes their characteristic side-to-side foraging maneuver (Quammen 1982).

An alternative to the plot-based methods is to record the precise location of individuals and use direct linear measures of nearest neighbors to quantify the spatial distribution. This approach is based on a complete enumeration of individuals in the population (Krebs 1999); thus, unlike the plot-based approach already described, density is known (total individuals in the study area). Briefly, if all individuals in a population are plotted using a geographic information system, then the distance between each individual and all conspecifics can be used to calculate the mean distance to the nearest neighbor (r_A). This value is compared with an

expected distance (r_E), which is based on the density of organisms in the study area. The ratio $r_A:r_E$ yields the INDEX OF AGGREGATION. Note that this index is related to spatial pattern in a manner opposite to the index of dispersion. If the ratio is 1, then the distribution is random. However, clumped patterns have values that approach 0, whereas even distributions exceed 1. Again, Krebs (1999) provides details on statistical tests to evaluate the extent to which a spatial pattern deviates from random.

The Long-billed Curlew provides another example of the nearest neighbor technique applied to dispersion patterns (Colwell et al. 2002; Mathis et al. 2006). Within Humboldt Bay, the Elk River estuary has the highest density of curlews (0.36 ± 0.09 birds ha^{-1}), and varying numbers of individuals defend low-tide feeding territories from mid-June through March. Using a spot-mapping technique (Bibby et al. 2000), observers mapped the location of individuals on 130 visits to the estuary during the course of a single nonbreeding season. During summer, when the number of resident curlews was highest and hence the sample size was adequate for statistical analysis, the nearest neighbor approach indicated that on most (73%) occasions curlews were uniformly distributed within the estuary. This observation suggests that territoriality caused birds to avoid one another and space out at rather uniform distances. On remaining occasions, however, birds were randomly distributed within intertidal habitats, and this pattern was significantly more likely when there were more nonterritorial curlews around. Mathis et al. (2006) used the nearest neighbor method to characterize the movement of individuals within their territories. In this case, the location of a focal-sampled curlew was plotted at 2-minute intervals, providing a sample of 60 observations per 2-hour focal sample. The diet of birds was recorded continuously during these observations (Colwell et al. 2002). At the scale of the territory, the nearest neighbor method showed that most (75%) of the time curlews were uniformly distributed within their territo-

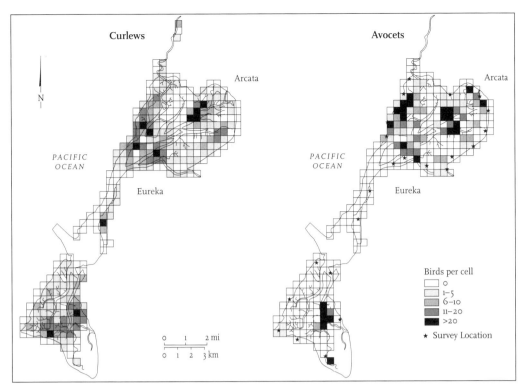

FIGURE 9.1. Low-tide spatial distributions of the Long-billed Curlew (*left*) and American Avocet (*right*) exhibiting an aggregated or clumped pattern in the intertidal habitats of Humboldt Bay, California. The areas of concentration correspond to underlying variation in habitat features such as elevation of tidal flats and proximity to high-tide roosts (curlews), or substrate particle size (avocets). After Danufsky and Colwell (2003).

ries and were less often distributed in a random (16%) or aggregated (8%) pattern. This variation in space use correlated with diet. Aggregated patterns were obtained from individual curlews feeding for extended periods on bivalves that were concentrated in "hotspots"; by contrast, the uniform distributions stemmed from curlews walking about their territories while searching visually for prey.

These examples illustrate the methods used to characterize the range of dispersion patterns of shorebirds during the nonbreeding season. It should be apparent that there is considerable variation within and among species in dispersion pattern. This variation is of interest, especially when coupled with basic ecological data such as food distribution and danger. In addition, the variation in dispersion pattern may depend on the spatial scale of analysis. For ex-

ample, across large spatial scales that span significant variation in the types and quality of habitats within an estuary, it is common for birds to aggregate in some habitats and be absent from others. This may stem from the variation in suitability of foraging habitats, as dictated by the various abiotic factors (such as substrate particle size or salinity) that influence the distribution of invertebrates. Moreover, at large spatial scales of analysis, strong aggregated patterns of dispersion stem, in part, from the inclusion of low-quality or unsuitable areas in the habitats sampled. For example, Black Oystercatchers were observed using only 2 of 40 study plots on sandy beaches in coastal California, but their aggregated distribution on beaches occurred adjacent to one of the main rocky intertidal headlands where they are permanent residents (Colwell and Sundeen 2000).

A RANGE OF SOCIAL ORGANIZATION

In some waterfowl (such as geese and swans), the young remain with their parents in cohesive family groups into their first winter. This is not known to occur in shorebirds. Thus, the social cohesion of nonbreeding flocks is not attributable to pairs or family groups. Rather, the dispersion patterns and social organization are an outcome of individual decisions to live alone versus in groups, under the influence of environmental factors. Myers (1984) and Goss-Custard (1985) provided excellent reviews of the range of sociality exhibited by shorebirds during the nonbreeding season, as well as a treatment of the selective forces that have shaped these patterns. The following summary details interspecific and intraspecific variation in the social organization of shorebirds when they are feeding. Later, I address the behaviors and circumstances of roosting, when large numbers of birds, often of many species, rest collectively at a location.

TERRITORIALITY

On a daily basis, a species' social system can be characterized as lying along a continuum spanning solitary to group living (Fig. 9.2). At the solitary end of the continuum are territorial species, in which individuals defend food resources within a fixed area. This defense may last for minutes to years and may span the lifetime of individuals. Nonbreeding territoriality occurs in many shorebirds, but there is considerable variation within and among species in patterns of resource defense (Myers 1984; Goss-Custard 1985; Colwell 2000). In total, nearly one-quarter of the 215 species of shorebird have been reported to defend territories. Territoriality has been reported most commonly for plovers and lapwings, calidridine sandpipers, tringine sandpipers, and curlews (Colwell 2000).

In many large-bodied shorebirds such as curlews, thick-knees, and some large plovers, territoriality is a regular occurrence; however, not all individuals in a population may be territorial. For example, studies from across the winter

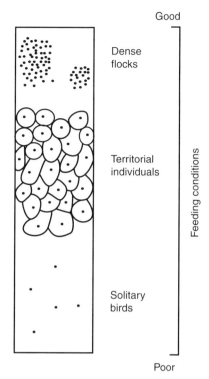

FIGURE 9.2. Highly variable shorebird dispersion patterns, from dense flocks to solitary feeding birds. Favorable feeding territories feature moderately rich and predictable food resources and a low density of competitors such that resource defense is economically feasible. After Goss-Custard (1985).

range of the Black-bellied Plover indicate that some individuals defend permanent feeding territories in intertidal habitats whereas others tend to feed in loose flocks (Townshend 1985; Turpie 1995). At the northern extent of the species' wintering range in coastal England, some individuals defend territories year after year. Interestingly, the territorial plovers were larger than the individuals that were nomadic (Townshend 1985). Territoriality has been well documented in other members of the genus *Pluvialis* (Byrkjedal and Thompson 1998). The intraspecific variation in territoriality of Black-bellied Plovers and other large shorebirds raises interesting questions regarding the extent to which the behavioral strategy is heritable versus environmentally determined. Specifically, to what extent are individuals capable of changing their behavior as they age and gain experience, and how might this be affected by the flock composition (the fluidity of varying numbers of conspecifics, each differing in these attributes)? Evidence from the Eurasian Oystercatcher, a species in which dominance interactions lead to individuals feeding in a dispersed (but not truly territorial) pattern, suggests that individual behaviors are established early in life and are somewhat inflexible (Ens and Cayford 1996).

Although territoriality has been observed in smaller shorebirds, in most cases it is an ephemeral phenomenon lasting hours to a few days. During fall migration (September 8 to 25), 8% to 19% of the populations of Semipalmated, Western, and Least Sandpipers defended territories. Territories occurred where food was particularly abundant along a few meters of the shores of salt ponds in Puerto Rico (Tripp and Collazo 1997). Sanderlings appear to be an exception to the transitory nature of territoriality in calidridine sandpipers, at least for the years that they were studied by Myers and his colleagues in coastal California (Connors et al. 1981; Myers 1984; Myers et al. 1979). In the 1970s, individual Sanderlings regularly defended 100 m stretches of ocean-fronting beach for weeks to months during the winter. At low tide, individuals fed in flocks on exposed tidal flats of the local estuary. When flooding tides inundated these areas, Sanderlings moved to beaches where they defended territories. Territoriality resulted from the profitability of feeding on rich invertebrate prey, principally the burrowing decapod *Emerita* and the isopod *Excirolana*, which occurred in the upper beach strand at intermediate and higher tide levels (Connors et al. 1981). Interestingly, the occurrence of territoriality varied annually during the 1970s and early 1980s with the presence of wintering Merlins (*Falco columbarius*). However, now it is uncommon to observe territoriality in Sanderlings along this same stretch of coastline. The benefits of living in groups have apparently increased with the recovery of falcon populations in North America (Myers 1984), probably due to the increased predation risk to solitary individuals compared with those in a flock.

A RANGE OF DISPERSION PATTERNS IN ONE SPECIES

The Eurasian Oystercatcher is an ideal subject for studying spatial ecology owing to its large size and the ease with which foraging observations can be made on individuals feeding on large prey. Ens and Cayford (1996) discussed the spatial patterns and behaviors of oystercatchers feeding on bivalves (such as *Cerastoderma* and *Mytilus*) at low tide. Although oystercatchers appear to aggregate in loose flocks at times, individuals in these flocks generally are not actively feeding. When feeding on mussel beds, however, the nearest neighbor distances of actively feeding oystercatchers show that spatial patterns are nonrandom, and they indicate that individuals avoid one another. Short nearest neighbor distances were underrepresented, suggesting avoidance of conspecifics, whereas intermediate distances were overrepresented. These observations suggest that aggregations of oystercatchers are scale dependent; in other words, they concentrate in actively feeding flocks in areas where their food supply is richest, but individuals avoid each other within these flocks. Studies of marked individuals confirm this. Long-lived adults are faithful to their wintering sites, feeding in a predictable pattern dictated by tides. Many oystercatchers confine their foraging to a restricted winter home range, whereas others range more widely. Repeated observations of marked individuals suggest that oystercatchers do not form feeding flocks that are stable assemblages of the same individuals, as may be the case for other wintering shorebirds (Whitfield 1985; Conklin and Colwell 2008). Rather, individual oystercatchers simply return day after day to the same feeding area, which happens to be used by other individuals that behave similarly. These individuals do roost, however, in tight flocks at high tide.

At the opposite end of the continuum describing sociality in nonbreeding shorebirds are the flocks typical of so many species throughout the winter and especially during migratory periods. In fact, virtually all the species that have been characterized as exhibiting some form of territoriality routinely forage in flocks. However, the size, density and cohesiveness of these flocks are highly variable (Myers 1984). Some species, such as the Black Turnstone and Semipalmated Plover, forage in loose-knit groups, often mixed with other shorebirds. For other shorebirds, large, dense, mixed-species assemblages are the norm. Still, within these high-density flocks there can be considerable aggressive interactions over food (Recher and Recher 1969).

Myers (1984) first used a comparative approach to address the question of flock cohesion. Cohesiveness refers to the extent to which individuals behave in an organized and coordinated manner to maintain the composition and integrity of flock membership from one moment to the next. Imagine a flock of several hundred Western Sandpipers in a local population that is much larger. The question of cohesiveness addresses the extent to which individuals in this flock are likely to associate with one another tomorrow, next week, and for the remainder of the winter. This question is of interest because it addresses the issues of the benefits of flocking per se versus being in a group of familiar individuals who may share similar interests and hence may behave altruistically or reciprocally (Myers 1984). Myers (1983) examined this question with color-marked Sanderlings. In a winter population of several hundred, the composition of flocks was so fluid from one day to the next that groups essentially consisted of random assemblages of individuals. Recent work on the Dunlin (Conklin and Colwell 2008) substantiates this view that sandpiper flocks were a random assemblage of individuals behaving opportunistically in their associations

with conspecifics. In this case, wintering sandpipers were radio-marked in a study of high-tide roost fidelity around Humboldt Bay, California. The association between individuals was judged by the shared use of high-tide roosts on consecutive high tides (separated by approximately 12 hours). There was little evidence that Dunlins occurred in cohesive associations in roosting flocks. In fact, associations (pairwise co-occurrences of marked birds) were rather ephemeral, barely lasting from one high tide to the next (1.1 consecutive high tides on average).

Two principal ecological factors shape the tendency for species to forage in dense or loose flocks versus alone on territories (Myers 1984): food and predators. Individuals track food resources and concentrate their foraging activity where food is readily available (see Chapters 7 and 8), as indicated by high intake rates (Goss-Custard 1984). As a consequence, shorebird densities and intake rates correlate positively with high densities of prey. But living with others entails costs. The main cost to an individual foraging in a flock is a reduction in intake rate associated with several facets of competition (Goss-Custard 1984). First, individuals may experience reduced intake rates owing to direct interference from others in close proximity, a pattern demonstrated for Common Redshanks foraging for *Corophium* on intertidal mudflats of The Wash in England (Goss-Custard 1977). A second cost of foraging in dense flocks is that the foraging activities of other birds may reduce intake rates by eliciting evasive responses of prey, which reduce the availability of food. A third cost is that conspecifics will consume prey and reduce food availability. These relationships between food supply and foraging have been addressed at length in the previous chapter.

The defense of feeding territories by shorebirds is also related to their food supply. Goss-Custard (1985) argued that territory defense was favored by rather specific conditions involving the density of food and competitors. Specifically, territoriality was more likely to occur under conditions when food was of moderate quality,

DOES FLOCKING DIFFER AMONG SPECIES THAT FORAGE IN DIFFERENT WAYS?

Some shorebirds routinely forage in loosely spaced groups with nearest neighbors at considerable distances from one another. By contrast, other species regularly forage in dense flocks, with birds less than a body length from several neighbors. This variation in dispersion probably stems from the competing costs of foraging together (interference lowering intake rates) and the benefits of predator vigilance. Over the years, several authors have commented on the differences in dispersion among shorebirds that feed using tactile versus visual methods. Specifically, Blick (1980) argued that visual feeders could more easily be vigilant for predators as they scanned surrounding areas with their head-up posture; tactile feeders were less capable of being vigilant because they foraged with their head down. Goss-Custard (1970, 1985) offered an alternative argument to the variation in flocking tendency among visual and tactile feeders. He suggested that visual species were more likely to interfere and reduce intake rates of conspecifics in close proximity, and hence that they were less likely to occur in dense flocks. Barbosa (1995,

1997, 2002) further subdivided shorebirds based on how birds moved about habitats and captured prey. CONTINUOUS-TACTILE species (such as godwits and many calidridines) feed almost exclusively by touch, with birds walking continuously while probing below the surface for prey. CONTINUOUS-VISUAL species (such as turnstones, stilts, and oystercatchers) walk steadily about habitats using visual cues to search for food in close proximity and then peck at prey from the surface. PAUSE-TRAVEL species (such as plovers and coursers) scan widely within habitats, and capture prey in a walk-stop-peck fashion. Barbosa (1995) argued that vigilance of these groups differed because pause-travel species could more easily be vigilant for predators when foraging owing to their head-up foraging method. By contrast, tactile and continuous-visual feeders more often have their heads down as they forage and are thus less able to be vigilant as they feed. Continuous-visual and pause-travel species tended to be solitary or formed smaller flocks than the continuous-tactile species, which confirms the relationship between foraging method and dispersion pattern. The vigilance of these species also differed, with pause-travel species scanning less frequently than the other two groups.

such that large numbers of individuals attracted to the resource would not overwhelm the defensive abilities of a single territorial bird. Evidence to support this hypothesis was provided by wintering Sanderlings (Myers et al. 1979). Sanderlings sometimes defend food resources along linear territories of 10 to 120 m in the wave-washed zone of ocean beaches. The size of Sanderling territories was negatively related to conspecific density. An inverse relationship between territory size and food density disappeared when researchers controlled statistically for competitor density. Relative to Goss-Custard's (1985) model, it is worth noting that Sanderlings abandon territories at particularly high conspecific

densities, presumably when the benefits of defense of food are outweighed by the time and energy costs of doing so.

There is more to the relationship between food and territory defense, however. Based on reviews of sociality (e.g., Myers 1984; Goss-Custard 1985; Colwell 2000), territory defense is variable in duration. Individuals of some species defend territories for intervals as short as a few minutes or hours during migration, before they move on to wintering areas (Tripp and Collazo 1997). On the other hand, long-term territory defense (of months' duration) has been reported in a variety of species, including some oystercatchers, tringine sandpipers, large plovers, and curlews. This variation in tenure requires additional explanation.

Defense of a feeding territory, regardless of how long, has been suggested to stem from feeding interference (Myers et al. 1979; Goss-Custard 1985). Specifically, interference favors territorial behavior when the benefits of increased energy intake (despite the costs of defense) outweigh the costs of lower intake rates stemming from tolerating conspecifics. Goss-Custard (1985) reported that the time spent by Common Redshanks defending territories was compensated by increased intake rate. This argument applies to territory defense regardless of duration, and it may be especially applicable when food is patchily distributed and variable in time. Individuals should defend territories whenever it is energetically economical, whether that period is for minutes, hours, or days. However, the permanency of feeding territories in some species requires additional comment. Specifically, why do individuals sometimes defend the same intertidal areas from conspecifics year after year, and what features of the underlying food resources favor long-term defense? Long-term territories may conserve food supplies over the course of a winter (Goss-Custard 1985). Given that shorebirds sometimes can significantly reduce the populations of their invertebrate prey over the course of the nonbreeding season (Goss-Custard 1984), it is plausible that territory defense acts to conserve food for the long term. However, prey populations were reduced to similar extents in areas defended by wintering Common Redshanks compared with habitats occupied by flocks (Goss-Custard et al. 1984). To date, there is no clear answer to this question.

The large, dense flocks formed during the nonbreeding season by so many shorebirds stand in sharp contrast to the relatively less common occurrence of territoriality. This suggests that group living has significant benefits that outweigh the costs of lowered intake rates owing to interference and prey reduction.

The benefits of flocking are many (Hamilton 1971; Vine 1971), and most or all involve reducing the risks of predation, principally by raptors (Stinson 1980; Myers 1984). Raptors, especially falcons and accipiters, are widely considered to be the main predators of shorebirds during the nonbreeding season. Early support for the selective advantages of flocking came from modeling flock size in relation to frequency of predation. Predation attempts do not have to be frequent for foraging in large flocks to be selectively advantageous (Stinson 1980). Goss-Custard (1985) offered several observations in support of the notion that predation was the main cause of flocking behavior in shorebirds. First, raptors have a significant effect on shorebird populations during the nonbreeding season. Second, flocking has not been shown to positively affect the foraging rates of shorebirds. Rather, intake rate typically declines with increasing bird density. Third, individuals feeding on territories regularly abandon this social strategy and form flocks for short duration upon the appearance of a raptor (Myers 1980; Whitfield 1988). Finally, considerable evidence demonstrates that individuals that form flocks benefit by becoming less vulnerable to predation.

There are several antipredator benefits accrued by an individual shorebird that joins a flock. First, an individual decreases its probability of being depredated, otherwise referred to as the *dilution effect* or safety-in-numbers argument. Second, individuals benefit from cumulative vigilance of flock-mates, which results in early detection of danger and higher intake rates. Third, predators may have difficulty locating flocks and may be injured when hunting flocks. These antipredator advantages are collectively summarized by a negative relationship between flock size and an individual's risk of predation. Roberts (1996) formulated this argument as the product of three components: (1) predator attack rate, (2) probability of detecting an attacking predator, and (3) the probability of escaping. The latter two components have a direct relationship to flock size, but disentangling the mechanisms by which group size affects the safety of individuals has been challenging (Elgar 1989; Roberts 1996).

The attack rate of predators, principally raptors, stems from the combined functional and numerical responses of predators to variation in the availability of shorebirds as prey. At first glance, there appears to be little that shorebirds can do to directly alter the attack rate of their predators. But shorebirds can choose to feed in habitats characterized by low predator density, thus reducing the attack rate and predation risk. The selection of habitats of varying predation risk can take the form of daily movements within the landscape of a wintering area, or it can be on a fine spatial scale on a short-term basis. Fine-scale adjustments in habitat use in response to predation risk and foraging have been shown for several shorebirds. Pomeroy et al. (2006) studied Western Sandpipers foraging along a tidal flat gradient of varying predation risk, as gauged by openness of habitat and proximity to cover that obstructed the view of approaching raptors. They erected artificial structures on tidal flats to further decrease the ability of Western Sandpipers to detect raptors. Using fecal counts on tidal flats as an index of use, they showed that the abundance of sandpipers was lower in obstructed habitats. In the same general area, Ydenberg et al. (2002) argued that the decline in Western Sandpipers staging at a sheltered estuary in coastal British Columbia was most plausibly explained by an increase in danger at this site, where habitats obstructed the view of approaching predators.

One clear advantage to being in a flock is a safety-in-numbers argument (Hamilton 1971). Simply stated, the probability of being a victim decreases with increasing flock size. Evidence supporting this idea comes from the success of individual predators hunting shorebirds in flocks of varying size. Page and Whitacre (1975) produced one of the first assessments of this risk. They studied in detail the foraging behavior of a female Merlin and the diets of other raptors wintering at a small estuary on the northern California coast. The likelihood of a single, small shorebird being the victim of predation compared with a shorebird in a flock was 3.2:1. These data were based on a single female falcon studied over one winter season. More recent analyses of wintering shorebirds in Scotland shed additional light on the benefits of flocking. Cresswell and Quinn (2004) examined the decisions of Eurasian Sparrowhawks hunting Common Redshanks in groups that varied in vulnerability. Their study area was the Tyninghame estuary where Common Redshanks feed on the tidal flats and adjacent saltmarsh, two habitats that differ in foraging profitability and predation danger. Tidal flats were safer places to feed when sparrowhawks hunted because the openness of habitat allowed early detection of a fast-approaching raptor. However, when poor weather made foraging increasingly difficult and heightened the risk of starvation, Common Redshanks shifted to feed in the saltmarsh. This shift in habitat occurred despite a much greater risk of predation by sparrowhawks (Yasue et al. 2003). Based on detailed observations of one to four resident sparrowhawks hunting along a stretch of estuary, Cresswell and Quinn (2004) estimated that the probability of a Common Redshank becoming a victim of predation was based on group size and the proximity to saltmarsh and forested area. The vegetation was viewed as cover that obstructed the view of redshanks of an approaching raptor and thus allowed a sparrowhawk greater surprise of attack. The probability that a Common Redshank was captured by a sparrowhawk decreased steadily with both flock size and distance from cover (Fig. 9.3). Furthermore, raptors maximized their probability of hunting success by preferentially attacking the most vulnerable group of redshanks (Cresswell and Quinn 2004). Because sparrowhawks appeared to be so effective at making decisions regarding the vulnerability of groups of varying size, redshanks should constantly evaluate their environment and change groups and habitats to reduce predation risk. If so, then one would expect shorebird flocks to be quite dynamic in composition as individuals make these moment-to-moment decisions.

Another advantage of group living is that individuals share vigilance with other flock members (Hamilton 1971; Myers 1984; Elgar 1989;

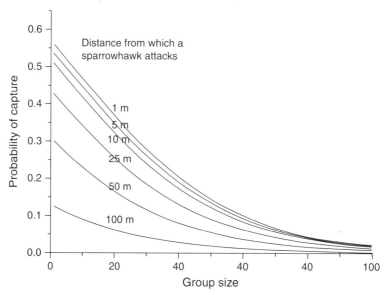

FIGURE 9.3. The probability that a Common Redshank is captured by a Eurasian Sparrowhawk decreases steadily with group size, and this relationship is also influenced by distance (shown in meters) to vegetative cover, which affords an approaching raptor greater surprise in its attack. After Cresswell and Quinn (2004).

Roberts 1996). This benefit relates to early detection of danger posed by an attacking raptor, which enhances escape probability. But reduced vigilance may also allow an individual to feed more efficiently. There is considerable evidence in support of the relationship between flock size and various measures of vigilance. For example, vigilance of Eurasian Curlews, estimated by the percentage time a bird had its head up, decreased rapidly and then leveled off in larger flocks (Abramson 1979). The relationships among flock size, vigilance, early detection of raptors (Eurasian Sparrowhawk and Peregrine Falcon), and risk of death were nicely worked out by Cresswell (1994). Flock sizes varied from 1 to 30 birds. An individual's vigilance, gauged by the interscan interval or time between successive head-up postures, decreased with increasing flock size, as did the interval to detection of an attacking raptor. Collectively, this yielded a higher mortality rate for Common Redshanks feeding alone or in small flocks. Because there was no difference in peck rate among birds feeding in different sized flocks,

Cresswell (1994) concluded that this was a clear case of predation favoring flock formation.

Another advantage to being in flocks stems from the challenge in finding them. Specifically, Vine (1973) suggested that flocks, especially large ones, were more difficult for a predator to locate amid the large landscape. This assumes that a raptor has limited knowledge of its home range and cannot accurately predict where to find prey. This seems unlikely to apply to shorebirds when considering that most studies have reported that high-tide roosts, for instance, occur predictably in the same locations over sometimes very long periods (Hale 1980).

A last consideration in arguments of costs and benefits of flocking addresses the predator's point of view. A predator attacking flocks (versus individuals) may suffer injury when doing so. It is noteworthy that when raptors attack shorebird flocks of varying size, they ultimately attempt to single out individuals as prey and often pursue individuals, not the flock. In fact, it is not uncommon to observe falcons give up pursuit rather quickly if they are unsuccessful in singling out prey. Otherwise, there is little empirical evidence to evaluate these ideas as they apply to shorebirds as prey of raptors.

In summary, selective forces associated with finding food and avoiding predation strongly influence the dispersion patterns and social organization of foraging shorebirds during the nonbreeding season. Species vary in their sociality, with larger, visual-feeding taxa often feeding solitarily or defending territories. By contrast, smaller, tactile-feeding species more commonly forage in flocks. However, these flock-mates often behave aggressively toward neighbors in response to feeding opportunities, and there is no indication that flocks consist of stable social groups. Additionally, there is considerable variation within species, which highlights the moment-to-moment decisions made by individuals regarding group versus solitary living. This dynamic environment is driven by ever-changing predation danger. It is not unusual to witness a territorial individual quickly abandoning its turf when a raptor suddenly appears and then reclaiming its territory when danger passes.

ROOSTS

The social nature of shorebirds becomes especially conspicuous when individuals coalesce into dense, mixed-species flocks at roosts. The selective factors that shape flocking are apparent when considering roosting shorebirds because birds generally do not forage while at roosts. When the flocking behavior of roosting birds can be interpreted without reference to the costs associated with foraging in groups (such as interference), it becomes clear that the flocking behavior observed at roosts is driven by the danger posed by predators. Furthermore, the advantages of roosting together are exemplified by species that regularly defend territories yet come together to roost when they are not feeding.

Several temporal and spatial elements characterize roosts and the behavior of roosting by shorebirds (Hale 1980; Myers 1984); there are, however, exceptions to these generalities. The seasonality and timing of roost formation is rather predictable. Roosts are almost exclusively a phenomenon of the nonbreeding season, when birds alternate between periods of feeding and relative inactivity. Early in the breeding season, however, Eurasian Curlews breeding in meadows near wintering areas come and go as pairs from roosts (Hale 1980). In coastal areas, roosts typically form on a regular schedule dictated by tides. Birds depart feeding areas and move to roosts when foraging opportunities become increasingly limited as flooding tides inundate tidal flats. These patterns often vary with the height of spring and neap tides. Spring tides always force shorebirds off their main estuarine feeding areas to roosts in surrounding habitats, whereas during neap tides shorebirds may continue to feed and often roost on high-elevation tidal flats (Rogers et al. 2006; Rosa et al. 2006). In habitats distant from the coast, where shorebird activity follows a circadian pattern typified by daytime feeding, individuals come and go from roosts at dusk and dawn, respectively (Hale 1980; Myers 1984).

Almost universally, roosts occur at a terrestrial location within a habitat, such as an island of salt marsh (Hale 1980; Conklin and Colwell 2007), a flooded salt pan (Rogers et al. 2006; Rosa et al. 2006), an unvegetated stretch of sandy beach (Myers 1984), or an elevated man-made structure (Burton et al. 1996). Roosts occur in habitats that are often comparatively open, which affords individuals the ability to scan the surrounding area for danger. One exception to this terrestrial-based description is the occurrence of large, localized flights of shorebirds. Although referred to as *aerial roosts*, the fact that birds are flying precludes preening and resting behaviors. These flights commonly occur over water and vary greatly in size, density, and energy expended by birds, as flocks may remain airborne for hours on end (Hale 1980; Dekker 1998; Hötker 2000; Conklin and Colwell 2007). The openness of habitats and occurrence of aerial roosts has been interpreted as resulting from disturbance by humans or raptors that forces birds off their main terrestrial roosts (Hale 1980; Dekker 1998; Hötker 2000). In this context, habitat is viewed in terms of a location safe from predators and humans. But shorebirds may select roosting sites based on energy advantages related to thermoregulation. For instance, the use of flooded salt pans in Western Australia by Red Knots and Great Knots is probably related to individuals managing excessive heat loads in tropical areas during the nonbreeding season (Rogers et al. 2006).

At a larger spatial scale, roosts can also be characterized by their location within the landscape, and this is often interpreted by proximity to feeding areas. Roosts typically form near important foraging areas. For instance, roosts often occur atop salt marsh adjacent to unvegetated tidal flats such that birds move at most a few kilometers to feeding areas. In some cases, however, individuals are known to travel round-trip 50 to 100 km on a daily basis, which may reflect limitation in high-quality roosting habitats (Hale 1980; van Gils et al. 2006). In this context, roost location and proximity to feeding habitats is important from the perspective of the energy costs of commuting to feeding areas (van Gils et al. 2006).

Lastly, the behaviors exhibited by shorebirds help to define roosting and to understand the selective forces shaping this phenomenon. Roosting conjures images of a dense flock, with birds standing idly in close proximity. Roosting implies relative inactivity, compared with other times of day when birds spend much of the time foraging. While they are at roosts, the main behaviors of shorebirds are resting and maintenance. The latter consists of preening activity, which is especially evident during periods of molt, when large numbers of shed feathers may be scattered about the site (Hale 1980). It is common, however, to observe a small percentage of birds feeding in habitats immediately adjacent to or amid a roost. Perhaps the most important activity observed among roosting shorebirds is vigilance. Resting and preen-

ing birds often stop these activities and scan their surroundings for predators. While the collective vigilance of the flock enhances early detection of predators, there are potential energy costs to being in a roost. For example, large flocks are more "nervous," with more frequent false alarms that result in individuals taking flight for varying durations (Conklin et al. 2008).

Why roost? With few exceptions, a roost is essentially a group of individuals, often of multiple species. These birds coalesce when the quality of local feeding opportunities diminishes as the tide rises or night falls. This simple observation hints strongly at the main reasons why individuals form roosts: to aid survival in the face of predation and enhance foraging opportunities by gaining knowledge about the quality and location of ephemeral food. A third explanation for roosting behavior draws on principles of group selection (Wynne-Edwards 1962). Briefly, this idea posits that individuals form roosts to evaluate conspecific population size and self-regulate their fitness. If this were the case, one would expect individual shorebirds to use information at roosts to gauge local population size in relation to food availability and then to adjust individual behavior to match the availability of food resources. For instance, individuals acting for the "good of the species" may choose to disperse from a wintering site to reduce competition among conspecifics for limiting food resources. Unfortunately, this seemingly altruistic behavior cannot be separated from an interpretation that an individual has dispersed to increase its own chances of survival. As pointed out by Hale (1980), there is little evidence to suggest that shorebirds in flocks alter their distribution in a local area for the collective good of the species.

There are two additional reasons as to why individuals join roosts, and these entail the same selective advantages of avoiding predation and enhancing foraging that have been addressed as benefits of joining a flock (Hale 1980; Myers 1984). First, individuals gain advantages from roosting together for the same three antipredator reasons they forage in flocks: (1) dilution effect, (2) enhanced vigilance, and (3) challenges to predators in finding and attacking flocks. It is noteworthy that several studies of Dunlin show that they regularly form large, dense aggregations at daytime roosts yet disperse to roost singly at night. This contrast in behavior has been interpreted as a response to differences in danger posed by diurnal and nocturnal predators (Mouritsen 1992, 1994; Conklin et al. 2007). Given that very little feeding goes on at roosts, it is clear that the antipredator benefits must be overwhelmingly responsible for the formation of flocks at roosts.

A second benefit of joining a roost concerns information shared among flock-mates regarding the location and quality of food resources. Specifically, the information center hypothesis (Ward and Zahavi 1973) postulates that birds join roosts during nonfeeding intervals (at high tide or at night) such that individuals can gain information on the location of ephemeral food sources. This argument is unlikely to apply to shorebirds because the food supplies they utilize are rather predictable from one day to the next. For instance, on the British Columbia coast, Black Oystercatchers roost nightly, and daily departures by small groups are dictated by the timing of the dropping tide relative to daylight hours. Each winter day, oystercatchers fly predictably between their roost and permanent mussel beds. In other cases, however, shorebirds may share information. For instance, van Gils et al. (2006) suggested that Red Knots roosting on the Dutch Wadden Sea share information about the quality of food resources.

HABITAT SELECTION

In describing roosting habitats and behaviors typical of shorebirds at roosts, it should be evident that certain features of the environment influence the use and quality of a roost. Therefore, habitat features may influence the behavioral process by which a bird selects a roost within surrounding habitats. Here, quality is judged by an individual's ability to manage three critical facets of the environment during the nonbreeding season: time, energy, and danger. From an energy perspective, a high-quality roost offers open habitat in close proximity to prime foraging areas. This combination affects an individual's daily energy budget by reducing maintenance costs while at the roost and travel costs to and from the roost. In the former case, birds wintering at northern latitudes may select roosts that afford some shelter from wind and rain (e.g., Peters and Otis 2007). However, the behavior of shorebirds while at roosts may ameliorate the energy costs of roosting in open habitats. For instance, birds may cluster closer together and face into the wind during inclement weather (Hale 1980). An exception to this cold-weather perspective comes from the tropics of northwestern Australia, where shorebirds contend with a hot climate and excessive heat loads generated by flights to and from roosts. At Roebuck Bay, Australia, Great Knots and Red Knots roost at just a few locations, where they are often disturbed by raptors and humans. Nevertheless, knots appear to use this site rather than an alternative roost because they would generate significant heat during an approximately 25 km flight from their main roost (Rogers et al. 2007).

The energy costs of travel between roosts and foraging areas may be important for some species in large estuaries, especially where human activity limits the number of high-quality roosts. In the Dutch Wadden Sea, the *islandica* subspecies of Red Knot roosts at just a few spots amid a large area of intertidal flats. The mollusks that comprise the diet of knots vary in availability from day to day with tides such that intake rates vary substantially among patches of feeding habitat at varying distances from the roost. Van Gils et al. (2006) modeled the dynamics of transit between roosts and foraging areas. They reported empirical data that suggested that knots managed starvation risk and predation danger by feeding in high-quality patches during the middle of the low-tide interval when intake rates were highest. This pattern resulted in birds departing roosts and overflying suitable foraging areas to reach higher quality sites. In some cases, however, the energy costs of transit flight between roosts and foraging areas appear to be trivial, given the short distances and abundance of options. Dunlin wintering on Humboldt Bay, California, can choose from hundreds of roost sites, all of which are located at most a few kilometers from foraging habitats (Conklin et al. 2008).

In addition to time and energy costs associated with roost-site selection, predation danger appears to be an important selective force shaping roost-site selection. A high-quality roost is one with open, unobstructed views of surrounding habitats such that birds can scan effectively for approaching danger, whether posed by raptors, mammals, dogs, or humans. Virtually all studies of shorebirds characterize their roosting habitats in this way. Unfortunately, no study has quantified variation in roost quality from the perspective of danger posed by predators. Conklin et al. (2008) sampled raptor attack rates at several of the most consistently used Dunlin roosts at Humboldt Bay and reported that predation did not differ among roost locations. However, Dunlins were more likely to depart roosts during a given high tide when raptor attacks were more frequent. It may be that the best option for managing predation is to join roosts in open habitat, remain vigilant for predators, and move with flock-mates among roosts when feasible to avoid predation. Given that raptors probably have knowledge of roost locations and can readily detect flocks from a great distance, the best a shorebird can do to manage predation is to remain in a flock wherever it may choose to roost.

ROOST QUALITY

If roosting habitat is in short supply, then one would expect that the composition of flocks among and within roosts would differ in some predictable way. For instance, with limited habitat, social interactions might push some birds into lower quality roosts. This is more likely to occur in species in which dominance influences access to resources. Additionally, at any one roost, individuals might be forced to use parts

of the roost that afford less thermal benefits or greater predation risk. There is some support for this from just a few species.

Swennen (1984) reported on the age and sex of Eurasian Oystercatchers at winter roosts in a portion of the Dutch Wadden Sea. Oystercatcher roosts were located in relatively close proximity to the main feeding areas in intertidal habitats. The abundance of oystercatchers at roosts correlated positively with the area of intertidal habitat available for feeding. Larger roosts formed adjacent to larger areas, and birds from different roosts generally fed near their roost, overlapping only slightly with birds from other roosts. The age structure of the roosts varied in a predictable manner, with some roosts occupied predominantly by adults, whereas other roosts were composed largely of juveniles and subadults. The incidence of disease and bill abnormalities varied in concert with these age differences among roosts. The roosts with more juveniles and subadults had a higher frequency of these infirm birds. Lastly, during a particularly severe cold snap that lasted 11 days, approximately 2% to 3% of the population perished, and the proportion of dead birds was greatest on the roosts with more juveniles and subadults. These findings indicate several things. First, the consistent differences among roosts in age composition, abnormalities, and mortality suggest that dominance played a role in relegating some birds to lower quality roosts. Second, these patterns indicate that roosting habitat was limited by the size of the local population supported by the area of intertidal foraging habitat.

These same arguments may apply to the structure of the flock at any one roost. If habitat is in short supply within a roost, then dominance interactions may relegate young birds or individuals in poor condition to disadvantageous positions within the roost. For example, subordinate birds may be forced to the periphery of the flock where they are exposed to the elements or greater predation risk. Ruiz et al. (1989) reported that the spatial structure of roosting Dunlin flocks was nonrandom, with more juveniles and heavier birds captured from the interior of the roost as opposed to the edges. They were unable to explain this pattern, in part because it contradicts expected patterns. Fine-scale structure to a roost may result from the energy costs of being exposed to the elements on the edge of the flock, where they may be exposed to higher predation rates. Testing the latter idea would be difficult in Dunlin owing to the frequent disturbance of flocks. It would be relatively straightforward to test the thermal advantage hypothesis using data loggers positioned both amid birds in the interior and at the periphery of the flock.

TRADITIONAL VERSUS EPHEMERAL USE

Roosts are often characterized as being traditional phenomena because they form predictably at certain locations day after day, and over much longer intervals. Other data, however, suggest that shorebirds exhibit low fidelity to roosts on a daily basis. For instance, Hale (1980) showed that there was substantial day-to-day variation in the abundance of shorebirds at roosts in the British Isles. Two recent studies using simple counts of birds at roosts confirm high variation in shorebird numbers, even for roosts considered to be traditional (Colwell et al. 2003; Peters and Otis 2007). This suggests that individuals move frequently among roosts. Traditional use implies that individuals exhibit high fidelity to locations within a nonbreeding season and among years. In the latter case, cultural transmission of knowledge about the location and quality of roosts may take place such that juveniles learn about these locations, including those roosts used for short intervals at staging areas during migration, early in their first nonbreeding season. Alternatively, juveniles may simply follow the lead of experienced flockmates in deciding where to roost. Interestingly, evidence from marked shorebirds indicates that juveniles range more widely than adults, whether examined from the perspective of the home range (Myers et al. 1988) or from the perspective of the average distance moved between

roosts (Rehfisch et al. 1996). This suggests that as inexperienced birds age they continue to learn about their environment and alter their behaviors, including refining their use of feeding and roosting habitats. The benefits of experience are also suggested by the restricted home range sizes of adults, who presumably are more efficient in their use of space.

Evidence to support the idea of traditional use comes from a variety of observations. First, studies of marked populations show that individuals exhibit rather restricted movements during the winter (Myers 1984). For a population at a particular estuary, the **FUNCTIONAL UNIT** describes the collective area used by individuals in a local population to satisfy their daily needs. In this case, the functional unit includes habitat for feeding and roosting within an estuary or complex of wetlands (Tamisier 1985; Luis and Goss-Custard 2005). Several studies have reported site fidelity of individuals to localized areas of an estuary, which strongly hints at fidelity to individual roosts. Rehfisch et al. (1996, 2003) used long-term data from banding groups at two large estuaries in Great Britain to assess fidelity to roosts. Individuals of several species of shorebird were initially captured at a roost and then recaptured in the same or subsequent years. Roost-site fidelity was defined as the percentage of individuals of a species that were recaptured within the same section of the estuary. The two estuaries, The Wash and Moray Firth, spanned many kilometers of shoreline, and fidelity to roosts was evaluated based on the proportion of recaptures within 10 to 15 km of estuary shoreline, not necessarily to a particular roost. Most species exhibited relatively high fidelity, with better than 75% of within-year movements within a section of the estuary. Many birds were often recaptured at the very roost where they were originally caught or within a few of kilometers of the capture site (Rehfisch et al. 1996, 2003). Of the nine species, only Red Knots wintering on the Moray Firth moved widely among roosts, with recaptures averaging approximately 15 km apart. Given these localized movements, it is not surprising to hear of historical records showing that some roosts have been used for as long as humans have been keeping track of shorebirds. For instance, the Crossens roost on the Ribble marshes in northwestern England has been used for over 100 years (Hale 1980). So we may tentatively conclude that shorebirds are traditional in that they limit their movements within estuaries and often occur at predictable roosts. However, there are certainly instances in which individuals have been shown to frequent many roosts over short intervals. This variation probably occurs within and among species of shorebird in response to landscape-level differences in the availability of alternative habitats, as well as site-specific variation in the proximity of roosts to foraging habitats, and danger posed by predators.

With the advent of radio telemetry, researchers have examined fidelity to roosts in finer detail both temporally and spatially. Warnock and Takekawa (1995) identified just five roosts used by 106 radio-marked Western Sandpipers at San Francisco Bay. In northwestern Australia, Rogers et al. (2006, 2007) used receivers at fixed locations to track the use of various roosts around Roebuck Bay by Great Knots and Red Knots. Both species roosted at just a few locations around the bay, suggesting strong fidelity to roosts. Variation in use of roosts was associated with the magnitude of spring and neap tides.

One telemetry study stands in contrast to others in its conclusion about roost-site fidelity. Conklin (Conklin and Colwell 2007; Conklin et al. 2008) tracked 55 radio-marked Dunlin over three consecutive winters on Humboldt Bay. From the perspective of the local population, Dunlin use of some roosts was somewhat predictable, with 75% of diurnal detections of radio-marked birds occurring at just 10 of 89 roosts. But this last number is a surprisingly high number of roosts used by just 55 individuals (out of a population of approximately 10,000), and it suggests transient use of roosts, at least on a short time interval. Details on individual movements among roosts confirm this impression. Individ-

ual Dunlin frequented many roosts, and they commonly switched roosts on consecutive high tides. By the time the study ended after three winters, observers were continuing to discover roosts used by the population.

What can we conclude about the notion of traditional use from these studies? It appears that some roosts are used consistently by a local population, and this may be the case when alternative habitats, especially those of high quality, are in short supply. Nevertheless, variation in shorebird numbers at these traditional roosts often varies, which suggests that individuals move widely among multiple roosts. In other instances, it is clear that roosts are more ephemeral phenomena, with use by individuals varying greatly from day to day. In this case, the ephemeral nature of roosts is apparent from year to year, with some roosts used consistently in one year and abandoned the next. This suggests that the quality of individual roosts may be ephemeral, changing with type and number of predators as well as flock composition.

However, these conclusions are confounded by several methodological differences among studies. First, it is clear that the methods of defining and studying roost use may influence conclusions about traditional use. It is important to recognize that conflicting evidence may arise from studies based on counts of birds representing a local population compared with those derived from tracking individuals. In some cases, traditional use is a fair descriptor when it is evident from extensive surveys and population counts that only a limited number of roosts are used by virtually all individuals of a species. This appears to be the case for the two species of knot wintering on Roebuck Bay. By extension, a large number of roosts suggests that individual fidelity to any one roost is low. For example, 14 shorebird species used 240 different roosts around Humboldt Bay during a single January to October interval (Fig. 9.4), and this ephemeral use of roosts by local populations was confirmed by tracking individuals (Conklin and Colwell 2007; Conklin et al. 2008).

A second feature of study design that may influence impressions about traditional use concerns the spatial scale used to define a roost site (Conklin and Colwell 2007). In the initial Humboldt Bay study (Colwell et al. 2003), birds were mapped into a geographic information system, and a roost was defined as distinct from one another based on a 50 m buffer placed around multiple observations of birds using a habitat over the course of a year. The fine-scale definition of a roost accounts, in part, for the large number of roosts (240) identified around this estuary. A similar approach was used with the telemetry locations derived from marked Dunlins at Humboldt Bay. By contrast, other researchers have defined a roost with less spatial precision, as dictated by expansive stretches of habitat that may span a few kilometers along a beach (Rogers et al. 2006) or estuary (Rehfisch et al. 1996, 2003). These differences in spatial scale used to define a roost likely account for some of the differences in estimates of roost-site fidelity: high levels of fidelity naturally result from coarse scale definitions of a roost, whereas lower fidelity corresponds to finer distinctions in the location of roosts.

ROOST AVAILABILITY AND POPULATION SIZE

During the winter, the functional unit consists of the foraging areas and roosts that provide the essential needs for a local population (Tamisier 1985; Luis and Goss-Custard 2005). An important question concerns the extent to which habitat limits shorebird population size. The effect of foraging habitat and food on local population size is relatively straightforward and draws on the principle of carrying capacity. When

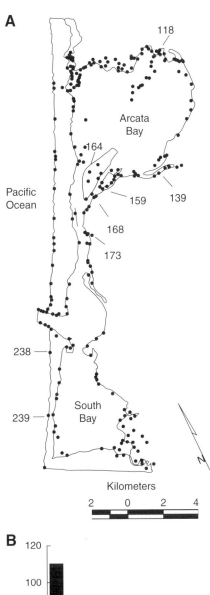

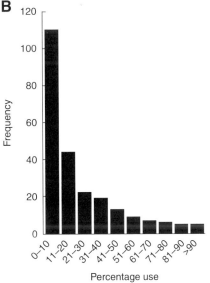

FIGURE 9.4. During the nonbreeding season, 14 species of shorebird used 240 different roosts (•) around Humboldt Bay, California (A). Shorebirds used most of these roosts infrequently, and they used only a few consistently. Numbers with lines show the location of the most consistently used roosts, which correspond to the histogram (>90% use) in the graph (B). From Colwell et al. (2003).

foraging habitat and food is in short supply, individuals cannot meet their daily energy requirements. Consequently, individuals either die or emigrate, and local population is reduced. In this way, the local population can be limited in size by availability and quality of food resources. Moreover, further loss or degradation of foraging habitats can have direct consequences for individual survival and therefore for local population size (Burton et al. 2006). Several recent papers have addressed the extent to which roosting habitat may limit local population size of shorebirds (van Gils et al. 2006; Rogers et al. 2006; Conklin et al. 2008). In this case, it is important to understand how the number and quality of roost sites may limit population size. First, it is possible that the area of habitat available to a local population is in short supply, such that individuals cannot find suitable roosts. Evidence in support of this comes from changes in local population size when roosts are either lost or created. In the Firth of Forth, for instance, creation of roosting habitat resulted in a dramatic increase in the local population of wintering shorebirds (Furness 1973): overall, shorebird numbers increased tenfold. But the increase appeared to stem from birds altering the travel distances and simply roosting nearer the estuary (Hale 1980). In other words, it is likely that the creation of this roosting habitat simply altered the distribution of wintering birds, concentrating them nearer the estuary. If the area of roosting habitat limits local population size, then loss or degradation of habitat, especially of high-quality sites, should result in a local decrease in population size. In northeast England, port development resulted in the loss of an important roost. Populations of several species of shorebird declined in the years immediately after the loss of this roost (Burton et al. 1996). At Humboldt Bay, the sheer number of roosts suggests that habitat does not limit local population size, and there was no evidence that the area of habitat at roosts was insufficient to support the local population of approximately 10,000 wintering Dunlins (Conklin et al. 2008).

Roosting habitat may limit local population size in more subtle ways, however. Specifically, individuals may incur added energy costs by travel to and from feeding areas or by increasing the energy spent in avoiding disturbance, whether natural or anthropogenic. For example, Rogers et al. (2007) argued that Great Knots and Red Knots at Roebuck Bay incurred significant energy costs of maintenance and flight, which limited their use of roosts to just a few locations. Similarly, Red Knots in the Dutch Wadden Sea incurred significant energy costs associated with travel to and from their roosts (van Gils et al. 2006).

CONSERVATION IMPLICATIONS

There are several important conservation and management implications that derive from understanding a species' dispersion patterns and social organization. First, a species' social organization may influence the extent to which the habitat of an estuary can support a given number of birds during the winter (Goss-Custard 1980, 1985). This is especially the case with territorial species, as interspecific aggression and competition for space may limit the number of individuals that can settle at an estuary. Moser (1988) suggested that this was the case for winter populations of the Black-bellied Plover at estuaries around Great Britain. Specifically, territorial plovers limited the number of conspecifics that could settle in an area, with the result that some estuaries appeared to have reached carrying capacity for the winter population and others were still adding birds. Second, the dispersion patterns and mobility of shorebirds have the potential to influence estimates of local population size. For any given estuary, the success of a coordinated survey effort, whether focused on counting birds at roosts or within foraging habitats, relies on a sampling regime that accounts for the spatial and temporal variation in shorebird numbers (Colwell and Cooper 1993). Estimates of abundance for territorial species, which are relatively predictable in their spatial distribution, are probably more reliable than estimates for flocking species that are patchily distributed in space and time. Survey protocols should specifically address variation that may arise owing to differences in the tendencies of species to occur as solitary individuals as opposed to moving in flocks of varying size and density. Finally, given the importance of roosts in the daily existence of nonbreeding shorebirds, it is imperative that the energy costs of natural and anthropogenic disturbance be quantified to elucidate the extent to which quality of the roosting habitat may limit local shorebird population size.

LITERATURE CITED

Abramson, M. 1979. Vigilance as a factor influencing flock formation among curlews *Numenius arquata*. *The Ibis* 121: 213–216.

Barbosa, A. 1995. Foraging strategies and their influence on scanning and flocking behaviour of waders. *Journal of Avian Biology* 26: 182–186.

———. 1997. The effects of predation risk on scanning and flocking behavior in Dunlin. *Journal of Field Ornithology* 68: 607–612.

———. 2002. Does vigilance always covary negatively with group size? Effects of foraging strategy. *Acta Ethologica* 5: 51–55.

Bibby, C. J., N. D. Burgess, D. A. Hill, and S. H. Mustoe. 2000. *Bird census techniques*. 2nd ed. London: Academic Press.

Blick, D. J. 1980. Advantages of flocking in some wintering shorebirds. Ph.D. dissertation, University of Michigan.

Bryant, D. M. 1979. Effects of prey density and site characters on estuary usage by over-wintering waders (Charadrii). *Estuarine and Coastal Marine Sciences* 9: 369–384.

Burton, N. H. K., P. R. Evans, and M. A. Robinson. 1996. Effects on shorebird numbers of disturbance, the loss of a roost site and its replacement by an artificial island at Hartlepool, Cleveland. *Biological Conservation* 77: 193–201.

Burton, N. H. K., M. M. Rehfisch, N. A. Clark, and S. G. Dodd. 2006. Impacts of sudden winter habitat loss on the body condition and survival on redshank *Tringa totanus*. *Journal of Applied Ecology* 43: 464–473.

Byrkjedal, I., and D. Thompson. 1998. *Tundra plovers: The Eurasian, Pacific and American Golden Plovers and Grey Plover*. London: T. & A. D. Poyser.

Colwell, M.A. 2000. A review of territoriality in non-breeding shorebirds. *Wader Study Group Bulletin* 93: 58–66.

Colwell, M. A., and R. J. Cooper. 1993. Estimates of coastal shorebird abundance: The importance of multiple counts. *Journal of Field Ornithology* 64: 293–301.

Colwell, M. A., T. Danufsky, N. Fox-Fernandez, J. E. Roth, and J. R. Conklin. 2003. Variation in shorebird use of diurnal high-tide roosts: How consistently are roosts used? *Waterbirds* 26: 484–493.

Colwell, M.A., T. Danufsky, R.L. Mathis, and S.W. Harris. 2001. Historical changes in the abundance and distribution of the American Avocet at the northern limit of its winter range. *Western Birds* 32: 1–15.

Colwell, M.A., and S.L. Landrum. 1993. Nonrandom shorebird distribution and fine-scale variation in prey abundance. *The Condor* 95: 104–114.

Colwell, M. A., R. L. Mathis, L. W. Leeman, and T. S. Leeman. 2002. Space use and diet of territorial Long-billed Curlews (*Numenius americanus*) during the non-breeding season. *Northwestern Naturalist* 83: 47–56.

Colwell, M. A., and K. D. Sundeen. 2000. Shorebird distributions on ocean beaches of northern California. *Journal of Field Ornithology* 71: 1–15.

Conklin, J. R., and M. A. Colwell. 2007. Diurnal and nocturnal roost site fidelity of Dunlin (*Calidris alpina pacifica*) at Humboldt Bay, California. *The Auk* 124: 677–689.

———. 2008. Individual associations in a wintering shorebird population: Do Dunlin have friends? *Journal of Field Ornithology* 79: 32–40.

Conklin, J. R., M. A. Colwell, and N. W. Fox-Fernandez. 2008. High variation in roost use by Dunlin wintering in California: Implications for habitat limitation. *Bird Conservation International* 18: 275–291.

Connors, P. G., J. P. Myers, C. S. W. Connors, and F. A. Pitelka. 1981. Interhabitat movements by Sanderlings in relation to foraging profitability and the tidal cycle. *The Auk* 98: 49–64.

Cresswell, W. 1994. Flocking is an effective antipredator strategy in redshanks, *Tringa totanus*. *Animal Behavior* 47: 433–442.

Cresswell, W., and J. L. Quinn. 2004. Faced with a choice, sparrowhawks more often attack the more vulnerable prey group. *Oikos* 104: 71–76.

Danufsky, T., and M.A. Colwell. 2003. Winter shorebird communities and tidal flat characteristics at Humboldt Bay, California. *The Condor* 105: 117–129.

Dekker, D. 1998. Over-ocean flocking by Dunlins, *Calidris alpine,* and the effects of raptor predation at Boundary Bay, British Columbia. *Canadian Field-Naturalist* 112: 694–697.

Elgar, M. A. 1989. Predator vigilance and group size in mammals and birds. *Biological Reviews* 64: 13–33.

Ens, B. J., and J. T. Cayford. 1996. Feeding with other oystercatchers. In *The oystercatcher,* ed. J. D. Goss-Custard, 77–104. Oxford: Oxford University Press.

Furness, R. W. 1973. Roost selection by waders. *Scottish Birds* 7: 281–287.

Goss-Custard, J. D. 1970. Feeding dispersion in some overwintering wading birds. In *Social behavior in birds and mammals,* ed. J. H. Crook, 3–35. London: Academic Press.

———. 1977. Predator responses and prey mortality in Redshank, *Tringa totanus* (L.), and a preferred prey, *Corophium volutator* (Pallas). *Journal of Animal Ecology* 46: 21–35.

———. 1980. Competition for food and interference among waders. *Ardea* 68: 31–52.

———. 1984. Intake rates and food supply in migrating and wintering shorebirds. In *Shorebirds: Migration and foraging behavior,* ed. J. Burger and B. L. Olla, 233–270. New York: Plenum Press.

———. 1985. Foraging behaviour of wading birds and the carrying capacity of estuaries. In *Behavioural ecology,* ed. R. M. Sibly and R. H. Smith, 169–188. Oxford: Blackwell.

Goss-Custard, J. D., A. M. Nicholson, and S. Winterbottom. 1984. Prey depletion inside and outside Redshank, *Tringa totanus,* territories. *Animal Behavior* 32: 1259–1260.

Hale, W. G. 1980. *Waders.* London: Collins.

Hamilton, W. D. 1971. Geometry for the selfish herd. *Journal of Theoretical Biology* 31: 295–311.

Hötker, H. 2000. When do Dunlins spend high tide in flight? *Waterbirds* 23: 482–485.

Krebs, C. J. 1999. *Ecological methodology.* 2nd ed. Menlo Park, CA: Addison Wesley Longman.

Luis, A., and J. D. Goss-Custard. 2005. Spatial organization of the Dunlin *Calidris alpina* L. during winter—the existence of functional units. *Bird Study* 52: 97–103.

Mathis, R.L., M.A. Colwell, L.W. Leeman, and T.S. Leeman. 2006. Long-billed Curlew distributions in intertidal habitats: Scale-dependent patterns. *Western Birds* 37: 156–168.

Moser, M. E. 1988. Limits to the numbers of grey plovers (*Pluvialis squatarola*) wintering on British estuaries: An analysis of long-term population trends. *Journal of Applied Ecology* 25: 473–485.

Mouritsen, K. N. 1992. Predator avoidance in night-feeding Dunlins *Calidris alpina*: A matter of concealment. *Ornis Scandinavica* 23: 195–198.

———. 1994. Day and night feeding in Dunlins *Calidris alpina*: Choice of habitat, foraging technique and prey. *Journal of Avian Biology* 25: 55–62.

Myers, J. P. 1980. Territoriality and flocking by Buff-breasted Sandpipers: Variations in non-breeding dispersion. *The Condor* 82: 241–250.

———. 1983. Space, time and the pattern of individual association in a group-living species: Sanderlings have no friends. *Behavioral Ecology and Sociobiology* 12: 129–134.

———. 1984. Spacing behavior of nonbreeding shorebirds. In *Shorebirds: Migration and foraging behavior,* ed. J. Burger and B. L. Olla, 271–321. New York: Plenum Press.

Myers, J. P., P. G. Connors, and F. A. Pitelka. 1979. Territory size in wintering Sanderlings: The effects of prey abundance and intruder density. *The Auk* 96: 551–561.

Myers, J. P., C. T. Schick, and G. Castro. 1988. Structure in Sanderling (*Calidris alba*) populations: The magnitude of intra- and inter-year dispersal during the nonbreeding season. In *Acta XIX congressus internationalis ornithologici,* ed. Henri Ouellet, 604–615. Ottawa: National Museum of Natural Science/University of Ottawa Press.

Page, G. W., and D. F. Whitacre. 1975. Raptor predation on wintering shorebirds. *The Condor* 77: 73–83.

Peters, K. A., and D. L. Otis. 2007. Shorebird roost-site selection at two temporal scales: Is human disturbance a factor? *Journal of Applied Ecology* 44: 196–209.

Pomeroy, A. C., R. B. Butler, and R. C. Ydenberg. 2006. Experimental evidence that migrants adjust usage at a stopover site to trade off food and danger. *Behavioral Ecology* 17: 1041–1045.

Quammen, M. L. 1982. Influence of subtle substrate differences on feeding by shorebirds on intertidal mudflats. *Marine Biology* 71: 339–343.

Recher, H. F., and J. A. Recher. 1969. Some aspects of the ecology of migrant shorebirds. II. Aggression. *Wilson Bulletin* 81: 140–154.

Rehfisch, M. M., N. A. Clark, R. H. W. Langston, and J. J. D. G. Greenwood. 1996. A guide to the provision of refuges for waders: An analysis of 30 years of ringing from the Wash, England. *Journal of Applied Ecology* 33: 673–687.

Rehfisch, M. M., H. Insley, and B. Swann. 2003. Fidelity of overwintering shorebirds to roosts on the Moray Basin, Scotland: Implications for predicting impacts of habitat loss. *Ardea* 91: 53–70.

Roberts, G. 1996. Why individual vigilance declines as group size increases. *Animal Behavior* 51: 1077–1086.

Rogers, D. I., P. F. Battley, T. Piersma, J. A. van Gils, and K. G. Rogers. 2006. High-tide habitat choice: Insights from modeling roost selection by shorebirds around a tropical bay. *Animal Behaviour* 72: 563–575.

Rogers, D. I., T. Piersma, and C. J. Hassell. 2007. Roost availability may constrain shorebird distributions: Exploring the energetic costs of roosting and disturbance around a tropical bay. *Biological Conservation* 133: 225–235.

Rosa, S., A. L. Encarnacao, J. P. Granadeiro, and J. M. Palmeirim. 2006. High water roost selection by waders: Maximizing feeding opportunities or avoiding predators. *The Ibis* 148: 88–97.

Ruiz, G. M., P. G. Connors, S. E. Griffin, and F. A. Pitelka. 1989. Structure of a wintering Dunlin population. *The Condor* 91: 562–570.

Stinson, C. H. 1980. Flocking and predator avoidance: Models of flocking and observations on the spatial dispersion of foraging winter shorebirds (Charadrii). *Oikos* 34: 35–43.

Swennen, C. 1984. Differences in quality of roosting flocks of oystercatchers. In *Coastal waders and wildfowl in winter,* ed. P. R. Evans, J. D. Goss-Custard, and W. G. Hale, 160–176. Cambridge: Cambridge University Press.

Tamisier, A. 1985. Some considerations on the social requirements of ducks in winter. *Wildfowl* 36: 104–108.

Townshend, D. J. 1985. Decisions for a lifetime: Establishment of spatial defence and movement patterns by juvenile Grey Plovers (*Pluvialis squatarola*). *Journal of Animal Ecology* 54: 267–274.

Tripp, K. J., and J. A. Collazo. 1997. Non-breeding territoriality of Semipalmated Sandpipers. *Wilson Bulletin* 109: 630–642.

Turpie, J. K. 1995. Non-breeding territoriality: Causes and consequences of seasonal and individual variation in the grey plover *Pluvialis squatarola* behaviour. *Journal of Animal Ecology* 64: 429–438.

van Gils, J. A., B. Spaans, A. Dekinga, and T. Piersma. 2006. Foraging in a tidally structured environment by Red Knots (*Calidris canutus*): Ideal, but not free. *Ecology* 87: 1189–1202.

Vine, I. 1973. Risk of visual detection and pursuit by a predator and the selective advantage of flocking behavior. *Journal of Theoretical Biology* 30: 405–422.

Ward, P., and A. Zahavi. 1973. The importance of certain assemblages of birds as information centres for food-finding. *The Ibis* 115: 517–534.

Warnock, S. E., and J. Y. Takekawa. 1995. Habitat preferences by wintering shorebirds in a temporally changing environment: Western Sandpipers in the San Francisco Bay estuary. *The Auk* 112: 920–930.

Whitfield, D. P. 1985. Social organisation and feeding behavior of wintering Turnstone (*Arenaria interpres*). Ph.D. dissertation, University of Edinburgh.

———. 1988. Sparrowhawks *Accipiter nisus* affect the spacing behavior of wintering turnstone *Arenaria interpres* and Redshank *Tringa totanus*. *The Ibis* 130: 284–287.

Wynne-Edwards, V. C. 1962. *Animal dispersion in relation to social behavior.* New York: Hafner.

Yasue, M., J. L. Quinn, and W. Cresswell. 2003. Multiple effects of weather on the starvation and predation risk trade-off in choice of feeding location in Redshank. *Functional Ecology* 17: 727–736.

Ydenberg, R. C., R. W. Butler, D. B. Lank, C. G. Guglielmo, M. Lemon, and N. Wolf. 2002. Trade-offs, condition dependence and stopover site selection by migrating sandpipers. *Journal of Avian Biology* 33: 47–55.

10

Population Biology

I N THE SIMPLEST AND TRADITIONAL view, the goal of conservation and management is to affect the size of wildlife populations such that the rare native species increase, the common ones maintain their abundance, and pest or introduced species are extirpated. To accomplish this task requires detailed knowledge of a species' population size, birth rate, and death rate, also known as vital rates to demographers. It is also helpful to understand the extent to which immigration and emigration contribute to population change. From an applied perspective, it is imperative that conservationists address the ecological factors that limit population size (Caughley and Gunn 1996). Limiting factors include competition for food (or habitat), predation, and disease. Does predation compromise the survival of adults such that they are not replaced by annual recruitment of young? Does a shortage of high quality foraging areas and food limit the overwinter survival of individuals? To what extent does annual variation in predation of eggs and young dictate annual variation in population size?

Ultimately, the success of conservation and management actions is judged by their impacts on populations. For example, the reproductive success of some threatened and endangered shorebirds is limited by predation on eggs and chicks. Consequently, lethal and nonlethal management of predators is an important tool to recovering these populations. Similarly, it is implicit in habitat management that manipulations of wetland hydrology provide more food during migration. As a result, this will increase survival during this energy-taxing segment of the annual cycle; it may also increase reproductive success during the subsequent breeding season. The response of populations to management and conservation efforts can only be judged by knowing something of a species' population size and trends. Hence, it is essential to fully understand the demographic characteristics of shorebirds and the ecological factors that limit population size. In this chapter, I review the salient features of shorebird demography: how biologists collect and analyze these demographic data and apply this information to population dynamics. Next, I address what is known about historical and contemporary population sizes of shorebirds, causes for population declines worldwide, and programs currently in place to monitor their trends. Finally, I review the evidence that ecological factors limit shorebird population sizes, and whether shorebird populations are limited in their breeding or nonbreeding areas.

DEMOGRAPHY

A useful first step in understanding shorebird population biology is to characterize the various elements that affect annual change in population size. Evans and Pienkowski (1984) provided a useful approach to understanding shorebird demography by subdividing the annual cycle into stages or components that influence population growth. Briefly, a population is largest early in the nonbreeding season when juveniles join adults on migration to wintering areas. With this starting point, population size diminishes as natural factors (such as predation, food shortages, disease, and weather) act with anthropogenic sources of mortality to affect overwinter survival. In the subsequent spring, adults and yearlings return to breeding areas where they establish themselves with varying success as breeders. The productivity of a local population, or the number of young produced, is the product of the mating system, local breeding density, clutch size, hatching success, and fledging success. For example, monogamous species (such as godwits and curlews) have lower reproductive potential than species practicing rapid multiple-clutch polygamy (for instance, the Sanderling and Temminck's Stint) simply because of the doubling of productivity stemming from biparental and uniparental care, respectively. Large annual increases in population arise when high densities of breeding adults produce many clutches that successfully hatch and many of these chicks survive. Similar arguments apply to clutch size, which varies from one to four eggs across shorebird taxa. Productivity is influenced by the same set of ecological factors (predation, resource availability, etc.) that influences overwinter survival.

SURVIVAL

Evans and Pienkowski (1984) provided a detailed summary of the survival of shorebirds, broken down into various segments of the annual cycle; Evans (1991) updated this information. These reviews relied on evidence compiled from the return rates of marked birds to both wintering and breeding areas. Sandercock (2003) updated this information and reviewed the strengths and weaknesses of various means by which survival could be estimated. Briefly, four approaches exist to quantify annual survival: (1) maximum longevity, (2) life-table methods, (3) return rates, and (4) mark-recapture analyses. It is clear that only the last method provides accurate estimates of survival, and even then rarely are estimates given for true (as opposed to apparent) survival owing to the failure to account for emigration.

LONGEVITY

Researchers have used the reciprocal of a species' longevity record as an index of adult survival (Sandercock 2003). Longevity records for shorebirds vary greatly (Table 10.1), but even some of the smallest species may live surprisingly long. The longevity record for the smallest North American shorebird, the Least Sandpiper, is 19 years. Maximum longevity of larger species is probably double this value. Eurasian Oystercatchers occasionally live in excess of 40 years (Exo 1993; Hockey 1996). Several Marbled Godwits observed at Humboldt Bay, California, were minimally 23 years old, as they had been aged as adults when they were first captured in 1969 (Colwell et al. 1995). Longevity records, however, provide an unsatisfactory approach to estimating survival for several reasons. Most importantly, the index is biased by sampling effort, including the number of birds marked as well as the duration of time over which data have been collected.

LIFE TABLES

In a population with a stable age distribution, it is possible to use life tables to estimate annual survival from a representative sample of adults and juveniles (Sandercock 2003). The proportion of adults to the total number of adults and juveniles yields an estimate of annual survival. Unfortunately, few studies of shorebirds have been conducted for sufficiently long periods

and with intensive monitoring to satisfy the conditions for the use of life tables to provide survival estimates. Two studies that used this method yielded similar estimates of adult survival: 0.709 for the Common Sandpiper (Holland and Yalden 1991) and 0.755 for the Shore Plover (Sandercock 2003 from data in Dowding and Kennedy 1993).

RETURN RATE

The annual return of a sample of individually marked birds to a site provides a minimum estimate of survival (Evans and Pienkowski 1984). Sandercock (2003) pointed out that this estimate represents the product of four independent probabilities: (1) true survival, (2) the tendency for an individual to return to a site (site fidelity), (3) the likelihood that an individual does not breed every year (breeding propensity), and (4) the probability that an individual is detected when it is present in the population (Sandercock 2003). The last three components all reduce the detection of birds that are, in fact, alive. That is, a bird may be alive and either not return to a study area, skip a breeding season, or not be detected by researchers. There is some evidence that high Arctic species may occasionally skip breeding in years when a late spring and heavy snow cover negatively affect the availability of breeding habitat and food (Green et al. 1977). However, it is probably a rare phenomenon for individuals of other shorebird species to skip years. The age composition of a sample of birds used to estimate return rate may also influence data. If a sample consists of a disproportionate number of first-winter birds, who tend to wander more widely than adults (e.g., Myers et al. 1988), then survival estimates based on return rates are likely to be biased low. Collectively, therefore, a return rate represents a minimum estimate of survival for a population.

The return of shorebirds to their breeding sites may provide a useful index of adult survival because the breeding season is often rather short, especially for Arctic-breeding species (Evans and Pienkowski 1984). However, it is apparent from various species discussed in

TABLE 10.1

Longevity record (year) and estimates of adult survival (ø) for selected species of shorebird

SPECIES	LONGEVITY	1/LONGEVITY	SURVIVAL	SOURCE
Northern Lapwing			0.705	Peach et al. (1994)
Black-bellied Plover	25	0.9600		Wash Wader Ringing Group [WWRG] (2008)
Pacific Golden-Plover	21	0.9524	0.75–0.85	Johnson et al. (2004)
Common Ringed Plover	19	0.9475		WWRG (2008)
Semipalmated Plover	8		0.71	Badzinski (2000)
Piping Plover	14		0.74	Larson et al. (2000)
Eurasian Oystercatcher	44	0.9722		Exo (1993)
Common Greenshank	16	0.9375		WWRG (2008)
Common Redshank	19	0.9474	0.74	Insley et al. (1997), WWRG (2008)
Black-tailed Godwit	23	0.9565	0.90	Gill et al. (2001), WWRG (2008)
Bar-tailed Godwit	32	0.9688		WWRG (2008)
Marbled Godwit	29	0.9655		Gratto-Trevor (2000)
Bristle-thighed Curlew	23	0.9565	0.85	Marks (1992), Marks and Redmond (1996)
Eurasian Curlew	29	0.9655		WWRG (2008)
Semipalmated Sandpiper	16	0.9375	0.56–0.73	Gratto-Trevor and Vacek (2001), Sandercock and Gratto-Trevor (1997)
Western Sandpiper			0.45–0.57	Sandercock et al. (2000), Fernandez et al. (2003)
Least Sandpiper	19	0.9474		Miller and McNeil (1988)
Dunlin	21	0.9524	0.74	Staav (1983), Warnock et al. (1997)
Red Knot	24	0.9583	0.84	Kew et al. (1999), Brochard et al. (2002)
Sanderling	17	0.9412		Toms and Clark (1998)
Ruddy Turnstone	19	0.9474		WWRG (2008)

Chapter 5 (see Table 5.1) that return rates to breeding sites often provide a poor estimate of adult survival. In many cases, especially for nonterritorial (such as phalaropes), lek-breeding (such as the Buff-breasted Sandpiper), and colonial breeders (such as stilts and avocets), return rates are so low that it is clear that they are unreliable estimates of survival. By contrast, higher return rates for many territorial species (see Evans and Pienkowski 1984) suggest that they are more representative of adult survival.

There exist considerably fewer data on site fidelity of shorebirds during the nonbreeding season. The return rates of several species wintering in northeast England suggest that survival can be rather high for species with very different ecologies (Evans and Pienkowsk 1984). The average annual return rate of the Sanderling, Ruddy Turnstone, and Eurasian Curlew over five winters was 83%, 85%, and 77%, respectively. For the first two species, estimates were 27% and 19% higher, respectively, than those based on the recovery of banded birds; estimates for curlews from the two approaches were similar. Interestingly, the lower estimates for curlew may stem from the fact that it continued to be hunted in Great Britain until 1981.

With detailed knowledge of marked birds, it is possible to partition mortality into that occurring at wintering sites versus occurring during migration and breeding intervals based on spring departure dates of marked individuals (Evans and Pienkowski 1984; Evans 1991). Using this approach, Evans (1991) estimated that disproportionately more mortality occurred during migration for the Sanderling and Ruddy Turnstone, whereas it appeared a greater percentage of mortality for the Eurasian Curlew happened on wintering grounds.

MARK-RECAPTURE ANALYSIS

Mark-recapture techniques represent a more sophisticated analytical approach to survival estimation, which has largely replaced early return rate analyses. Mark-recapture analyses make use of essentially the same data as those presented in the return rate approach: the resighting of individuals in subsequent years. As already stated, individuals may be recorded as absent from a population from one year to the next for three reasons other than their death: site fidelity, breeding propensity, and detection probability. One advantage to the mark-recapture analysis is that it enables one to quantify the detection probability or the likelihood that an individual is detected (or missed) when it is, in fact, alive and present in the population. This allows for improved estimates of survival.

Survival estimates for adults based on mark-recapture analyses yield appreciable variation among species, which is correlated with mass. Adult survival commonly varies from 0.50 to 0.70 in small species, such as calidridine sandpipers and small plovers. By contrast, survival is higher in some tringines, godwits, and curlews, with estimates ranging from 0.70 to 0.90 (Sandercock 2003). In virtually all species, survival of juveniles is low compared with adults. In most instances, juvenile survival is roughly half that of adults. These age differences probably occur owing to the inexperience of young birds in feeding and avoiding predation in the short interval that encompasses their first migration and establishing residency at a wintering site. Evidence to support this comes from comparable survival of juveniles and adults once the juveniles had established themselves in late autumn on their wintering sites (Pienkowski and Evans 1985).

Nearly all studies that used mark-recapture to estimate survival have reported apparent survival because they cannot account for permanent emigration. Only one study to date has reported true survival for a shorebird, and it was for juvenile survival in a population of Snowy Plovers breeding along the central California coast. Stenzel et al. (2007) reported on a 16-year effort and a large sample (>1,000) of plovers tracked through their first year by an extensive network of observers. Researchers individually color-marked plovers as chicks and followed them for their first year of life, regardless of whether they bred locally or emigrated hundreds of kilometers to breed elsewhere (Stenzel et al. 1994). The sample consisted of 66,000 live encounters of juveniles and 35 dead birds. Because they had included these dispersing birds in their analyses, Stenzel et al. (2007) argued that they had accounted for virtually all individuals that would otherwise have gone undetected in a study with a limited study area. Hence, they provided a value nearer the true survival. It is important to note that the population of Snowy Plovers that they studied is a partial migrant population, consisting of a mix of birds that migrate short distances and other birds that are permanent residents in one location where they both breed and winter (Stenzel et al. 1994). Overall, the survival of juveniles from fledging until early in April of the next breeding season (6.5 to 10.5 months later) averaged 46%. There was, however, considerable annual variation (28% to 58%) in survival, and this variation was not correlated with indices of winter weather or the interval during which mammalian predators were managed.

In summary, shorebirds exhibit relatively high survival, with most adults surviving well from one year to the next. There are, however, occasional years of particularly low survival that correspond to winters of especially severe weather in northern latitudes. As in all birds, juvenile shorebirds experience comparatively lower survival, with highest mortality occurring in the few months after they become independent and coincident with their first migration and when they establish themselves on wintering grounds. Most importantly, Sandercock (2003) concluded that variation in adult survival has a greater effect than productivity on the rate of change in shorebird populations. Koivula et al. (2008) reported a very similar result from a long-term study of Temminck's Stint at the southern limit of its breeding range. Adult survival was so low that the stint population was predicted to go extinct in a few decades and even moderate gains in survivorship would not offset this result. This conclusion regarding the importance of adult survivorship has strong applied value, considering that management is often directed at increasing productivity rather

than adult survival. For example, predator control (see Chapter 12) is often used to increase breeding success. Such efforts are fruitless without attention paid to management that increases survivorship. By contrast, the impetus behind management during the nonbreeding season (such as wetland manipulations) often is aimed at improving feeding habitats, which may lead to greater overwinter survival.

PRODUCTIVITY

The reproductive potential of a population varies under the influence of a number of elements, including age of first reproduction, breeding density, clutch size, and the success with which individuals hatch eggs and fledge young (Evans and Pienkowski 1984). There is a large and growing literature based on intensive monitoring of breeding populations, which provides insights into these facets of reproduction in shorebirds.

AGE OF FIRST REPRODUCTION

For nearly a century, ornithologists have noted the presence of sometimes large numbers of shorebirds in wintering areas during the breeding season (Bent 1927, 1929; Loftin 1962). These observations suggested that some subset of a population did not return to breeding areas, and it was widely thought that these individuals were likely young birds. Evidence indicates that these flocks are often composed of first-year individuals that do not return to breeding grounds (Summers et al. 1995). Additional information from breeding areas indicates that even when yearlings return they tend to arrive later than older individuals (e.g., Oring and Lank 1982; Summers and Nicoll 2004), and they may not acquire a breeding territory or mate. Other studies in breeding areas indicate that some individuals do not breed when they are physiologically ready.

Overall, smaller species breed for the first time when 1 or 2 years old, whereas larger species often are slightly older when they first breed (Evans and Pienkowsk 1984; Summers et al. 1995). A small number of color-marked Piping Plovers remained on wintering areas during their first breeding season (Haig and Oring 1988). In the Pacific Golden-Plover, some individuals oversummer in the tropics as yearlings and 2-year olds (Johnson and Johnson 1983). Yearling Western Sandpipers commonly oversummer in Panama but not in Mexico, suggesting that the age of first reproduction depends on the latitude of the wintering area. Female Western Sandpipers tend to winter farther south than males (Nebel et al. 2002; Fernández et al. 2004).

On breeding grounds, additional evidence indicates that delayed breeding is common in shorebirds. Approximately 50% of Common Redshanks breed as yearlings (Thompson and Hale 1989). This life history trait suggests that, before human influence, breeding opportunities were limited in some species. It is difficult to know, however, whether these individuals are truly unable to breed unless they remain and are monitored in a study area for an entire breeding season. This is especially true for nonterritorial species. For example, a small percentage of female Wilson's Phalaropes in a color-marked population were not observed to obtain a mate, but most of these females left the study area after relatively short periods and may have bred elsewhere (Colwell and Oring 1988). There is, however, evidence in some species that some individuals occasionally do not breed. This is more likely in territorial species in which competition and defense of a nesting site and the food resources therein may limit settlement of some percentage of the population. Several studies (Holmes 1966; Harris 1970; Ens et al. 1996) experimentally removed males from their territories and documented that they were quickly replaced by "floaters," males that must have been waiting for the opportunity to breed. This suggests that there is a segment of the population that does not breed. It is possible, however, that these floaters could have bred elsewhere, given enough time (Evans and Pienkowski 1984). On the breeding grounds, the presence of nonbreeding birds also suggests that habitat

or availability of mates may limit the number of breeding adults in an area.

In exceptional cases, individuals of some large-bodied shorebirds may wait for more than a decade to breed. In the Eurasian Oystercatcher, for instance, data from several intensively studied populations offer fascinating details on the extent to which individuals delay breeding (Ens et al. 1995, 1996). First, nonbreeding flocks persist on wintering areas during the summer, and plumage characters indicate that these flocks are dominated by 1- and 2-year-old oystercatchers. Second, data from marked birds on breeding grounds show that most oystercatchers do not breed until they are 3 or 4 years old; females tend to breed first, at an earlier age than males. Third, oystercatchers in full alternate plumage are present in flocks in breeding areas. The percentage of birds in these flocks appears to decrease with latitude and distance from the coast. In other words, nonbreeders are present in greater proportions in regions near the center of the species' distribution. Finally, removal experiments indicate that nonbreeders are present on the breeding grounds and they rapidly replace individuals that were removed from a territory.

One additional pattern stands out in the data concerning the prevalence of oversummering birds, and it relates to the latitude of the wintering area. In general, a greater percentage of younger birds oversummer in southern compared with northern estuaries (Evans and Pienkowski 1984; Summers et al. 1995; O'Hara et al. 2005). This pattern suggests that delayed breeding is a conservative life-history strategy that balances the cost of long-distance migration against the probability of breeding successfully (Summers et al. 1995). Others have proposed that internal parasites may play a role in the tendency of young birds to oversummer (McNeil et al. 1995), but these explanations seem unlikely as an evolutionary argument because they do not account for the size-dependent pattern across species (Summers et al. 1995). Collectively, these observations indicate that delayed breeding is common among shorebirds, especially oystercatchers and other large-bodied shorebirds. Furthermore, delayed breeding suggests that populations may be limited by habitat availability or opportunities to acquire a territory and mate on breeding grounds.

BREEDING DENSITY

A universal ecological pattern is that a species is patchily distributed across its range. This observation certainly applies to shorebirds during the breeding season. Some areas support particularly high densities, whereas other locales with seemingly suitable habitat lack breeding adults. This is especially true in northern latitudes, where considerable habitat remains unaltered by humans in boreal forests and tundra; this is probably not the case for habitats in temperate and tropical regions. Holmes (1971) reported that breeding densities of Western Sandpipers in Subarctic Alaska varied annually from 33 to 49 pairs ha^{-1}. Brown et al. (2007) documented considerable variation in densities of breeding shorebirds across a broad area of the wet tundra of the coastal plain of the Arctic National Wildlife Refuge. Low densities of Long-billed Curlews were recorded across much of their temperate North American breeding range (Jones et al. 2008). These patchy distributions make it challenging to characterize breeding density, to extrapolate findings to population estimates, and to address the extent to which habitat or breeding opportunities limit a population.

In essence, variation in breeding density and especially the presence of unoccupied habitats emphasizes the degree to which availability or quality of breeding habitat limits population size. Few studies have explicitly approached this question from the perspective of the factors that limit a population probably because of the logistical difficulties of conducting such a program in remote northern regions and across the entirety of a species' range. Nevertheless, recent attempts to document variation in the density of breeding shorebirds are increasingly aimed at estimating population size, either locally (Childers and Dinsmore 2008), regionally (Brown et al. 2007), or across a species' range

(Jones et al. 2008). This is an important first step in attempts to monitor population trends.

CLUTCH SIZE

Shorebirds are conservative breeders in that they have small fixed clutch sizes (see Chapter 5). Therefore, this life-history feature is less important in its effect on population growth (Evans and Pienkowski 1984). Any increase in productivity associated with this component comes from variation in the mating system. For example, species exhibiting rapid multiple-clutch polygamy (Little and Temminck's Stints), where males and females split incubation duties between two separate clutches laid in rapid sequence, can double the productivity of a population. In some cases, clutch size for a population declines slightly with date of the breeding season, suggesting that females that have already nested and failed to hatch chicks are unable to muster the energy or calcium to complete a full replacement clutch.

HATCHING SUCCESS

Abundant evidence from long-term studies that have intensively monitored breeding populations shows that hatching success, often measured as the percentage of nests that hatch at least one chick, varies greatly over space and time. There is a hint that clutches survive better in northern latitudes compared with temperate and tropical regions (Skutch 1949; Jehl 1971). But even in the Arctic there can be years in which nesting success is very low. It is clear that in some years local populations experience no recruitment of young because nearly all eggs fail to hatch and the few chicks that do hatch die. In other years, hatching success can approach 100%. For example, annual variation in nesting success was 0 to 100% for a small, island-breeding population of the Spotted Sandpiper in Minnesota. The island was located in the middle of a large lake, and total reproductive failure occurred in the 2 years when an American mink (*Mustela vison*) wintered on the island and ate virtually all clutches or the few broods that hatched (Oring et al. 1983).

Although annual variation may not be as extreme for Arctic-breeding species, variation in hatching success can be large. Annual variation in egg predation of Western Sandpipers ranged from 40% to 90% over 4 years (Niehaus et al. 2004). Nesting success can vary substantially at a much finer spatial scale, making inferences about the causes of clutch loss problematic (McCaffery and Ruthrauff 2003).

Predation is widely recognized as the most important cause of clutch failure for birds in general (Martin 1993), and this is especially true for ground-nesting shorebirds (Evans and Pienkowski 1984). The main predators of shorebird eggs tend to be mammals, especially foxes, mustelids, raccoons, and ground squirrels. The most commonly documented avian predators of eggs are corvids and gulls. Still, some species can experience appreciable clutch loss owing to other environmental factors, such as flooding in tidal areas (Hale 1980).

FLEDGING SUCCESS

There is less information on the success of shorebirds in fledging chicks, probably because their mobility often makes them difficult to monitor in open habitats. Fledging success appears to parallel annual variation in hatching success: when predators consume a large percentage of clutches, the few chicks that hatch often succumb to the same predators. Oring et al. (1991) provided support for this notion in their analysis of components of lifetime reproductive success in the Spotted Sandpiper. In a 20-year study, there was positive covariation in hatching and fledging success of individuals. Some sandpipers experienced successive years of poor reproduction owing to high predation of both eggs and chicks, whereas others had high reproductive success because they bred in years when predators had less of an impact on the island population (Oring et al. 1983, 1991).

MONITORING POPULATION GROWTH

Accurate estimates of survival and productivity are essential for predicting population growth. Demographers characterize the growth of a population (stable, increasing, or decreasing) using a variety of methods. The most widely used approach combines survival estimates for adults and juveniles with productivity data. Simplistically, a population remains stable when adults survive and produce sufficient numbers of young each year to replace the adults that die. Populations grow when a combination of high adult survival and high productivity enhance numbers; declines occur when productivity is insufficient to replace dead individuals. Estimating population growth requires intensive monitoring of per capita reproductive success and long-term monitoring of individually marked birds to get at survival.

The demographic parameters that influence population growth include sex ratio, density, survival, productivity, immigration, and emigration. Several of these parameters, also referred to as vital rates, can be used to quantify growth by estimating a population's intrinsic growth parameter or lambda (λ). In a stable population, $\lambda = 1.0$; increasing and decreasing populations are characterized by $\lambda > 1.0$ and $\lambda < 1.0$, respectively. Lambda is the annual growth rate, which can be estimated in several ways. First, lambda can be estimated as the ratio of consecutive annual estimates of population size, $\lambda = N_{t+1}/N_t$, where N is the number of breeding individuals in successive years (N_t, N_{t+1}, etc.). In a population of unmarked individuals, however, it is impossible to distinguish between productivity and immigration as the cause for an increase in population size. An increase in breeding population size may stem from high productivity the previous year with subsequent recruitment of yearlings into the breeding population, or it may stem from immigration. Similar arguments apply to distinguishing between mortality and emigration in the cause of a population's decline. A second method algebraically estimates λ as the product of per capita fledging success and juvenile survival, summed with adult survival. In short, when the production of young and their survival is high enough to replace the adults that die, the population remains stable ($\lambda = 1.0$). A final method of estimating population growth makes use of individually marked animals and sophisticated mark-recapture analytical techniques that estimate λ from the tenure of individuals in a population. Specifically, Pradel models use the probability that a marked individual was alive in the population in a prior year (also referred to as "seniority") to estimate growth (Sandercock 2003).

Several papers have examined population growth for localized shorebird populations. Pearce-Higgins et al. (2009) analyzed long-term data for two populations of the Common Sandpiper, a trans-Saharan migrant, to evaluate the extent to which global warming may be responsible for population change. Adult survival for a population in England was 0.688. Annual variation in adult survival was correlated with the North Atlantic Oscillation—yearling and older adults survived less well during winters that were warm and wet in Europe and cool and dry in Africa. Productivity (here, the percentage of adults tending broods) correlated positively with warmer June temperatures. Overall, the population decreased by 59% over 28 years, and much (40%) of this decline was attributable to variation in adult survival. However, a similar decline over a 12-year interval for a population in Scotland was not correlated with the North Atlantic Oscillation, which suggests that the climate effect was not universal. The population decline for Common Sandpipers is mirrored by some, but not all, Afro-Palearctic migrants (Sanderson et al. 2006).

The Semipalmated Sandpiper has a broad distribution across tundra habitats of the Nearctic. Hitchcock and Gratto-Trevor (1997) used a stochastic matrix population model to diagnose the causes of a precipitous population decline that occurred in the 1980s near the southern limit of the species' breeding range at La Perouse Bay, Manitoba. The most influential demographic factor affecting the population decline was adult survivorship; a small reduction in adult survival can cause a stable population to decline. Only high immigration rates could counter the effects of low adult survival.

The population of the Snowy Plover, listed as threatened under the U.S. Endangered Species Act, breeds along the Pacific Coast from Washington south through California. Mullin et al. (2010) used the three previously described methods to estimate growth (λ) for a small population in coastal northern California. Algebraic estimates showed a population in rapid decline (λ = 0.71 – 0.79 annually), whereas averaged estimates from consecutive annual surveys (λ = 0.98 ± 0.10) and Pradel modeling (λ = 0.96 ± 0.26) suggested that the population was nearly stable or declining slowly. The discrepancy between methods was explained by high rates of immigration into the population, which was determined from plovers marked elsewhere along the Pacific Coast.

POPULATION SIZES AND TRENDS

With few exceptions, notably threatened and endangered species, the size of shorebird populations is poorly or imprecisely known (Morrison et al. 2001; Brown et al. 2001; Delany et al. 2009). In fact, attempts to estimate population sizes have sometimes represented best guesses by experts, which have prompted at least one rather pointed exchange regarding a North American species, the Long-billed Curlew (Farmer 2008; Lanctot et al. 2008). Importantly, early summaries of shorebird conservation (Senner and Howe 1984; Evans et al. 1984) did not address shorebird population sizes because they simply were not available. This shortcoming has been partly redressed with the publication of several recent compendia (e.g., Delany et al. 2009) and conservation plans (Brown et al. 2001) with population size estimates (see Appendix I). Still, knowledge of current population sizes is limited at best for many species.

Worldwide, shorebird populations have been shown to be declining dramatically. Roughly half (52%) of the world's 237 populations with trend data are declining, whereas a much smaller percentage (8%) are increasing (Wetlands International 2006). Regional trends are no less encouraging. In the western Palearctic, Delany et al. (2009) report on 230 biogeographic populations, of which 59 are known or suspected to be in decline; the status of many other populations is unknown. In Australia, Nebel et al. (2008) drew attention to the fact that population trend data exist for only a third of shorebirds. Of these, most migratory (73%) and resident (81%) shorebirds in southeastern Australia have declined over roughly the quarter century spanning 1983 to 2006. For North America, the U.S. Shorebird Conservation Plan (Brown et al. 2001) has acknowledged the imprecision of most estimates of population size. For example, 63% of the 72 estimates for population size of Nearctic taxa (including subspecies) were categorized as either poor or low quality. By contrast, only 18% of taxa were characterized as having good or high-quality population estimates. Only one North American species, the Piping Plover, was categorized as having high-quality information on population size. The reason for the lack of knowledge about shorebird population sizes stems from the general absence of well-organized and geographically broad monitoring programs until relatively recently. The absence of these programs may stem, in part, from the nongame status of all but two shorebirds in North America, the exceptions being Wilson's Snipe and the American Woodcock. Still, the quality of data for the snipe is categorized as poor. Because virtually all Nearctic shorebirds have not been hunted under regulations of the Migratory Bird Treaty Act, there has been little need to monitor the changes in their populations in relation to hunting pressure in the manner of waterfowl. In the past two decades, however, biologists confronted with the perceived declines of shorebird populations have become more active in conservation, and they have recognized the need for scientifically defensible estimates and the ability to monitor changes. Moreover, conservationists recognize the value of reliable population estimates, as these estimates, regardless of their recency, provide the benchmark against which to compare future population change.

Population estimates for 72 species and subspecies of shorebird breeding in North America derive from surveys conducted during the breeding and nonbreeding seasons by a host of researchers representing government and nongovernmental organizations (Brown et al. 2001; Morrison et al. 2001). It is noteworthy that estimates for some taxa (such as the Piping Plover and Short-billed Dowitcher) have been subdivided by subspecies or at least by geographic region, recognizing that separate populations use very different flyways and often experience different threats. For example, of the three subspecies of the Red Knot that occur in the Nearctic, population sizes and trends are especially problematic for *Calidris canutus rufa*, which winters in Argentina and migrates through Delaware Bay. This population has experienced a dramatic decline in a very short period owing to the harvest of horseshoe crabs along the Atlantic seaboard of the United States (Baker et al. 2004).

Population estimates often are presented on a logarithmic scale owing to substantial variation among species. At the low end of the scale is a group of temperate-breeding plovers and the Hawaiian Stilt, most of which are listed under the U.S. Endangered Species Act as threatened or endangered, with population sizes of <10,000. The precision of these estimates are all quite good because of concerted survey efforts mandated by their federal status (e.g., Haig et al. 2005). By contrast, estimates are much higher for most other taxa, especially those breeding in boreal and Arctic regions: the precision of these estimates, however, is generally much lower (Brown et al. 2001). Overall, population size correlates negatively with body size, and this relationship is driven by the especially large populations of small calidridines and in spite of the low populations of several small plovers (Morrison et al. 2001). Nevertheless, two groups are outliers: one consists of species "at risk" or that are difficult to count, and the second is composed of taxa with higher population sizes than expected. In this latter group are the two North American shorebirds that continue to be hunted (Morrison et al. 2001).

MONITORING PROGRAMS

Successful conservation requires knowledge of population sizes and trends. From a practical standpoint, monitoring programs allow an assessment of effectiveness of actions at local, regional, and global scales. If funding for conservation is effective in ameliorating ecological factors that limit populations, then on-the-ground management actions should translate into population increases. Consequently, population monitoring is essential to evaluating conservation and management practices. Worldwide, a variety of programs exist to monitor shorebird populations or track their distribution and abundance during the nonbreeding season. Principal among these efforts are the Wetland Bird Survey (WeBS), Canadian Maritime Shorebird Survey (MSS), International Shorebird Survey (ISS), and Pacific Flyway Project (PFP). There are also efforts to survey breeding populations. These programs include the Breeding Bird Survey (BBS), Program for Regional and International Shorebird Monitoring (PRISM), and several examples in which concerted efforts are made to count threatened or endangered species. Most of these programs have relied heavily on the cooperation of volunteers to collect data; consequently, they have traded off quality in their survey methods for willing participants. Recently, however, population monitoring has increased in rigor with development of double-sampling methods to estimate breeding densities and total population sizes. The following is an overview of monitoring programs, beginning with wetland-based surveys during the nonbreeding season.

WETLAND BIRD SURVEY

Coordinated counts of waterbirds at estuaries around Great Britain have been ongoing for over 60 years. Early efforts to count wildfowl were formalized in 1947 with waterfowl counts of the International Wildfowl Inquiry. In 1969, the Birds of Estuaries Enquiry initiated monitoring of shorebirds at estuaries. These programs were combined in 1993 under the Wetland Bird Survey (WeBS), a cooperative effort by the British Trust for Ornithology, Wildfowl and Wetlands Trust, Royal Society for the Protection of Birds, and Joint Nature Conservation Committee. The objectives of WeBS were to estimate population sizes, monitor trends, understand distributions, and facilitate research on shorebirds during the nonbreeding season. To accomplish this, WeBS uses mostly volunteers to collect data around the British Isles. The program consists of Core Counts, conducted monthly at approximately 2,000 wetlands from September through March. Observers visit a predefined area, identify all species, and estimate their abundances. For shorebirds, these counts often are done at high-tide roosts. Each winter, from November to February, a fewer number of observers conduct Low Tide Counts at a subset of estuaries to understand patterns of distribution and abundance in relation to their foraging habitats and food. Originally, an attempt was made to conduct Low Tide Counts at 59 estuaries around the United Kingdom, where shorebird populations numbered at least 50,000 individuals. For a variety of reasons, however, coverage has been reduced to fewer estuaries. Low Tide Counts were coordinated to occur in the morning, within 2 hours of low tide, on specific dates to facilitate data interpretation and avoid double counting of birds.

Cayford and Waters (1996) used data from WeBS combined with several other special survey efforts to estimate the total number of shorebirds wintering in the United Kingdom and to estimate population trends over a short interval. Results indicated that coastal sites around Great Britain supported approximately 1.65 million shorebirds during the winter and that populations were increasing for most species over approximately the 10-year interval between successive estimates. By contrast, Browne et al. (1996) reported significant declines (28% to 53%) over a 10-year period for wintering populations of the Ruddy Turnstone, Purple Sandpiper, and Common Ringed Plover, but not the Sanderling. These species were surveyed using slightly different methods, and habitats they occupied are not those typically covered by WeBS. Atkinson et al. (2006) used WeBS data to evaluate the usefulness of an alert system for monitoring population trends in shorebirds.

INTERNATIONAL SHOREBIRD SURVEY/ MARITIMES SHOREBIRD SURVEY

Several programs exist in North America to monitor shorebirds during the nonbreeding season. These programs were initiated to obtain information on the distribution and abundance of shorebirds at wetlands throughout the continent. They were not originally designed to provide data for population monitoring, although several papers have made use of the data for this purpose. The Manomet Bird Observatory (now the Manomet Center for Conservation Sciences) initiated the International Shorebird Survey (ISS) in 1974. In Canada, the Canadian Maritimes Shorebird Survey (MSS) began at the same time under the auspices of the Canadian Wildlife Service. The methods used to survey shorebirds in each of these programs vary somewhat. For example, MSS/ISS guidelines ask that observers count shorebirds at selected wetlands with a fixed study area on particular dates (for example, on days 5, 15, and 25 of each month) or at approximately 10-day intervals during migration windows. It is expected that the same observers will survey sites to reduce interobserver variability in counts. The timing of surveys varies depending on whether the site is coastal or inland. For example, at some coastal sites shorebirds are surveyed at high-tide roosts; at other estuaries surveys are conducted at lower tide levels.

A challenge in using MSS/ISS data to detect population trends arises from considerable variation in species' abundances stemming from a variety of factors, including interobserver variability and year-to-year variation in the timing of movement of large numbers of migrants. Despite the shortcomings, two early papers analyzed MSS and ISS data from the Atlantic seaboard. Howe et al. (1989) reported

that three of nine species sampled by ISS observers had significant downward trends in abundance over the first 12 years of the program. Morrison et al. (1994) reported similar results for three of 13 species sampled over the first 18 years of the MSS. Bart et al. (2007) expanded the geographical and temporal extent of these early analyses and reported slightly different results for 30 species in the north Atlantic region and 29 species in the U.S. Midwest. In the Atlantic region, 73% of 30 species declined over 25 years, and nine of these trends were significant. In the Midwest, however, few significant declines were detected. Bart et al. (2007) evaluated several hypotheses to explain the different results between the two regions. They suggested several plausible explanations for the regional differences, including (1) sampling problems associated with fewer and more recent coverage in the Midwest, (2) a shift in the migratory pathways from Atlantic to Midwest, (3) a decrease in detection rate because birds are migrating more rapidly, and (4) a true population decline.

PACIFIC FLYWAY PROJECT

In western North America, the Pacific Flyway Project (PFP) was established in the late 1980s by biologists at Point Reyes Bird Observatory (now PRBO Conservation Science). In a manner similar to the ISS/MSS, this project used "citizen science" to amass data on the distribution and abundance of shorebirds at coastal and interior wetlands from Alaska to Baja California, Mexico (Page et al. 1999); it was not designed to monitor populations. The PFP surveys were fewer in number (than ISS/MSS), occurring four times (August, November, February, and April) during the nonbreeding season, but they were conducted in a coordinated fashion across the region. The methods were similar to those used in MSS and ISS surveys. Coordinated groups of observers conducted counts at important estuaries and interior wetland habitats. Still, survey methods varied somewhat among locations. For instance, observers sometimes surveyed high-tide roosts, often conducted a single observation at a site, or, rarely, coordinated a series of four sequential counts by multiple observers on a rising tide (Colwell and Cooper 1993).

Nevertheless, the PFP has provided some data that may be useful in monitoring shorebird populations, and it has been resurrected recently with a one-time annual count conducted. These efforts have resulted in significant contributions to understanding shorebird distributions along the Pacific Flyway (e.g., Shuford et al. 1998; Page et al. 1999). This information has been used to categorize the relative importance of wetlands along the flyway to shorebirds under the Western Hemisphere Shorebird Reserve Network and Ramsar Convention.

PROGRAM FOR REGIONAL AND INTERNATIONAL SHOREBIRD MONITORING

The Program for Regional and International Shorebird Monitoring, or PRISM, represents a recent concerted effort to improve the methods by which shorebird population sizes are estimated and monitored in Canada and the United States. The objectives of PRISM are (1) to estimate the size of breeding populations in North America; (2) to describe species' distributions, abundances, and habitat relationships; (3) to monitor trends in population size; (4) to monitor numbers at stopover sites; and (5) to assist local managers in meeting conservation goals (Bart et al. 2005). PRISM was conceptualized in 2001 to further the goals of the U.S. and Canadian Shorebird Conservation plans. From a practical perspective, it was apparent to those in shorebird conservation that any successes in gaining support for their work would require more rigorous and scientifically defensible methods of population estimation. Specifically, they reasoned that shorebirds required a similar approach as used in waterfowl management in which species had annual monitoring linked to a long-term maintenance or recovery of populations. When a population was below its recovery objective, management actions would be increased to affect growth, and funding could

be directed at those species that remained below a target population size.

To estimate population sizes in breeding areas, PRISM uses a system of double sampling (Bart and Earnst 2002), with extensive sampling coupled to intensive observations of rectangular plots throughout the Arctic. Briefly, the protocol calls for the establishment of a large number of plots that are spread extensively across a species' range, and these plots are sampled a few times during the breeding season. Subsets of these plots are sampled more intensively such that population densities are better known. This double sampling is used to calculate detection indices for species. A species' index is the ratio of birds detected by naïve surveyors visiting a plot once compared with results obtained from more intensive surveys conducted repeatedly during the breeding season. The detection ratio ranges from 0.0 to 1.0. Detection indices are then used to estimate species' densities in an area by multiplying the number of birds detected during surveys of a sample of habitat as represented by the extensive plots. Assuming that species are distributed within suitable habitats similarly across a species' range, these numbers are then extrapolated to produce estimates of total population size.

PRISM has started to produce meaningful results. Estimates of abundance and population size will serve as the foundation for population trend analyses. For example, Brown et al. (2007) used this approach to estimate shorebird population sizes in the Arctic National Wildlife Refuge and to argue that these relatively large populations (approximately one quarter of a million birds of 14 species) justified designation of the refuge as a site of international importance under the Western Hemisphere Shorebird Reserve Network (see Chapter 11).

BREEDING BIRD SURVEY

The Breeding Bird Survey (BBS) was initiated in 1966 and is coordinated by the U.S. Geological Survey and the Canadian Wildlife Service. The survey method is based on a 40-km roadside survey of secondary or tertiary roads. Observers drive along roads and stop every 0.8 km to conduct a point count. During the 3-minute point count, observers tally all individuals seen or heard within a 0.4-km radius circle. Observers conduct surveys during the morning, and each route is sampled once annually during the peak of the breeding season, typically June (Robbins et al. 1986). Several features of the BBS make it a difficult method to apply to monitoring breeding populations of shorebirds, including poor precision, potential observer bias, and road biases. The roadside method of the survey probably does not adequately sample the habitats used by most temperate-breeding shorebirds, which tend to be aggregated at wetlands that may or may not be well distributed along the route. Another shortcoming of the BBS is that the network of roads in North America precludes sampling the breeding areas of most Nearctic species. The exceptions to this may be a handful of species that breed in wetlands and associated upland habitats across the midcontinent and intermountain west.

Only a few studies have made use of BBS data to assess shorebird population trends. Sanzenbacher and Haig (2001) analyzed data for the Killdeer, the most ubiquitous Nearctic plover. Among temperate-breeding shorebirds, the Killdeer is well suited for the BBS owing to their extensive range, use of diverse habitats, and the ease with which they are detected when they vocalize. Killdeer populations had declined significantly in some regions, such as in Canada and the western United States where populations had declined by roughly 2.5% annually from 1966 to 1996. Morrison et al. (2001) assessed the population trends of 35 North American species, including data on 15 from the BBS. In addition to the Killdeer, significant negative trends were recorded for two other species, the Lesser Yellowlegs (−8.2% annual decline) and Wilson's Phalarope (−0.2%). The Upland Sandpiper exhibited a significant positive population trend (+1.0%).

Recently, the roadside transect method of the BBS was adapted in a continentwide effort to estimate the population size of the Long-billed Curlew (Stanley and Skagen 2007; Jones et al. 2008). Early "best guesses" indicated that the population size was 20,000 (Brown et al. 2001), a number that was adjusted upward to 40,000 (Morrison et al. 2007). Researchers used a stratified random-sampling design to establish 32-km road transects within four habitats that differed in quality based on the percentage of grassland cover. In 2004 and 2005, observers surveyed transects by conducting 5-minute point counts at 0.8 km intervals along roads; observers recorded the number of curlews within a 400-m radius circle. These data were used to estimate density along routes and then were extrapolated to the amount of area in each of the four habitat types across the western U.S. and Canada. The population estimates for these 2 years averaged 161,181. Although this population estimate has wide confidence intervals, it is substantially higher than the previous estimates based on expert opinion. Two recent opinion pieces highlight the controversy of these estimates (Farmer 2008; Lanctot et al. 2008).

Finally, population trend data exist for the American Woodcock, one of two shorebirds that are hunted in North America (Kelly 2004). Singing-ground and wing-collection surveys indicate long-term population declines since 1968 (Kelly 2004). Low recruitment (based on juvenile to adult female ratios) appears to have been low throughout much of the period of population decline. Hunters harvested approximately 250,000 woodcock during one recent hunting season (2003 to 2004).

CHRISTMAS BIRD COUNT

The Christmas Bird Count, organized by the National Audubon Society, was initiated in 1900 as a nonconsumptive alternative to outings in which hunters sought to shoot large numbers of game birds in a single day. The CBCs take place each year from December 14 through January 5. Each local CBC is conducted by multiple observers in one 24-hour period. The effort strives to tally as many species as possible within a

24-km radius circle, which is often positioned so as to maximize habitat diversity and, hence, the number of species detected. The survey protocol consists of groups of observers attempting to estimate the abundance of all species encountered. Despite its shortcomings, CBC data are useful in understanding range shifts, especially given the century-long history of the program. Using CBC data, Sanzenbacher and Haig (2001) determined that there was no rangewide decline in the Killdeer population across North America, although there were regions in which both significant negative and a few positive trends were detected. The CBC data provided early (1960s to 1980s) evidence that the Pacific Coast population of the Snowy Plover was in decline in southern California (Page et al. 1986; Butcher and Lowe 1990). The American Avocet winters along the Pacific Coast of North America as far north as Humboldt Bay, California, and the CBC data were one of several sources that indicated the size of the local (wintering) population had increased over several decades until declining slightly in the latter years of the 1990s (Colwell et al. 2001). Buchanan (1999) examined CBC data and showed that Rock Sandpipers' abundance decreased in Oregon, Washington, and British Columbia coincident with a strong El Niño event in the 1982 to 1983 period. Finally, at least one recent conservation plan (Fernández et al. 2008) used CBC data to examine population trends for a wintering sandpiper.

SINGLE SPECIES CENSUSES

The temperate or tropical breeding distributions of some shorebirds make population surveys logistically more feasible compared with remote areas (such as Arctic tundra or boreal forest). This is the case for several North American species of plover that are listed as threatened or endangered under the U.S. Endangered Species Act. The Piping Plover was listed in 1985, with the Atlantic Coast and Great Plains populations designated as threatened, and the Great Lakes population receiving en-dangered status. Beginning in 1991, coordinated surveys of breeding areas have been conducted every 5 years to document population size and trends (Table 10.2). These efforts have yielded a high-quality estimate of population size owing to the efforts of a vast network of observers surveying virtually all suitable breeding habitats during a short period during the breeding season. The overall trend in the population has been increasing for the Atlantic and Great Lakes regions but decreasing in the Great Plains. The regional trends cancel one another out such that the total population has remained relatively stable since 1996. It is unlikely that the regional differences in trends stem from large-scale shifts in the breeding distribution (that is, an increase in Atlantic Coast birds coming from individuals emigrating from the Great Plains) because there has been only limited movement of color-marked birds between populations. The maximum dispersal distance reported was approximately 1,500 km for a bird that had hatched in northern Minnesota and bred on Lake Erie (Haig and Oring 1988).

In some instances, researchers have used a combination of censuses and habitat modeling to estimate population size based on an extrapolation of suitable habitat to population size. Long et al. (2008) undertook this approach with the Madagascar Plover, a threatened shorebird that breeds in wetlands difficult to access along the western coast of Madagascar. Observers found 236 plovers during several months of fieldwork over several years. Using habitat modeling, they determined the habitat preferences of plovers and then established the amount of suitable habitat based on remotely sensed images. Finally, they calculated that approximately 139 km^2 of suitable habitat supported 3,100 ($\pm$ 396) plovers. This approach was based on presence (not abundance) data and may fail to account for the social nature of plovers, which would contribute to a patchy distribution across the landscape. Its effect on population size estimate is unknown.

TABLE 10.2
Population estimates for the Piping Plover (Charadrius melodus) *based on range-wide surveys*

GEOGRAPHIC REGION/SUBSPECIES	1991	1996	2001	2006
Atlantic coast (*C. m. melodus*)	1,971	2,591	2,911	3,312
Canada	509	422	481	457
United States	1,462	2,169	2,430	2,855
Great Lakes (*C. m. circumcincutus*)	40	48	72	110
Great Plains (*C. m. circumcincutus*)	3,469	3,286	2,953	4,662
Canada	1,437	1,687	972	1,703
United States	2,032	1,599	1,981	2,959
Total	5,480	5,925	5,936	8,084

Sources: Eliot-Smith et al. 2009; Haig et al. 2005.

LIMITING FACTORS

The capacity of a population to grow is influenced by both intrinsic and extrinsic factors (Newton 1998). **INTRINSIC FACTORS** are those features of a species' biology that affect birth rate, death rate, immigration, and emigration and hence dictate the rate at which a population changes. They are also considered demographic characteristics of a population. As reviewed earlier, shorebirds are rather conservative in their reproductive biology (that is, they have small clutch sizes and delayed breeding). This means that their populations, especially small ones, are capable of comparatively slow growth. Fortunately, however, the survival rates of shorebirds are relatively high. As a result, populations may experience a mix of good and bad years on the breeding grounds and still persist in abundance. Problems arise, however, when survival rates are compromised or when chronic reproductive failure continues for long intervals. In fact, changes in survival rate have a stronger effect on population growth than changes in productivity (Hitchcock and Gratto-Trevor 1997; Sandercock 2003). The effect of compromised survival has been shown to have a strong effect on the viability of populations of some threatened or endangered shorebirds.

EXTRINSIC FACTORS are those features of the environment that affect population growth via their impacts on birth and death rates. Several extrinsic factors act on populations: competition, predation, and disease. These biotic factors are also viewed as the important ecological processes that have shaped the life histories of birds (Wiens 1989). The term **LIMITING FACTOR** is synonymous with extrinsic factor; specifically, limiting factors are those that keep a population from growing either by acting on the breeding grounds to negatively affect reproduction or by compromising survival at any time of year. A few examples illustrate the point. Ecologist studying shorebirds wintering at northern latitudes have long been concerned with evidence that competition for food resources limits a local population. Earlier treatment (see Chapter 7) of the carrying capacity of an estuary specifically addressed the issue of competition for food and population limitation. In this case, competition may limit a wintering population by keeping some individuals, most likely inexperienced juveniles, from feeding effectively (Goss-Custard et al. 2002). The negative effects of competition would be especially apparent during prolonged periods of cold, inclement weather. Clark (2004) used recoveries of banded shorebirds in Great Britain to show that mortality was unusually high during severe as compared with mild winters.

For competition to limit a population, food resources must be in short supply, such that individuals cannot sustain themselves and they experience poor survival or emigrate. It is also possible that competition may act more subtly on individuals such that they are in poor condition at the end of winter and experience low reproductive success in the subsequent spring. Extrinsic factors may also limit populations on the breeding grounds. A second example concerns the high levels of reproductive failure sometimes observed in shorebirds. Researchers working in Arctic regions have often been impressed by the strong annual variation in reproductive success stemming from predation of eggs and chicks. In this case, predation is the factor that acts to limit population size via its negative impact on the recruitment of young each year. In temperate latitudes, there is growing concern that predation may limit shorebird populations, especially when small populations breed within refugia (nature reserves) where predators may have a strong impact (Rönkä et al. 2006).

Ecologists have long been intrigued by the relative strengths that extrinsic factors play in affecting population growth, and the literature abounds with reference to limiting factors. The same factors that limit populations may also regulate them. Population regulation means that the same extrinsic factors that limit a population may vary in the strength of their effects with population density. So when extrinsic factors regulate a population, it specifically means that either per capita birth or death rates vary with population density. Hence, ecologists often refer to density-dependent phenomena. For example, if competition regulates a population on the wintering grounds, it must be the case that the mortality rate increases with population density. This may arise because individuals experience lower intake rates owing to greater interference or lower prey availability; prolonged cold weather may exacerbate this situation. Alternatively, nest predation may act to regulate a breeding population if predators vary in their functional and numerical response to the density of breeding shorebirds and consume a higher percentage of clutches as shorebird population density increases. As a result, per capita fecundity decreases.

The magnitude of the effect that a limiting factor has on a population may depend on the geographic area and the duration of time over which it acts. That is, predation may act locally to compromise reproduction, but it cannot have a large effect on population size unless it acts over the full extent of a species' range. Similarly, repeated years of cold winter weather in nonbreeding areas will have a stronger impact on population size than a single, localized cold snap of short duration. Consequently, it is essential to define the space and time over which studies are made when discussing limiting factors (Newton 1998).

It is clear that the abundances of natural populations are limited by competition, predation, and disease. If they weren't, then populations would grow exponentially. So what are the relative strengths of extrinsic factors in shaping shorebird population sizes? Here, I address evidence that these factors affect shorebird populations. Next, I turn to the effect of severe weather on shorebird populations and address its importance in affecting mortality and population sizes. The challenge in summarizing the information is that the migratory nature of shorebirds causes their populations to be affected throughout the annual cycle by factors operating on breeding grounds, at stopover sites, and in wintering areas. This issue is especially important when considering that anthropogenic effects magnify the demographic challenges that shorebirds face in an ecologically dynamic world.

PREDATION

If predation limits shorebird populations on the breeding grounds, it does so by compromising productivity or the number of young produced annually. Predation of eggs and chicks is the single most important factor contributing to substantial annual variation in the productivity of shorebirds (Evans and Pienkowski 1984; Piersma and Lindström 2004; MacDonald and Bolton 2008). It is common for researchers to report that predation has been the main cause of near total reproductive failure. Occasionally, this phenomenon occurs across a broad geographic area. In 1992, for instance, there was widespread reproductive failure among Arctic-breeding shorebirds, and predation was implicated as the cause of much of the low productivity (Ganter and Boyd 2000). These bad years are, however, generally offset by successive years of high reproductive success. It is generally perceived that predation exerts a stronger effect on year-to-year variation in productivity in temperate and tropical latitudes as compared with the Arctic (Ricklefs 1969; Jehl 1971). Alternatively, predation of adults during the breeding season may limit population size. Although there is no evidence for this in shorebirds, raptors occasionally have been shown to limit gamebird populations in Europe (Valkama et al. 2005).

The importance of predation in driving shorebird population dynamics is readily apparent from a variety of studies of Arctic-breeding shorebirds. In tundra habitats, annual variation in breeding success of ground-nesting birds has been shown to exhibit a cyclical pattern, typically with a 3-year periodicity (Summers and Underhill 1987; Blomqvist et al. 2002). This pattern in breeding performance has been linked to the population fluctuations of lemmings, which are the main prey of Arctic fox (*Alopex lagopus*) and several jaegers (*Stercorarius* spp.). Summers and Underhill (1987) first reported this cyclical phenomenon using a 33-year data set showing the percentage of juveniles in nonbreeding flocks of Brent Geese, Sanderlings, and Curlew Sandpipers (Fig. 10.1). A similar cyclical pattern of reproductive performance was reported for Red Knots and Curlew Sandpipers, using a 50-year dataset (Blomqvist et al.

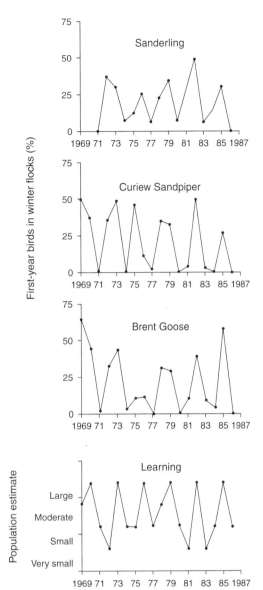

FIGURE 10.1. The percentage of juveniles in nonbreeding flocks of two species of shorebird and one goose vary on a predictable 3-year cycle. Sanderlings, Curlew Sandpipers, and Brent Geese all breed in tundra habitats of the Taimyr Peninsula, Russia. Estimates of percentage juveniles in flocks come from the East Atlantic Flyway. Annual variation in lemming abundance in the Taimyr region is shown also. After Summers and Underhill (1987).

2002). In both instances, there was a highly significant 3-year periodicity to the long-term annual variation, with years of complete reproductive failure (few or no juveniles) followed by years with high percentages of juveniles in nonbreeding flocks. Summers and Underhill (1987) pointed out that these species have a common breeding area on the Taimyr Peninsula, Russia. What is notable is that the data come from observations collected in northwest Europe and South Africa. Given the very different ecologies and habitats of these species during the nonbreeding season, it is difficult to reason that the pattern of similar productivity across species is somehow associated with factors acting during the nonbreeding season. Rather, the species all breed in tundra, and it is here that it is argued that predators strongly influence annual variation in productivity.

The predator-prey relationships that drive the cyclical variation in Arctic shorebird productivity are relatively straightforward. The mechanism involves the functional and numerical responses of Arctic foxes, and to a lesser extent gulls and jaegers, to the alternating availability of prey (e.g., Wilson and Bromley 2001). The alternative prey hypothesis posits that predators switch from eating lemmings to preying on eggs and chicks of ground-nesting birds in a manner driven by the cyclical pattern of lemming abundance. Cyclical patterns of small mammal populations have been reported from throughout the Arctic, but they seem to be stronger across broad regions of the Palearctic compared with the Nearctic (Lindén 1988; Steen et al. 1990). When lemmings are abundant, shorebirds experience high productivity; when lemming populations crash, foxes and jaegers switch to birds and consume most clutches and the few chicks that hatch. It is during these years of declining or low lemming abundance that shorebird populations have especially low productivity. Evidence from a variety of Arctic (e.g., Underhill et al. 1993; Small et al. 1993) and temperate regions (e.g., Šálek et al. 2004) lend some support to the alternative prey hypothesis.

Outside the relatively simple ecosystems of far northern latitudes (Ims and Fuegli 2005), more complex trophic interactions between diverse predators and a diverse array of prey species make it difficult to generalize the Arctic patterns to shorebirds that breed in temperate and tropical latitudes. Nevertheless, it is widely perceived that reproductive success is generally lower at more southerly latitudes, and that predation is again the main cause of reproductive failure (MacDonald and Bolton 2008). Some have even gone so far as to suggest that the southern limit of a species' range is set by predation, as was the case with the Common Ringed Plover (Pienkowski 1984; Evans and Pienkowski 1984).

The production of large numbers of young in a year does not necessarily equate to a substantial increase in population size the next breeding season. This is because juveniles may not survive well, for a variety of reasons related to the fact that inexperienced birds do not forage well and are often victims of predation. Insley et al. (1997) analyzed a long-term banding data set for Common Redshanks and found substantial annual variation in the survival of adults and juveniles. Survival increased from 43% in juveniles to 67% and 74% in 2-year olds and older birds, respectively. There was no relationship between annual survival and local population density. However, an 11-year study of Common Redshanks reported density-dependent mortality of both adults and juveniles owing to predation by a small population of Eurasian Sparrowhawks (Whitfield 2003). In this case, predation rates increased in both age classes as local population size increased. Juveniles were especially vulnerable because at high population density they were forced to feed in areas of salt marsh where predation danger was higher. Unfortunately, there are few other long-term studies that have quantified the effects of predation on wintering shorebird populations, especially over large geographic areas that are required to understand predation as a limiting factor. This latter condition is necessary because it is possible that high predation danger may be localized to particular sites where prey (the shorebirds) are abundant. Moreover, individuals may alter their distributions owing to predation, which would result in a population decrease.

In summary, predation can act to limit a population during the nonbreeding season via decreased survival of juveniles and adults. Given that most shorebirds spend up to 8 months of the year away from breeding areas, it is not surprising that mortality can be high during the nonbreeding season. Furthermore, there is some evidence that predators occasionally remove a large percentage of individuals from local populations during winter (Page and Whitacre 1975). It is important to recognize that the study that detailed appreciable effects of a single Merlin on wintering shorebirds was conducted during a period in which raptor populations were recovering from low numbers associated with a variety of anthropogenic effects, including pesticides and persecution. It is almost certain that falcons are more abundant now than several decades ago. Hence, the effects on shorebird populations are probably stronger today than several decades ago. However, knowledge of the extent and variation in predation on nonbreeding shorebirds remains poor.

DISEASE AND PARASITES

A diverse array of pathogens affects the survival and productivity of birds and thus has the potential to strongly impact populations. Pathogens can be categorized into two main groups: microparasites and macroparasites (Anderson and May 1986; Newton 1998). MICROPARASITES are characterized by their small size, rapid reproduction within hosts, and short life cycles. Examples include viruses, bacteria, fungi, and protozoa. The transmission of microparasites in a population acts in a density-dependent manner: transmission increases with population density. As a result, a greater proportion of individuals die. However, surviving individuals may develop immunity that results in resistance to subsequent reinfection. By contrast, MACROPARASITES often do not multiply within hosts but rather produce infective stages that pass out of the host with fecal material, for instance. Macroparasites

are comparatively long-lived and often produce only a limited immune response in infected birds. The resulting infections tend to persist for the life of the host and may sometimes reach levels that cause mortality. Examples of macroparasites include helminth and nematode worms, tapeworms, and flukes. ECTOPARASITES such as lice, hippoboscid flies, Hemiptera, and ticks also infest shorebirds.

Disease is generally considered to act in a density-dependent manner in its effects on populations. That is, the rate of transmission and level of infections increase with population density (Anderson and May 1986; May 1997). Despite widespread agreement that disease plays a dramatic role in the population dynamics of many species of bird (Newton 1998), there are surprisingly few documented examples of shorebird populations being affected by disease. In most instances, effects of disease are often reported as a side note to estimates of mortality in other taxa, notably waterfowl. Massive dieoffs from botulism (type C) are irregular occurrences at interior wetlands of North America, especially late in summer when wetland conditions favor the spread of the disease amid dense concentrations of birds. The National Wildlife Health Center of the U.S. Geological Survey maintains a Web site with information on avian cholera outbreaks. Two of 17 outbreaks in California during 1999 included shorebirds, but most incidents involved waterfowl. In a review of the occurrence of avian disease in the Salton Sea, California, Friend (2002) reported that shorebirds were increasingly victims of both avian botulism (type C) and avian cholera.

One exception to the general lack of information on diseases in shorebirds involves the effects of a neurotoxin from marine dinoflagellates that produces paralytic shellfish poisoning in many species of marine birds (and humans). The African Black Oystercatcher has a total population size of approximately 5,000. Oystercatchers are resident in coastal habitats where they feed on shellfish and other intertidal invertebrates. One episode of poisoning reduced the world's population by 6% (Hockey and Cooper 1980). In another instance, avian cholera killed roughly 4% of the population (Hockey 1996). There is some evidence that other species of oystercatcher have resistance to the toxins (such as domoic acid) that produce paralytic shellfish poisoning in other coastal regions of the world (Falxa 1992). Recent data suggest that some shorebirds alter their prey preferences when the seasonal risk of poisoning increases, suggesting that they sense concentrations of toxins (Kvitek and Bretz 2005).

In his review of the role of disease in avian population dynamics, Newton (1998) gives only one example of a shorebird. The paucity of information on disease effects on shorebirds is somewhat remarkable given the fact that shorebirds share strong affinities for freshwater habitats with many waterfowl and other waterbirds, and examples of avian diseases are replete in these groups. Piersma (1997, 2003) and Mendes et al. (2005) have suggested that most shorebirds have evolved patterns of habitat use in high Arctic and marine systems to minimize the impact of microparasites. By contrast, species using freshwater wetlands are prone to higher levels of infection from some parasites, such as hematozoa.

COMPETITION

At some times of year, resources may be in short supply. By definition, a resource must be used (but not necessarily consumed) by an individual such that others cannot use it. As a result, individuals may be affected and populations limited by a shortage of the resource (Wiens 1989). For shorebirds, resources include food and habitat. During winter, food may limit a population, especially at northern latitudes during long cold spells. It is also possible that competition for habitat (or space) occurs. Specifically, territoriality may limit access to foraging habitat for some individuals in a local population. Both of these scenarios address the notion of carrying capacity of a habitat, a topic discussed in Chapter 8. During the breed-

ing season, competition for territories may limit a population if there is insufficient habitat to support all individuals.

Competition is considered to act in a density-dependent fashion: as population increases, the negative effects of food shortage are measured in decreased reproduction and survival. It is highly unlikely that competition for food resources exerts a strong and consistent impact in breeding areas owing to the seasonal flush of invertebrate food that typifies most breeding habitats, especially in northern latitudes. However, several pieces of evidence suggest that competition for space or breeding opportunities may limit some populations. First, as pointed out earlier, a proportion of birds of a variety of species forego breeding early in life and oversummer on their wintering grounds. Additionally, removal experiments suggest that some adult males may not acquire a breeding territory and remain as floaters in the population, waiting for the opportunity to breed. These phenomena may be more common under conditions of high population density, although no data have been collected to evaluate this notion.

Beyond this, however, there is comparatively little information to suggest that competition limits shorebird populations on the breeding grounds. However, these arguments apply to boreal and Arctic regions where breeding habitats are comparatively pristine and unaffected by humans. This is probably not the case in temperate and tropical areas, where habitat loss has been substantial. For example, it is clear that prairie-breeding shorebirds of the Nearctic have declined, in part owing to habitat loss. The range of the Long-billed Curlew once extended well into the eastern tallgrass prairie region of the Midwest. Today, however, curlews are restricted to breeding in short-grass prairie and shrub-steppe habitats of the Great Plains and Intermountain West (Jones et al. 2008). Similar arguments apply to populations of the Willet, Marbled Godwit, and Upland Sandpiper that breed in prairie habitats that fringe interior wetlands of the midcontinent. The effects of habitat loss may also apply to some populations of temperate breeding plovers. Populations of the Piping, Snowy, and Wilson's Plovers that breed in coastal habitats are listed as threatened or endangered. Anthropogenic habitat loss and degradation is thought to be an important factor contributing to the decline of these populations. A shortage of high-quality breeding habitat limits the populations. If this were true, then one would predict several things. First, suitable breeding habitat should be occupied by individuals at densities that preclude settlement of conspecifics. Second, the creation of suitable breeding habitat via restoration should result in the rapid colonization of these sites by individuals that would otherwise not be able to find a territory.

Evidence for food limitation is much stronger during the nonbreeding season (Newton 1998) and especially at northern estuaries during the winter. As mentioned earlier, an abundance of literature has addressed the question of food limitation and carrying capacity of estuaries (see Evans 1984; Goss-Custard 1980, 1985; Goss-Custard et al. 2002). Several important points are relevant. First, a latitudinal gradient in shorebird density (higher in southern than northern estuaries; Hockey et al. 1992) suggests food varies in availability on a global scale and that food limitation may not act as strongly on wintering populations in southern latitudes. Second, shorebirds (and other predators) often reduce the standing crop of invertebrate food during short and long intervals (Goss-Custard 1985). Third, much of what food is present may not be available to birds (Evans and Dugan 1984). Fourth, the energy costs of maintenance are higher for birds wintering at northern estuaries. Fifth, occasionally large percentages of a local population die of starvation during prolonged periods of cold weather (Clark 2004). These observations indicate that competition for food is strongest during winter at northern latitudes, and it may interact with density-independent factors, specifically weather, to limit population size.

Abundant evidence indicates that food supplies limit shorebird populations on wintering areas. I have already addressed this idea by considering carrying capacity of estuaries (see Chapter 8), but it is worth revisiting. Briefly, competition may limit a population of wintering shorebirds by increasing mortality owing to starvation when food is in short supply. Furthermore, competition may act in a density-dependent manner if the per capita death rate increases with local population density. In reality, the role of competition in limiting shorebird populations and the interpretation of the notion of carrying capacity are challenging (Goss-Custard et al. 2002).

During winter, habitat may be in short supply such that intraspecific competition for space limits the number of individuals that can settle on an estuary. This has been demonstrated most clearly for territorial species, especially the Black-bellied Plover. Moser (1988) analyzed a long-term data set on the numbers of plovers wintering at 45 estuaries around the British Isles. Overall, the national population was increasing; however, plover numbers had leveled off at certain estuaries, notably those on the south and west coasts of England, which have warmer winter conditions. By contrast, populations were still increasing at other estuaries in the east and north, where winter weather is typically more severe. Moser (1988) suggested that this pattern of sequential filling of sites was evidence that plovers had reached carrying capacity at some estuaries and not others. The mechanism that promotes this pattern is the territorial behavior of wintering Black-bellied Plovers. Defense of feeding territories limited the settlement of additional plovers at some estuaries and not others. The fates of individuals that were excluded from estuaries and the effect on populations remain unknown. For instance, it is not known whether birds (probably juveniles) arriving in their first autumn at filled estuaries but unable to establish themselves simply dispersed to other locations or suffered higher mortality. In the first case, the global population of plovers would not be affected; in the latter case, it would be reduced.

Although the clearest examples of habitat limitation on shorebirds occur in territorial species, it is also possible that habitat may limit species that form feeding aggregations, this is readily apparent when habitat is lost and populations decline. The introduction of cordgrass (*Spartina anglica*) into coastal estuaries of the British Isles reduced the amount of intertidal foraging habitats for many shorebirds. Goss-Custard and Moser (1988) showed that Dunlin populations decreased substantially in Great Britain, especially in those estuaries where cordgrass had colonized and displaced large amounts of intertidal foraging habitats. Elsewhere in the world, similar scenarios of cordgrass invasion have been shown to alter shorebird foraging habitats (Buchanan 2003; Patten and O'Casey 2007; Nehring and Hesse 2008), with potential impacts on the local populations.

In summary, there is considerable evidence that competition for food and occasionally space limits local shorebird populations. Competition for food probably exerts its strongest effect on individuals and populations at the limits of a species range, and this occurs at northern latitudes in the depths of winter when cold weather exacerbates the energy demands of daily maintenance. It is unlikely that competition for food exerts as strong an impact on shorebird populations during the breeding season because a strong seasonal flush of invertebrate food occurs in most breeding areas during the spring and summer. Some evidence suggests that availability of territories may limit some populations during both the breeding and nonbreeding seasons.

WEATHER

Extrinsic factors may act in a density-dependent fashion to regulate a population. Weather-related effects on per capita birth and death rates are considered to act in a density-independent manner. Specifically, the effect of extreme weather, such as an intense storm during the breeding season or a prolonged cold snap in winter, alters birth and death rates in a manner independent of population density. In other words, the percentage of birds that fail to reproduce or die does not vary with local population density. In reality, however, it is likely that the influence of weather on population size is linked to the effects of extrinsic factors. For example, inclement weather may make it difficult for shorebirds to feed effectively and may increase the energy costs of maintenance, which results in poor body condition. If this weather persists, individuals in poor condition may become more vulnerable to predation and disease. Clark (2004) showed that winter mortality was higher during severe winters in England.

On northern breeding areas, weather occasionally has a strong impact on shorebird populations, either by causing reproductive failure or, rarely, by resulting in high adult mortality. For example, cold weather subtly affects the availability of food and, hence, the timing of breeding in shorebird populations, the length of the breeding season, and the opportunity for renesting. Sometimes weather affects chick mortality. Depending on latitude and length of the breeding season, inclement weather may delay breeding. In some years, extensive snow cover may persist well into summer such that nearly all adults forego breeding (Green et al. 1977). At northern latitudes, weather commonly affects shorebird populations during the nonbreeding season. There are occasionally large losses of adults and juveniles during migration (Newton 2007) or upon arrival on breeding grounds. For example, Ruddy Turnstones and Red Knots experienced high mortality when a spring snowstorm overwhelmed birds that had recently arrived on their high Arctic breeding grounds after a long-distance migration (Morrison 1975). Similar reports of high mortality in early spring have been reported for some temperate breeding species (e.g., Marcström and Mascher 1979; Holland et al. 1982).

Northern Lapwings breed in arable lands and meadow habitats of the British Isles. Estimates of lapwing abundance in British farmlands suggest that the species has decreased. Peach et al. (1994) analyzed a long-term data set (>50 year; beginning in 1930) of lapwings first marked as chicks (before fledging) to obtain survival estimates. Annual adult survival was relatively high (0.705), with a noticeable increase (0.752) in recent years (1961 to 1988). Survival of first-year birds (0.60) was lower than adults but not to an extent that it would explain the population decline. There were several years in which survival of both adults and juveniles was especially low, and these corresponded with severe winters of prolonged cold and snow (e.g., 1946 to 1947). By contrast, a review of productivity based on published accounts from the British Isles indicated that in most cases lapwings were not reproducing sufficiently well to replace themselves. As a result, the population was decreasing.

The Eurasian Golden-Plover breeds across a broad range in the Palearctic. Its southern breeding range extends south into the British Isles where it nests in heathland. During the winter, plovers occupy estuaries and arable lands throughout the British Isles and elsewhere along the East Atlantic Flyway. Parr (1992) compiled detailed demographic data for a local population of plovers over an 18-year (1973 to 1990) interval, which ended with extirpation. The population breeding in the northeast Scotland moorlands was not large, numbering at most 110 birds during peak numbers. Its decline was associated with a series of winters in which prolonged cold weather contributed to high mortality of adults and juveniles. This was evidenced by few birds returning to breed in subsequent springs. During the breeding season, adult mortality was negligible, as only 3 of 70 color-marked adults were depredated. Predators were responsible for most losses of eggs and chicks, and this was especially problematic in the later years when the population was declining. Habitat loss was not a cause of the decline, as local land-use practices were largely unchanged over the 18-year study.

HUMAN IMPACTS

It should come as no surprise that humans have had a strong negative effect on shorebird populations (Senner and Howe 1984). Chief among these impacts are direct mortality related to unregulated hunting, and reduced survival and productivity stemming from loss and degradation of habitats. Additionally, shorebirds suffer mortality owing to collisions with manmade structures, and they die from pesticide poisoning and pollution. These last two impacts are probably greater than we know because in many cases the events leading to die-offs often go unnoticed.

HUNTING

Humans have hunted shorebirds for millennia. In recent times, it is clear that the negative impacts on some shorebird populations can be dramatic (Senner and Howe 1984; Warnock et al. 2001). Historically, widespread market hunting during the 1800s probably led to the decline of many large shorebirds such as the Whimbrel, Upland Sandpiper, American Golden-Plover, and Eskimo Curlew. John James Audubon referenced the daily tally of shorebirds taken by market hunters in the Louisiana parish where he lived in the early 1800s: he reported that hunters shot approximately 48,000 plovers in one day during the peak of autumn migration. Given this magnitude of harvest, it is not surprising that some species never recovered. The Eskimo Curlew is almost certainly extinct (Roberts et al. 2009), although there is some hope that a small isolated population still exists somewhere in Alaska or northwestern Canada. A similar scenario holds for the Slender-billed Curlew, which is perilously close to extinction.

Shorebirds continue to be hunted worldwide. In North America, substantial hunting mortality occurs in some species (such as Wilson's Snipe and the American Woodcock) that are hunted legally. Kelly (2004) reported a long-term (1968 to 2004) population decline for the American Woodcock, for which the annual harvest was approximately 337,350 birds in a single, recent season. Legal and illegal hunting of shorebirds continues throughout much of Europe today. In some years, mortality amounts to tens of thousands of birds, especially of larger species (such as curlews, golden-plovers, and oystercatchers). Lambeck et al. (1996) detailed the annual hunter-based mortality on Eurasian Oystercatchers. Shooting mortality normally ranges between 1,500 and 2,000 birds in France alone, where the season typically extends from July into February. However, in cold winters when birds are pushed south from northern regions, the toll is estimated to be 10 times this level. In 1987, for instance, various lines of evidence indicated that hunters in France killed 10,000 to 27,000 oystercatchers during a prolonged cold snap. The French government closed the hunting season in January of that year owing to the high mortality. At Morecambe Bay, England, approximately 16,000 oystercatchers were killed over a 14-year period (1956 to 1969) as part of a program to reduce predation on commercial cockles, and the program failed to arrest the decline in cockles (Prater 1981). Eurasian Golden-Plovers were hunted throughout much of Europe, and according to Byrkjedal and Thompson (1998), up to 100,000 were killed annually in France alone.

Elsewhere in the world, unregulated hunting probably has a large but known impact on overwintering populations. Senner and Howe (1984) reported instances of 15,000 Lesser Yellowlegs killed annually by sport hunters in Barbados. Along the Australasian Flyway, it is estimated that hunting kills 250,000 to 1,500,000 shorebirds annually, which may amount to 5% to 30% of some species' flyway population (Warnock et al. 2001). Bamford (1992) estimated that hunting accounted for 7% to 18% of annual mortality in the Great Knot. The most serious hunting pressure comes from China (Warnock et al. 2001).

HABITAT LOSS AND DEGRADATION

If, under natural conditions, shorebirds are limited by food or habitat at any time of year, then any anthropogenic loss or degradation of these areas is likely to have negative consequences for their populations. In essence, this is the critical issue faced by conservationists, not merely those concerned with declining shorebird numbers. Anthropogenic loss and degradation of foraging habitat is a principal concern of shorebird conservation efforts, and it rests on the assumption that food or habitat limits a population. If this is true, then any additional reduction in habitat is likely to have consequences for local population size. Several recent papers link declining regional populations to various forms of anthropogenic land conversion and habitat loss. For instance, Nebel et al. (2008) argued that the population declines of several resident shorebirds in southeastern Australia were correlated with loss and modification of wetlands accompanying agricultural intensification. Elsewhere, Sanderson et al. (2006) hinted that habitat loss was responsible for the more precipitous decline of long-distance migrants (several of which were shorebirds) that wintered in open, dry habitats of Africa versus resident birds.

One recent paper demonstrated that the condition and survival of wintering shorebirds decreased after human-caused loss of intertidal habitat. Burton et al. (2006) showed that construction of a tidal barrage in Cardiff Bay, U.K., resulted in the loss of important intertidal foraging habitats for approximately 300 Common Redshanks. Birds displaced by the project had difficulty maintaining weight in winter and suffered a 44% increase in overwinter mortality compared with conspecifics in other nearby habitats that were unaffected. These effects are likely to have population consequences (a reduction in size) if they are not offset by increased productivity on the breeding grounds. Historically, loss of breeding habitats probably impacted temperate and tropical breeding shorebirds more severely than those breeding in more remote areas of the boreal and Arctic regions. Wetland loss in the conterminous United States has been substantial in the midcontinent, where prairie wetlands support a diverse waterbird community.

In Asia, reclamation of the Saemangeum estuary of South Korea represents the largest (40,100 hectares) one-time project converting intertidal to terrestrial habitats, and the implications for shorebirds are great (Moores et al. 2008). Shorebirds stage at the Saemangeum estuary in large numbers during northward and southward migrations. Recent monitoring suggests that some species have experienced significant local population declines since 2006 when the constructed seawall was closed and had its first impacts on intertidal habitats. For instance, the Great Knot population declined by 20%; it is unknown, however, whether these birds have simply moved to other estuaries along the Korean Peninsula (Moores et al. 2008).

PESTICIDES/POLLUTION

There are scattered accounts of shorebirds suffering mortality in association with the use of chemicals to control agricultural or other pest species. For instance, Warnock and Schwarzbach (1995) reported the deaths of 36 Dunlins and 2 Killdeer from eating strychnine-laced grain that was meant to kill rats. Pesticides also may subtly affect reproduction in birds. White et al. (1983) reported that 95% of shorebirds sampled during the winter in Texas showed elevated levels of various organochlorine pesticides (DDE, Dieldren, and toxaphene). In several species, birds were contaminated with levels known to inhibit normal reproduction. In some regions of the world, agricultural lands provide important feeding areas for wintering and breeding shorebirds. Given the prevalence of agricultural chemical use in this landscape, pesticides probably have a larger impact on shorebird survival and productivity than is appreciated.

OTHER SOURCES OF MORTALITY

Erickson et al. (2005) estimated that up to 1 billion birds are killed annually in the United States owing to anthropogenic causes other than habitat loss. Some of this mortality comes from collisions with various man-made structures, including power lines. There are only a few reports of mortality associated with shorebirds hitting power lines or fence lines. Conklin and Colwell (2007) opportunistically recorded the deaths of Dunlins in the vicinity of a winter roost where a small diameter electrical wire strung across a brackish slough killed a total of 30 birds over 28 hours of observation. They projected that the observed mortality extrapolated to the local area would have accounted for roughly 4% of the Dunlin population. Shorebird mortalities associated with expanding wind turbines are generally low compared with other avian taxa (e.g., Johnson et al. 2002).

CONSERVATION IMPLICATIONS

Population biology is the foundation of applied ecology because it provides the data to evaluate the success of conservation and management actions. These data include estimates of a species' (1) densities across landscapes of varying habitat types, (2) population sizes and trends, and (3) vital rates (such as survival and reproductive success) of local populations. A few examples illustrate the point.

Protected areas are an important conservation tool, but reserves vary greatly in the degree to which nature is protected. The coastal plain of the Arctic National Wildlife Refuge, Alaska, has been the center of considerable conservation debate for several decades. Despite a rich biological diversity of its tundra ecosystems and protection afforded by its status as a national wildlife refuge in the U.S. system, prospects for considerable oil and gas resources continue under provisions of the Alaska National Interest Lands Conservation Act (1980). Brown et al. (2007) used PRISM methodology to estimate the densities of 14 species and regional population sizes in the refuge. Using this information, they proposed that the coastal plain, where the highest shorebird densities occurred, was worthy of designation as a site of international importance under both the Ramsar Convention and Western Hemisphere Shorebird Reserve Network (see Chapter 11).

Population monitoring is integral to gaining protection of species by demonstrating that populations are low and declining. Arguably the best example of this is the population decline of the Red Knot in the New World (Niles et al. 2008). Concerted efforts to census the Nearctic subspecies (*C. c. rufa, C. c. roselarii*) on their wintering grounds and during spring migration, especially when knots stage at Delaware Bay, has produced evidence of a catastrophic population decline in just over three decades. The Red Knot population using the Atlantic Americas Flyway declined from 100,000 to 150,000 in the 1970s and 1980s to 60,000 by 1999. In the past decade, the rate of this decline has increased, such that population estimates are 18,000 to 30,000 birds. These declines have been accompanied by low productivity in high Arctic breeding areas and substantial reductions in adult survival. The main cause of these changes in population size and vital rates appears to be a reduction in the quality of food resources during the nonbreeding season, especially at Delaware Bay. Commercial harvest of horseshoe crabs has compromised the food necessary for Red Knots to increase their weights to successfully complete migration to Canadian breeding areas. The case was so dire for Red Knots that an emergency listing under the U.S. Endangered Species Act was initiated in March 2005 by several conservation groups, including various chapters of the Audubon Society, American Bird Conservancy, and Defenders of Wildlife. This petition was denied by the U.S. Fish and Wildlife Service in January 2006. However, the USFWS placed the species on the candidate list for emergency action.

The success of local management actions taken to increase the quality of habitats also can be informed by information garnered from monitoring populations. For example, biologists have used a variety of lethal and nonlethal methods to reduce the negative impacts of egg predators on shorebird productivity. In this case, the success of predator management is evaluated by an increase in hatching and fledging success. In the case of habitat management, enhanced habitat quality is indexed by an increase in the survival of individuals, presumably stemming from greater availability of food and conditions that diminish the risk of predation.

LITERATURE CITED

Anderson, R. M., and R. M. May. 1986. The invasion, persistence and spread of infectious diseases within animal and plant communities. *Philosophical Transactions of the Royal Society of London, Series B* 314: 533–570.

Atkinson, P. W., G. E. Austin, M. M. Rehfisch, H. Baker, P. Cranswick, M. Kershaw, J. Robinson, et al. 2006. Identifying declines in waterbirds: The effects of missing data, population variability and count period on the interpretation of long-term survey data. *Biological Conservation* 130: 549–559.

Badzinski, D. S. 2000. Population dynamics of Semipalmated Plovers (*Charadrius semipalmatus*) breeding at Churchill, Manitoba. M.Sc. thesis, Trent University.

Baker, A. J, P. M. González, T. Piersma, L. J. Niles, I. de L. S. do Nascimento, P. W. Atkinson, N. A. Clark, C. D. T. Minton, M. K. Peck, and G. Aarts. 2004. Rapid population decline in Red Knots: Fitness consequences of decreased refuelling rates and late arrival in Delaware Bay. *Proceedings of the Royal Society of London, Series B* 271: 875–882.

Bamford, M. 1992. The impact of predation by humans on waders in the Asia/Australasian flyway: Evidence from the recovery of bands. Stilt 20: 38–40.

Bart, J., B. Andres, S. Brown, G. Donaldson, B. Harrington, V. Johnston, S. Jones, G. Morrison, and S. Skagen. 2005. The Program for Regional and International Shorebird Monitoring (PRISM).

U.S. Department of Agriculture, Forest Service General Technical Report PSW-GTR-191: 893–901.

Bart, J., S. Brown, B. Harrington, and R. I. G. Morrison. 2007. Survey trends of North American shorebirds: Population declines or shifting distributions? *Journal of Avian Biology* 38: 73–82.

Bart, J., and S. L. Earnst. 2002. Double sampling to estimate density and population trends in birds. *The Auk* 119: 36–45.

Bent, A. C. 1927. *Life histories of North American shore birds: Order Limicolae (Part 1)*. Bulletin of the U.S. National Museum 142. Washington, DC: U.S. Government Printing Office.

———. 1929. *Life histories of North American shore birds: Order Limicolae (Part 2)*. Bulletin of the U.S. National Museum 146. Washington, DC: U.S. Government Printing Office.

Blomqvist, S., N. Holmgren, S. Åkesson, A. Hedenström, and J. Pettersson. 2002. Indirect effects of lemming cycles on sandpiper dynamics: 50 years of counts from southern Sweden. *Oecologia* 133: 1432–1439.

Brochard, C., B. Spaans, J. Prop, and T. Piersma. 2002. Use of individual color-ringing to estimate annual survival in male and female Red Knot *Calidris canutus islandica*: A progress report for 1998–2001. *Wader Study Group Bulletin* 99: 54–56.

Brown, S., J. Bart, R. B. Lanctot, J. A. Johnson, S. Kendall, D. Payer, and J. Johnson. 2007. Shorebird abundance and distribution on the coastal plain of the Arctic National Wildlife Refuge. *The Condor* 109: 1–14.

Brown, S., C. Hickey, B. Harrington, and R. Gill. 2001. *United States Shorebird Conservation Plan*. 2nd ed. Manomet, MA: Manomet Center for Conservation Sciences.

Browne, S. J., G. A. Austin, and M. M. Rehfisch. 1996. Evidence of decline in the United Kingdom's non-estuarine coastal waders. *Wader Study Group Bulletin* 80: 25–27.

Buchanan, J. B. 1999. Recent changes in the winter distribution and abundance of Rock Sandpipers in North America. *Western Birds* 30: 193–199.

———. 2003. Spartina invasion of Pacific Coast estuaries in the United States: Implications for shorebird conservation. *Wader Study Group Bulletin* 100: 47–49.

Burton, N. H. K., M. M. Rehfisch, N. A. Clark, and S. G. Dodd. 2006. Impacts of sudden winter habitat loss on the body condition and survival of redshanks *Tringa totanus*. *Journal of Applied Ecology* 43: 464–473.

Butcher, G. S., and J. D. Lowe. 1990. *Population trends of twenty species of migratory birds as revealed by Christmas Bird Counts, 1963–87*. Final Report, Cooperative Agreement No. 14-16-009-88-941. Washington, DC: U.S. Fish and Wildlife Service, Office of Migratory Bird Management.

Byrkjedal, I., and D. Thompson. 1998. *Tundra plovers: The Eurasian, Pacific and American Golden Plovers and Grey Plover*. London: T & AD Poyser.

Caughley, G., and A. Gunn. 1996. *Conservation biology in theory and practice*. Cambridge, MA: Blackwell Science.

Cayford, J. T., and R. J. Waters. 1996. Population estimates for waders Charadrii wintering in Great Britain, 1987/88–1991/92. *Biological Conservation* 77: 7–17.

Childers, T. M., and S. J. Dinsmore. 2008. Density and abundance of Mountain Plovers in northeastern Montana. *Wilson Journal of Ornithology* 120: 700–707.

Clark, J. A. 2004. Ringing recoveries confirm higher wader mortality in severe winters. *Ringing and Migration* 22: 43–50.

Colwell, M. A., and R. J. Cooper. 1993. Estimates of coastal shorebird abundance: The importance of multiple counts. *Journal of Field Ornithology* 64: 293–301.

Colwell, M. A., T. Danufsky, R. L. Mathis, and S. W. Harris. 2001. Historical changes in the abundance and distribution of the American Avocet at the northern limits of its winter range. *Western Birds* 32: 1–15.

Colwell, M. A., R. H. Gerstenberg, O. E. Williams, and M. G. Dodd. 1995. Four Marbled Godwits exceed the North American longevity record for scolopacids. *Journal of Field Ornithology* 66: 181–183.

Colwell, M. A., and L. W. Oring. 1988. Variable female mating tactics in a sex-role reversed shorebird, Wilson's Phalarope (*Phalaropus tricolor*). *National Geographic Research* 4: 426–432.

Conklin, J. R., and M. A. Colwell. 2007. Interactions of predator and manmade object causes potentially significant mortality in a winter shorebird population. *Wader Study Group Bulletin* 112: 57–59.

Delany, S., D. Scott, T. Dodman, and D. Stroud, eds. 2009. *An atlas of wader populations in Africa and western Eurasia*. Wageningen, the Netherlands: Wetlands International.

Dowding, J. E., and E. S. Kennedy. 1993. Size, age structure and morphometrics of the Shore Plover population on South East Island. *Notornis* 40: 213–222.

Elliot-Smith, E., S. M. Haig, and B. M. Powers. 2009. Data from the 2006 International Piping

Plover Census. U.S. Geological Survey Data Series 426.

Ens, B. J., K. B. Briggs, U. N. Safriel, and C. J. Smit. 1996. Life history decisions during the breeding season. In *The oystercatcher*, ed. J. D. Goss-Custard, 186–218. Oxford: Oxford University Press.

Ens, B. J., F. J. Weissing, and R. H. Drent. 1995. The despotic distribution and deferred maturity: Two sides of the same coin. *American Naturalist* 146: 625–650.

Erickson, W. P., G. D. Johnson, and D. P. Young, Jr. 2005. A summary and comparison of bird mortality from anthropogenic causes with an emphasis on collisions. *U.S. Department of Agriculture, Forest Service General Technical Report* PSW-GTR-191: 1029–1042.

Evans, P. R. 1991. Seasonal and annual patterns of mortality in migratory shorebirds: Some conservation implications. In *Bird population studies: Relevance to conservation and management*, ed. C. M. Perrins, J-D. Lebreton, and G. J. M. Hirons, 346–359. Oxford: Oxford University Press.

———. 1984. Introduction. In *Coastal waders and wildfowl in winter*, ed. P. R. Evans, J. D. Goss-Custard, and W. G. Hale, 2–5. Cambridge: Cambridge University Press.

Evans, P. R., and P. J. Dugan. 1984. Coastal birds: Numbers in relation to food resources. In *Coastal waders and wildfowl in winter*, ed. P. R. Evans, J. D. Goss-Custard, and W. G. Hale, 8–28. Cambridge: Cambridge University Press.

Evans, P. R., J. D. Goss-Custard, and W. G. Hale. 1984. *Coastal waders and wildfowl in winter*. Cambridge: Cambridge University Press.

Evans, P. R., and M. W. Pienkowski. 1984. Population dynamics of shorebirds. In *Shorebirds: Breeding behavior and populations*, ed. J. Burger, and B. L. Olla, 83–123. New York: Plenum Press.

Exo, K.-M. 1993. Höchstalter eines beringten Austernfischers (*Haematopus ostralegus*): 44 Jahre. *Die Vogelwarte* 37: 144–148.

Farmer, A. H. 2008. "Anchoring" and research priorities: Factors that depress bird population estimates? *The Auk* 125: 980–983.

Falxa, G. A. 1992. Prey choice and habitat use by Black Oystercatchers: Interactions between prey quality, habitat availability, and age of bird. Ph.D. dissertation, University of California, Davis.

Fernández, G., J. B. Buchanan, R. E. Gill, Jr., R. Lanctot, and N. Warnock. 2008. Conservation plan for the Dunlin (*Calidris alpina arcticola*, *C. a. pacifica*, *C. a. hudsonia*) with breeding populations in North America. Version 1.0. Manomet, MA: Manomet Center for Conservation Sciences.

Fernández, G., H. de la Queva, N. Warnock, and D. B. Lank. 2003. Apparent survival rates of Western Sandpipers (*Calidris mauri*) wintering in northwest Baja California, Mexico. *The Auk* 120: 55–61.

Fernández, G., P. D. O'Hara, and D. B. Lank. 2004. Tropical and subtropical Western Sandpipers (*Calidris mauri*) differ in life history strategies. *Ornitologia Neotropical* 15: 385–394.

Friend, M. 2002. Avian disease at the Salton Sea. *Hydrobiologia* 473: 293–306.

Ganter, B., and H. Boyd. 2000. A tropical volcano, high predation pressure, and the breeding biology of arctic waterbirds: A circumpolar review of breeding failure in the summer of 1992. *Arctic Biology* 53: 289–305.

Gill, J. A., K. Norris, P. M. Potts, T. G. Gunnarsson, P. W. Atkinson, and W. J. Sutherland. 2001. The buffer effect and large scale population regulation in migratory birds. *Nature* 412: 436–438.

Goss-Custard, J. D. 1980. Competition for food and interference among waders. *Ardea* 68: 31–52.

———. 1985. Foraging behaviour of wading birds and the carrying capacity of estuaries. In *Behavioural ecology*, ed. R. M. Sibly and R. H. Smith, 169–188. Oxford: Blackwell.

Goss-Custard, J. D., and M. E. Moser. 1988. Rates of change in the numbers of Dunlin, *Calidris alpina*, wintering in British estuaries in relation to the spread of *Spartina anglica*. *Journal of Applied Ecology* 25: 95–109.

Goss-Custard, J. D., R. A. Stillman, A. D. West, R. W. G. Caldow, and S. McGrorty. 2002. Carrying capacity in overwintering migratory birds. *Biological Conservation* 105: 27–41.

Gratto-Trevor, C. L. 2000. Marbled Godwit (*Limosa fedoa*). In *The birds of North America*, ed. A. Poole and F. Gill, no. 492. Philadelphia/Washington, DC: The Academy of Natural Sciences/The American Ornithologists' Union.

Gratto-Trevor, C. L., and C. M. Vacek. 2001. Longevity record and annual adult survival of Semipalmated Sandpipers. *Wilson Bulletin* 113: 348–350.

Green, G. H., J. J. D. Greenwood, and C. S. Lloyd. 1977. The influence of snow conditions on the date of breeding of wading birds in north-east Greenland. *Journal of Zoology* 183: 311–328.

Haig, S. M., C. L. Ferland, D. Amirault, F. Cuthbert, J. Dingledine, P. Goossen, A. Hecht, and N. McPhillips. 2005. The importance of complete species censuses and evidence for regional declines in Piping Plovers. *Journal of Wildlife Management* 69: 160–173.

Haig, S. M., and L. W. Oring. 1988. Distribution and dispersal in the Piping Plover. *The Auk* 105: 630–638.

Hale, W. G. 1980. *Waders*. London: Collins.

Harris, M. P. 1970. Territoriality limiting the size of the breeding population of the Oystercatcher (*Haematopus ostralegus*)—a removal experiment. *Journal of Animal Ecology* 39: 707–713.

Hitchcock, C. L., and C. Gratto-Trevor. 1997. Diagnosing a shorebird local population decline with a stage-structured population model. *Ecology* 78: 522–534.

Hockey, P. A. R. 1996. *Haematopus ostralegus* in perspective: Comparisons with other oystercatchers. In *The oystercatcher*, ed. J. D. Goss-Custard, 251–285. Oxford: Oxford University Press.

Hockey, P. A. R., and J. Cooper. 1980. Paralytic shellfish poisoning: A controlling factor in Black Oystercatcher populations? *The Ostrich* 51: 188–190.

Hockey, P. A. R., R. A. Navarro, B. Kalejta, and C. R. Velasquez. 1992. The riddle of the sands: Why are shorebird densities so high in southern estuaries? *American Naturalist* 140: 961–979.

Holland, P. K., J. E. Robson, and D. W. Yalden. 1982. The breeding biology of the Common Sandpiper *Actitis hypoleucos* in the Peak District. *Bird Study* 29: 99–110.

Holland, P. K., and D. W. Yalden. 1991. Population dynamics of Common Sandpipers *Actitis hypleucos* breeding along an upland river system. *Bird Study* 38: 151–159.

Holmes, R. T. 1966. Breeding ecology and annual cycle adaptations of the Red-backed Sandpiper *Calidris alpina* in northern Alaska. *The Condor* 68:3–46.

———. 1971. Density, habitat, and the mating system of the Western Sandpiper (*Calidris mauri*). *Oecologia* 7: 191–208.

Howe, M. A., H. Geissler, and B. A. Harrington. 1989. Population trends of North American shorebirds based on the International Shorebird Survey. *Biological Conservation* 49: 185–199.

Ims, R. A., and E. Fuegli. 2005. Trophic interaction cycles in tundra ecosystems and the impact of climate change. *BioScience* 55: 311–322.

Insley, H., W. Peach, B. Swann, and B. Etherridge. 1997. Survival rates of Redshank *Tringa totanus* wintering on the Moray Firth. *Bird Study* 44: 277–289.

Jehl, J. R., Jr. 1971. Patterns of hatching success in subarctic birds. *Ecology* 52: 169–173.

Johnson, G. D., W. P. Erickson, M. D. Strickland, M. F. Shepherd, D. A. Shepherd, and S. A. Sarappo. 2002. Collision mortality of local and migrant birds at a large-scale wind-power development on Buffalo Ridge, Minnesota. *Wildlife Society Bulletin* 30: 879–887.

Johnson, O. W., P. L. Bruner, P. M. Johnson, and A. E. Bruner. 2004. A new longevity record for the Pacific Golden-Plover. *Journal of Field Ornithology* 75: 134–135.

Johnson, O. W., and P. M. Johnson. 1983. Plumage-molt-age relationships in "over-summering" and migratory Lesser Golden-Plovers. *The Condor* 85: 406–419.

Jones, S. L., C. S. Nations, S. D. Fellows, and L. L. McDonald. 2008. Breeding abundance and distribution of Long-billed Curlews (*Numenius americanus*) in North America. *Waterbirds* 31: 1–14.

Kelly, J. R., Jr. 2004. *American Woodcock population status*. Laurel, MD: U.S. Fish and Wildlife Service.

Kew, A., S. Wakeham, and J. Clark. 1999. Species information: Section 5. In *1997–98 report: 40 years of ringing on the Wash*, ed. A. Kew, 59–77. Thetford, Norfolk, United Kingdom: Wash Wader Ringing Group.

Koivula, K., V.-M. Pakanen, A. Rönkä, and E.-J. Belda. 2008. Steep past and future population decline in an arctic wader: Dynamics and viability of Baltic Temminck's Stints *Calidris temminckii*. *Journal of Avian Biology* 39: 329–340.

Kvitek, R., and C. Bretz. 2005. Shorebird foraging behavior, diet, and abundance vary with harmful algal bloom toxin concentrations in invertebrate prey. *Marine Ecology Progress* Series 293: 303–309.

Lambeck, R. H. D., J. D. Goss-Custard, and P. Triplet. 1996. Oystercatchers and man in the coastal zone. In *The oystercatcher*, ed. J. D. Goss-Custard, 289–326. Oxford: Oxford University Press.

Lanctot, R. B., A. Hartman, L. W. Oring, and R. I. G. Morrison. 2008. Limitations of statistically derived population estimates, and suggestions for deriving national population estimates for shorebirds. *The Auk* 125: 983–985.

Larson, M. A., M. R. Ryan, and B. G. Root. 2000. Piping Plover survival in the Great Plains: An updated analysis. *Journal of Field Ornithology* 71: 721–729.

Lindén, H. 1988. Latitudinal gradients in predator-prey interactions, cyclicity and synchronism in voles and small game populations in Finland. *Oikos* 52: 341–349.

Loftin, H. 1962. A study of boreal shorebirds summering on Apalachee Bay, Florida. *Bird-Banding* 33: 21–42.

Long, P. R., S. Zefania, R. H. ffrench-Constant, and T. Székely. 2008. Estimating the population size of an endangered shorebird, the Madagascar Plover, using a habitat suitability model. *Animal Conservation* 11: 118–127.

MacDonald, M. A., and M. Bolton. 2008. Predation on wader nests in Europe. *The Ibis* 150 (Suppl.): 54–73.

Marcström, V., and J. W. Mascher. 1979. Weights and fat in lapwings and oystercatchers starved to death during a cold spell in spring. *Ornis Scandinavica* 10: 235–240.

Marks, J. S. 1992. Longevity record for the Bristle-thighed Curlew: An extension. *Journal of Field Ornithology* 63: 309–310.

Marks, J. S., and R. L. Redmond. 1996. Demography of Bristle-thighed Curlews *Numenius tahitiensis* wintering on Laysan Island. *The Ibis* 138: 438–447.

Martin, T. E. 1993. Nest predation and nest sites: New perspectives on old patterns. *BioScience* 43: 523–532.

May, R. M. 1997. Disease and the abundance and distribution of bird populations: A summary. *The Ibis* 137 (Suppl.): S85–S86.

McCaffery, B. J., and D. R. Ruthrauff. 2003. Spatial variation in shorebird nest success: Implications for inference. *Wader Study Group Bulletin* 103: 67–70.

McNeil, R., M. T. Diaz, B. Casanova, and A. Villeneuve. 1995. Trematode parasitism as a possible factor in over-summering of Greater Yellowlegs (*Tringa melanoleuca*). *Ornitologia Neotropical* 6: 57–65.

Mendes, L., T. Piersma. M. Lecoq, B. Spaans, and R. E. Ricklefs. 2005. Disease-limited distributions? Contrasts in the prevalence of avian malaria in shorebirds using marine and freshwater habitats. *Oikos* 109: 396–404.

Miller, E. H., and R. McNeil. 1988. Longevity record for the Least Sandpiper: A revision. *Journal of Field Ornithology* 59: 403–404.

Moores, N., D. Rogers, R.-H. Kim, C. Hassell, K. Gosbell, S.-A. Kim, and M.-N. Park. 2008. *The 2006–2008 Saemangeum shorebird monitoring program report.* Busan, South Korea: Birds Korea.

Morrison, R. I. G. 1975. Migration and morphometrics of European Knot and Turnstone on Ellesmere Island, Canada. *Bird-Banding* 46: 290–301.

Morrison, R. I. G., Y. Aubry, R. W. Butler, G. W. Beyersbergen, G. M. Donaldson, C. L. Gratto-Trevor, P. W. Hicklin, V. H. Johnston, and R. K. Ross. 2001. Declines in North American shorebird populations. *Wader Study Group Bulletin* 94: 34–38.

Morrison, R. I. G., C. Downes, and B. Collins. 1994. Population trends of shorebirds on fall migration in eastern Canada, 1974–1991. *Wilson Bulletin* 106: 431–447.

Morrison, R. I. G., R. E. Gill, Jr., B. A. Harrington, S. Skagen, G. W. Page, C. L. Gratto-Trevor, and S. M. Haig. 2000. Population estimates of Nearctic shorebirds. *Waterbirds* 23: 337–352.

———. 2001. *Estimates of shorebird populations in North America.* Occasional paper No. 104. Ottawa: Canadian Wildlife Service.

Morrison, R. I. G., B. J. McCaffery, R. E. Gill, Jr., S. K. Skagen, S. L. Jones, G. W. Page, C. L. Gratto-Trevor, and B. A. Andres. 2007. Population estimates of North American shorebirds, 2006. *Wader Study Group Bulletin* 111: 66–84.

Moser, M. E. 1988. Limits to the numbers of grey plovers (*Pluvialis squatarola*) wintering on British estuaries: An analysis of long-term population trends. *Journal Applied Ecology* 25: 473–485.

Mullin, S. M., M. A. Colwell, S. E. McAllister, and S. J. Dinsmore. 2010. Apparent survival and population growth of Snowy Plovers in coastal northern California. *Journal of Wildlife Management,* forthcoming.

Myers, J. P., C. T. Schick, and G. Castro. 1988. Structure in Sanderling (*Calidris alba*) populations: The magnitude of intra- and interyear dispersal during the nonbreeding season. In *Acta XIX congressus internationalis ornithologici,* ed. Henri Ouellet, 604–615. Ottawa: National Museum of Natural Science/University of Ottawa Press.

Nebel, S., D. B. Lank, P. D. O'Hara, G. Fernández, B. Haase, F. Delgado, F. A. Estela, et al. 2002. Western Sandpipers (*Calidris mauri*) during the nonbreeding season: Spatial segregation on a hemispheric scale. *The Auk* 119: 922–928.

Nebel, S., J. L. Porter, and R. T. Kingsford. 2008. Long-term trends of shorebird populations in eastern Australia and impacts of freshwater extraction. *Biological Conservation* 141: 971–980.

Nehring, S., and K.-J.Hesse. 2008. Invasive alien plants in marine protected areas: The *Spartina anglica* affair in the European Wadden Sea. *Biological Invasions* 10: 937–950.

Newton, I. 1998. *Population limitation in birds.* New York: Academic Press.

———. 2007. Weather-related mass mortality events in migrants. *The Ibis* 149: 453–467.

Niehaus, A. C., D. R. Ruthrauff, and B. J. McCaffery. 2004. Response of predators to Western Sandpiper nest exclosures. *Waterbirds* 27: 79–82.

Niles, L. J., H. P. Sitters, A. D. Dey, P. W. Atkinson, A. J. Baker, K. A. Bennett, R. Carmona, et al. 2008. *Status of the Red Knot (*Calidris canutus rufa*) in the Western Hemisphere,* ed. C. D. Marti. Studies in Avian Biology No. 36. Camarillo, CA: Cooper Ornithological Society.

O'Hara, P. D., G. Fernández, F. Becerril, H. de la Cueva, D. B. Lank. 2005. Life history varies with migratory distance in Western Sandpipers *Calidris mauri. Journal of Avian Biology* 36: 191–202.

Oring, L. W., M. A. Colwell, and R. M. Reed. 1991. Lifetime reproductive success in the Spotted Sandpiper (*Actitis macularia*): Sex differences and variance components. *Behavioral Ecology and Sociobiology* 28: 425–432.

Oring. L. W., and D. B. Lank. 1982. Sexual selection, arrival times, philopatry and site fidelity in the polyandrous Spotted Sandpiper. *Behavioral Ecology and Sociobiology* 10: 185–191.

Oring, L. W., D. B. Lank, and S. J. Maxson. 1983. Population studies of the polyandrous Spotted Sandpiper. *The Auk* 100: 272–285.

Page, G. W., F. C. Bidstrup, R. J. Ramer, and L. E. Stenzel. 1986. Distribution of wintering Snowy Plovers in California and adjacent states. *Western Birds* 17: 145–170.

Page, G. W., L. E. Stenzel, and J. E. Kjelmyr. 1999. Overview of shorebird abundance and distribution in wetlands of the Pacific coast of the contiguous United States. *The Condor* 101: 461–471.

Page, G. W., and D. F. Whitacre. 1975. Raptor predation on wintering shorebirds. *The Condor* 77:73–83.

Parr, R. 1992. The decline to extinction of a population of Golden Plover in north-east Scotland. *Ornis Scandinavica* 23: 152–158.

Patten, K., and C. O'Casey. 2007. Use of Willapa Bay, Washington, by shorebirds and waterfowl after *Spartina* control efforts. *Journal of Field Ornithology* 78: 395–400.

Peach, W. J., P. S. Thompson, and J. C. Coulson. 1994. Annual and long-term variation in the survival rates of British Lapwings *Vanellus vanellus*. *Journal of Animal Ecology* 63: 60–70.

Pearce-Higgins, J. W., D. W. Yalden, T. W. Dougall, and C. M. Beale. 2009. Does climate change explain the decline of a trans-Saharan Afro-Palaearctic migrant? *Oecologia* 159: 649–659.

Pienkowski, M. W. 1984. Breeding biology and population dynamics of Ringed Plovers *Charadrius hiaticula* in Britain and Greenland: Nest predation as a possible factor limiting distribution and timing of breeding. *Journal of Zoology* 202: 83–114.

Pienkowski, M. W., and P. R. Evans. 1985. The role of migration in the population dynamics of birds. In *Behavioural ecology: Ecological consequences of adaptive behavior*, ed. R. M. Sibly and R. H. Smith, 331–352. Oxford: Blackwell.

Piersma, T. 1997. Do global patterns of habitat use and migration strategies co-evolve with relative investments in immunocompetence due to spatial variation in parasite pressure? *Oikos* 80: 623–631.

———. 2003. "Coastal" versus "inland" shorebirds species: Interlinked fundamental dichotomies between their life and demographic histories. *Wader Study Group Bulletin* 100: 5–9.

Piersma, T., and Å. Lindström. 2004. Migrating shorebirds as integrative sentinels of global environmental change. *The Ibis* (Suppl. 1): 61–69.

Prater, A. J. 1981. *Estuary birds of Britain and Ireland.* Carlton, United Kingdom: T & AD Poyser.

Ricklefs, R. E. 1969. *An analysis of nesting mortality in birds.* Smithsonian Contributions to Zoology 9. Washington, DC: Smithsonian Institution Press.

Robbins, C. S., D. Bystrak, and P. H. Geissler. 1986. *The Breeding Bird Survey: Its first fifteen years, 1965–1979.* Resource Publication No. 18. Washington, DC: U.S. Department of the Interior, Fish and Wildlife Service.

Roberts, D. L., C. S. Elphick, and J. M. Reed. 2009. Identifying anomalous reports of putatively extinct species and why it matters. *Conservation Biology* 23: 1–10.

Rönkä, A., K. Koivula, M. Ojanen, V.-M. Paknen, M. Pohjoismäki, K. Rannikko, and P. Rauhala. 2006. Increased nest predation in a declining and threatened Temminck's Stint *Calidris temminckii* population. *The Ibis* 148: 55–65.

Šálek, M., J. Svobodová, V. Bejček, and T. Albrecht. 2004. Predation on artificial nests in relation to the numbers of small mammals in the Krušné hory Mts, the Czech Republic. *Folia Zoologica* 53: 312–318.

Sandercock, B. K. 2003. Estimation of survival rates for wader populations: A review of mark-recapture methods. *Wader Study Group Bulletin* 100: 163–174.

Sandercock, B. K., and C. L. Gratto-Trevor. 1997. Local survival in Semipalmated Sandpipers *Calidris pusilla* breeding at La Pérouse Bay, Manitoba. *The Ibis* 139: 305–312.

Sanderson, F. J., P. F. Donald, D. J. Pain, I. J. Burfield, and F. P. J. van Bommel. 2006. Long-term population declines in Afro-Palearctic migrant birds. *Biological Conservation* 131: 93–105.

Sanzenbacher, P. M., and S. M. Haig. 2001. Killdeer population trends in North America. *Journal of Field Ornithology* 72: 160–169.

Senner, S. E., and M. A. Howe. 1984. Conservation of Nearctic shorebirds. In *Shorebirds: Breeding behavior and populations*, ed. J. Burger and B. L. Olla, 379–421. New York: Plenum Press.

Shuford, D., G. W. Page, and J. E. Kjelmyr. 1998. Patterns and dynamics of shorebird use of California's Central Valley. *The Condor* 100: 227–244.

Skutch, A. F. 1949. Do tropical birds rear as many young as they can nourish? *The Ibis* 91: 430–455.

Small, R. J., V. Marcstrom, and T. Willebrand. 1993. Synchronous and nonsynchronous population fluctuations of some predators and their prey in central Sweden. *Ecography* 16: 360–364.

Staav, R. 1983. Åldersrecord för fåglar ringmärkta i Scerige. *Fauna och Flora* 78: 265–276.

Stanley, T. R., and S. K. Skagen. 2007. Estimating the breeding population of Long-billed Curlew in the United States. *Journal of Wildlife Management* 71: 2556–2564.

Stenzel, L. E., G. W. Page, J. C. Warriner, J. S. Warriner, D. E. George, C. R. Eyster, B. A. Ramer, and K. K. Neuman. 2007. Survival and natal dispersal of juvenile Snowy Plovers (*Charadrius alexandrinus*) in central coastal California. *The Auk* 124: 1023–1036.

Stenzel, L. E., J. C. Warriner, J. S. Warriner, K. S. Wilson, F. C. Bidstrup, and G. W. Page. 1994. Long-distance breeding dispersal of Snowy Plovers in western North America. *Journal of Animal Ecology* 63: 887–902.

Steen, H., N. G. Yoccoz, and R. A. Ins. 1990. Predators and small rodent cycles: An analysis of 79-year time series of small rodent population fluctuations. *Oikos* 59: 115–120.

Summers, R. W., and M. Nicoll. 2004. Geographical variation in the breeding biology of the Purple Sandpiper *Calidris maitima*. *The Ibis* 146: 303–313.

Summers, R. W., and L. G. Underhill. 1987. Factors related to breeding production of Brent Geese *Branta b. bernicla* and waders (Charadrii) on the Taimyr Peninsula. *Bird Study* 34: 161–171.

Summers, R. W., L. G. Underhill, and R. P. Prys-Jones. 1995. Why do young waders in southern Africa delay their first return migration to the breeding grounds? *Ardea* 83: 351–357.

Thompson, D. B. A., and W. G. Hale. 1989. Breeding site fidelity and natal philopatry in the Redshank *Tringa totanus*. *The Ibis* 131: 214–224.

Toms, M. P., and J. A. Clark. 1998. Bird ringing in Britain and Ireland in 1995. *Ringing and Migration* 19: 95–168.

Underhill, L. G., R. P. Prŷs-Jones, E. E. Syroechkovski, Jr., N. M. Groen, V. Karpov, H. G. Lappo, M. W. J. Van Roomen, et al. 1993. Breeding of waders (Charadrii) and Brent Geese *Branta bernicla bernicla* at Pronchishcheva Lake, northeastern Taimyr, Russia, in a peak and a decreasing lemming year. *The Ibis* 135: 277–292.

Valkama, J., E. Korpimäki, B. Arroyo, P. Beja, V. Bretagnolle, E. Bro, R. Kenward, et al. 2005. Birds of prey as limiting factors of gamebird populations in Europe: A review. *Biological Reviews* 80: 171–203.

Warnock, N., C. Elphick, and M. Rubega. 2001. Shorebirds in the marine environment. In *Biology of marine birds,* ed. E. A. Schreiber and J. Burger, 581–615. Boca Raton, FL: CRC Press.

Warnock, N., G. W. Page, and B. K. Sandercock. 1997. Local survival of Dunlin (*Calidris alpina*) wintering in California. *The Condor* 99: 906–915.

Warnock, N., and S. E. Schwarzbach. 1995. Incidental kill of Dunlin and Killdeer by strychnine. *Journal of Wildlife Diseases* 31: 566–569.

Wash Wader Ringing Group. 2008. British wader longevity records. Available at: freespace.virgin.net/holme.vale/WaderLongevity.htm (accessed May 06, 2010).

Wetlands International. 2006. *Waterbird population estimates.* 4th ed. Wageningen, the Netherlands: Wetlands International.

White, D. H., C. A. Mitchell, and T. E. Kaiser. 1983. Temporal accumulation of organochlorine pesticides in shorebirds wintering on the south Texas coast, 1979–80. *Archives of Environmental Contamination and Toxicology* 12: 241–245.

Whitfield, D. P. 2003. Predation by Eurasian sparrowhawks produces density-dependent mortality of wintering redshanks. *Journal of Animal Ecology* 72: 27–35.

Wiens, J. A. 1989. *The ecology of bird communities.* Vol. 1. Cambridge: Cambridge University Press.

Wilson, D. J., and R. G. Bromley. 2001. Functional and numerical responses of predators to cyclical lemming abundance: Effects on loss of goose nests. *Canadian Journal of Zoology* 79: 525–532.

PART THREE

Management and Conservation

11

Habitat Conservation and Management

CONTENTS

FOR CENTURIES, HUMANS HAVE sought to manipulate wildlife populations by protecting, altering, and managing habitats. In this way, we have increased or sustained the abundance of desirable species and decreased populations of others considered to be pests (Leopold 1933). From an ecological perspective, habitat management attempts to manipulate the factors that limit a population (Caughley and Gunn 1996). For example, a principal underlying assumption of efforts to protect and enhance wetlands in wintering areas is that food is in short supply. In principle, increasing the amount and quality of habitat would increase the availability of food, which has positive consequences for survival and reproduction. Alternatively, enhanced cover may act to decrease the thermal demands on birds during inclement winter weather. For shorebirds, the absence of obstructive cover enhances vigilance and allows individuals to evade predators more effectively. In either case, enhanced habitat acts to decrease the effects of limiting factors and increase productivity or survival, which positively affects population growth.

Wetlands are among the most threatened habitats worldwide, and they are some of the most productive ecosystems (Mitsch and Gosselink 2007). The principal means by which land managers alter wetland habitats to benefit shorebirds is by increasing food availability; a similar approach underlies waterfowl management. In this chapter, I address wetland conservation from several fronts. First, I review various international treaties and programs that protect and recognize wetland habitats. For conservation of migratory shorebirds, these efforts are based on a system of protected areas that conserve breeding and nonbreeding areas in a chain of sites, each of which may be critical to maintaining viable populations (Myers et al. 1987). Second, I provide an overview of manipulations of human-constructed wetlands that offer the opportunity to increase food availability to shorebirds via a program of water level manipulation that need not compromise the interests of waterfowl. Lastly, I examine shorebird use of agricultural lands, including rice paddies, fallow agricultural fields, and pastures for livestock; a special case concerns salt pans used for the commercial production of salt. Each of these habitats is widespread around the world, and they often lie adjacent to and appear to supplement the foraging of shorebirds that rely on seminatural wetlands. As such, these human-altered habitats are increasingly viewed as valuable foraging habitat for nonbreeding shorebirds, although there are potential negative aspects to shorebird use of these habitats (such as pesticides and other agricultural chemicals).

DECISION MAKING IN WILDLIFE MANAGEMENT

Any successful management approach proceeds with a well-thought out scheme of monitoring coupled with clearly identified objectives and expected outcomes (Keeney 1992; Clemen 1996; Lyons et al. 2008). In a well-organized decision-making process, there exist two types of objective. FUNDAMENTAL OBJECTIVES are the specific outcomes or performance measures that are used to evaluate success of management actions; MEANS OBJECTIVES are the steps sufficient to achieve fundamental objectives (Lyons et al. 2008). For example, a fundamental objective of conservation actions directed at endangered

and threatened species is to increase population size and reduce the risk of extinction. The various management actions (for instance, ameliorate the effects of predation on poor reproductive success, improve habitat, or reduce human disturbance) that are recommended to accomplish this goal are the means objectives. In effect, means objectives are outlined in detail in published recovery plans. In the context of management of wetland habitats for shorebirds (or any group of wildlife), fundamental objectives are often coupled with outcomes that include other taxa. For example, the notion of integrated wetland management (Laubhan and Fredrickson 1993; Fredrickson and Laubhan 1996; Erwin 2002) articulates a fundamental objective that strives to maximize the diversity of wildlife using a particular habitat or complex of wetlands. The means objectives to achieve this goal may include (1) physical alteration of wetland bathymetry to increase the diversity of microhabitat of varying water depth over which waterbirds forage; (2) manipulation of water levels directly via the delivery and removal of water through a system of water control structures; and (3) integration of these practices across a complex of wetlands that differ in size, proximity, and underlying vegetative communities.

Lyons et al. (2008) presented an excellent example of a well-structured approach to wetland management with clearly articulated fundamental and means objectives. Moreover, their research effort addressed adaptive management and the different types of uncertainty inherent in all management decisions. The project was conducted across a large spatial scale encompassing wetlands on 23 refuges of the U.S. National Wildlife Refuge system throughout the Midwest and Northeast. Lyons et al. (2008) identified a fundamental objective of their wetland manipulation experiment as maximizing the number of waterbirds using managed wetlands. In this case, the fundamental objective was a quantitative measure of performance; success was evaluated by monitoring the density of waterbirds using wetlands that were experiencing water level manipulation during the late summer and autumn. The following sections, which focus on different facets of wetland habitat management, vary in their fundamental and means objectives.

WETLAND CONSERVATION

Throughout history, humans have undertaken progressive steps to conserve wildlife, including measures to restrict hunting and control predators (Leopold 1933). A common next step in conservation is the establishment of protected areas where species are given refuge from humans to varying degrees and where management of habitats is intensified to ameliorate limiting factors. The principal mechanism by which wetland habitat can be protected for shorebirds is via international treaty, which binds signatory nations to recognize and manage habitats. At the same time, nongovernmental organizations can act to increase conservation of wetland habitats, including nonbinding programs to recognize sites of importance in the migratory cycle of shorebirds.

RAMSAR CONVENTION

The international treaty enacted in 1971 in Ramsar, Iran, protects wetlands of international significance based on unique features of ecology, botany, zoology, limnology, or hydrology. Birds, especially shorebirds, are one ecological feature that can be used to identify wetlands of significance under the convention. From the perspective of shorebirds, wetlands are often judged to be significant using criteria that measure their importance to a population. Under the Ramsar treaty, if a wetland supports ≥1% of a biogeographical (flyway) population, it is judged to be of zoological significance. In general, the Ramsar treaty seeks to recognize, protect, and enhance management of wetlands in order to facilitate the wise use of resources. In this case, one fundamental objective of the Ramsar convention is maintenance of shorebird population size, and the means objective is the recognition and protection of individual wetlands. These objectives are accompanied by an overarching anthropocentric perspective of sustainable development. In essence, implementation of the Ramsar treaty can be viewed as a multiple objective decision process that includes objectives related to both sustainable development and viable shorebird populations.

As of October 2009, there were 159 signatory nations, and these countries collectively had recognized 1,855 sites covering nearly 183,319,787 hectares. The United States became party to the Ramsar Convention in 1987, at which time it added the 610,497 hectares of Everglades National Park to the Ramsar List. Other U.S. wetlands on the list include Bolinas Lagoon, Tomales Bay, and the Grasslands Ecological Area in California; Delaware Bay along the Atlantic Coast; Cheyenne Bottoms in Kansas; and Izembek Lagoon in Alaska. As of October 2003, 24 sites encompassing 1,312,319 hectares had been recognized within the U.S. Virtually all of the Ramsar sites designated by the U.S. are protected as national parks, marine protected areas, or state and federal wildlife refuges. The significance of this is that adding a site to the Ramsar List does not necessarily afford additional protection to a wetland beyond that granted by federal or state ownership, although protection under existing federal and state programs is often substantial. Even so, designating a wetland to the Ramsar List may serve to increase public recognition of the significance of a site even though it may not add additional protective measures.

WESTERN HEMISPHERE SHOREBIRD RESERVE NETWORK

WHSRN (pronounced "whissern") is a nonbinding, international program coordinated since 1985 by the Manomet Center for Conservation Sciences (formerly the Manomet Bird Observatory). The goal of WHSRN is to increase recognition and ranking of wetland sites as shorebird staging, wintering, and breeding habitats in North, Central, and South America. As of October 2009, WHSRN had recognized 77 sites, covering about 10,000,000 hectares in

12 countries, stretching from the high Arctic tundra to Tierra del Fuego. The first WHSRN site was established in 1986 at Delaware Bay to protect spring staging areas for shorebirds moving along the Atlantic Americas Flyway from South America to breeding habitats in the Canadian Arctic. Ironically, recent evidence (Niles et al. 2008) shows that the recognition afforded Delaware Bay was insufficient as a conservation action to alleviate the dramatic population decline of Red Knots. This observation highlights one shortcoming that WHSRN shares with Ramsar: the program affords recognition for a site but does not necessarily effect conservation actions for anthropogenic factors causing the population decline. Specifically, the cause of the Red Knot population decline was the harvest of horseshoe crabs in Delaware Bay for bait, which is in turn used to harvest mollusks for human consumption. The crab harvest compromised the availability of a high-quality food resource that was used to fuel spring migration and subsequent breeding by Red Knots. Consequently, the subspecies is at risk of going extinct within the next 10 years (Baker et al. 2004; Niles et al. 2008). In the context of structured decision making, the current management approach to Red Knots involves multiple objectives comprising a sustainable crab harvest and a viable Red Knot population (Breese et al. 2007).

Similar to the Ramsar Convention, WHSRN benefits wetland conservation efforts by increasing the recognition and profile of wetland sites. In the United States, however, most if not all WHSRN sites are already protected under the system of protected areas administered by state and federal agencies. Like the Ramsar Convention, WHSRN merely recognizes sites and does little to afford additional protection to these wetlands. A useful next step in growing the WHSRN program would be to facilitate habitat management at protected sites, coupled with monitoring, to advance conservation science aimed at maintaining viable shorebird populations. Specifically, understanding how man-

agement at protected areas impacts shorebird populations will be critical in the face of anthropogenic loss and degradation of wetland habitats as well as climate change.

Much like the Ramsar Convention, the criterion used by WHSRN to recognize wetlands is based on the percentage of a flyway population that uses a site (Table 11.1); unlike Ramsar, however, there are three levels of recognition. Wetlands of *regional* importance harbor at least 20,000 shorebirds annually or 1% of any species' flyway population. *International* sites host 100,000 shorebirds or 10% of a flyway population. Lastly, sites of *hemispheric* importance support 500,000 shorebirds or 30% of a flyway population.

The use of percentages of a flyway population as a criterion for listing under both the Ramsar Convention and WHSRN is not without difficulties and challenges. For instance, the quality of shorebird survey data available to evaluate the importance of a site may be poor. Another problem concerns the population benchmarks (for instance, 1% of a population). Although they are useful, these standards suffer from at least two problems. First, linking a site's importance to the number of shorebirds (and hence the percentage of a flyway population) sidesteps meaningful interpretation of ecological data concerning bird densities. For instance, a large estuary may support large numbers of shorebirds that occur at high densities; hence, it is clearly important. But what can be said about an even larger area that supports similar numbers of birds, albeit at lower densities? As an example, consider the recent designation of the northern Sacramento Valley, which includes state and federal refuges combined with private lands dedicated to rice agriculture. This WHSRN site satisfies the criteria for total numbers and for percentage of the flyway population simply because it covers a large expanse of habitats.

A second problem with the flyway population criterion concerns the effect of long-term variation in population size and the shifting distributions of species. Specifically, local wet-

TABLE 11.1

Categories of importance and recognition under the Western Hemisphere Shorebird Reserve Network and Ramsar Convention

	WESTERN HEMISPHERE SHOREBIRD RESERVE NETWORK			RAMSAR
	REGIONAL	INTERNATIONAL	HEMISPHERIC	ZOOLOGICAL SIGNIFICANCE
Criterion				
Shorebird abundance	20,000	100,000	500,000	—
% Flyway population	1	10	30	1
EXAMPLES OF CRITICAL AREAS				
Waterbody	Ensenada de La Paz	Lagoa do Peixe	Bay of Fundy	Banc d'Arguin
Location	Baja California Sur	SE Brazil	SE Canada	Mauritania
Size (ha)	194	34,400	62,000	1,200,000
Year recognized	2008	1990	1987–88	1999
Jurisdiction	Federal	Federal	Provincial, Federal	Federal
Shorebird diversity	~25 species	27 species	23 species	~25 species
Important species	Snowy Plover, Semipalmated Plover, Marbled Godwit, Whimbrel	Hudsonian Godwit, Red Knot, Buff-breasted Sandpiper, Rufous-chested Dotterel	Semipalmated Sandpiper	~2.3 million migrants

lands may vary in their importance as regional populations fluctuate. Moser (1988) was the first to address this issue when he showed that during a period of regional population growth in the western Palearctic, the numbers of wintering Black-bellied Plover at a subset of 45 estuaries around the British Isles had leveled off while numbers at other estuaries were still increasing. Consequently, the *proportion* of the regional population at any of these preferred estuaries (those estuaries that filled to carrying capacity first) was originally large, but it was now decreasing as the population grew. Consequently, any measure of importance of wetland based on the percentage of a flyway population that uses a site is subject to change with population growth.

In summary, the Ramsar Convention and WHSRN have done much to increase recognition by the general public of the importance of wetlands to biodiversity and sustainable development. These programs have certainly increased understanding of the value of individual staging and wintering habitats to conservation of shorebird populations. Moreover, they have heightened awareness among the public as to the conservation challenges facing shorebirds and other wetland-dependent species. To the extent that this recognition increases the conservation, enhancement, and management of wetlands within a site suggests that these programs are successful. It is unclear, however, whether designation has resulted in enhanced habitat management, which has affected increases in shorebird population sizes. At the very least, these two programs have heightened the awareness of the public to the plight of shorebirds, which almost certainly facilitates other wetland conservation efforts.

CONSERVATION PLANNING AND IMPLEMENTATION

Effective conservation of highly migratory species such as shorebirds requires cooperation among governments that often are separated geographically by considerable distances. At the global scale, countries must work together to protect wetlands and other habitats necessary to sustain populations. Ideally, this is accomplished in some coordinated fashion with explicit conservation criteria used to identify, acquire, protect, and manage reserves.

PRIORITIZATION OF PROTECTED AREAS

A valuable first step in the conservation process is to evaluate the habitats (wetlands) that serve as the background of the system of protected areas. This is often accomplished by comparing the relative value of different sites based on multiple conservation objectives and goals, such as maximizing diversity or protecting a population of a rare species. As stated, these objectives are similar to the fundamental objectives of wetland management reviewed earlier. A rich literature exists regarding the most efficient approaches to sequentially selecting reserves as part of a system that maximizes protection of diversity (Pressey et al. 1993). These algorithms often approach the design of a system of protected areas using the principle of complementarity, which strives to set aside a system of protected areas that maximizes areawide diversity in the most efficient manner. Complementarity analysis begins by identifying the most diverse habitat and ranking it highest for protection. The identification and ranking of subsequent reserves for protection is based on how each successive choice adds new species and increases total diversity within the system of protected areas. By the end of the exercise, a system of protected wetlands is identified that maximizes nature protection while minimizing the costs associated with land acquisition. Alternative approaches and adjustments to criteria of complementarity analysis exist. These include maximizing protection of rare or endemic species and various methods that weigh the importance of sites based on abundance or rarity. From the perspective of decision making in wetland management, these adjustments amount to multiple objectives.

Only a few attempts evaluated and ranked a group of wetlands from the perspective of prioritizing sites for conservation of shorebirds. Two recent papers have addressed how various methods differed in ranking a collection of wetlands for protection of shorebirds. One evaluated estuaries of South Africa (Turpie 1995) and the other coastal wetlands of the Pacific Coast of the United States (Page et al. 1999). Both approaches compared complementarity analysis and several alternatives in their ranking of wetlands during the nonbreeding season.

South Africa's coastline extends more than 2,500 km and consists of rocky intertidal habitat and long stretches of sandy, ocean-fronting beach interspersed with estuaries and lagoons. Turpie (1995) collated information from published reports in regional journals on the distribution and abundance of 88 waterbird species at 42 estuaries and lagoons. She included sites if they had greater than 500 individuals of all species combined, excluding skulking groups that were difficult to survey. Shorebirds comprised approximately one-third of the species, and many were Palearctic breeders that wintered in large numbers in the region during the austral summer. By contrast, many of the local resident species were less abundant. Turpie (1995) evaluated the relative merits of five different scoring systems for ranking the estuaries, including two measures of diversity (species richness, and H'=Shannon-Wiener index), endemism, species' abundances, the conservation status of individual species (based on the IUCN Red List), and complementarity analysis. The results derived from different criteria were remarkably similar in ranking estuaries. A consistent group of 10 sites were identified as necessary to affect the conservation of waterbird diversity. Turpie discussed one shortcoming of the complementarity analysis—that it treats species as present or absent and ignores abundances. This becomes important because sites may rank high in complementarity analysis owing to the presence of a small number of individuals of a species. Other sites with larger numbers of species may be ignored when they have lower diversity. Turpie suggested that complementarity analysis could be amended to include abundance criteria that would place greater emphasis on viable populations. Turpie's analysis was confined to the boundaries of South Africa, and it is important to note that such constraints do not address natural variation in the abundance of species across their range. Specifically, some species in the analysis of South African wetlands may be quite abundant elsewhere on the continent, which would alter the rankings of estuaries based on regionally rare species.

A similar approach to evaluating the importance of estuaries along the Pacific Coast of North America was presented by Page et al. (1999). They summarized the seasonal abundance of 21 shorebirds at 56 estuaries, compiled as part of the Pacific Flyway Project. Next, they determined the minimum number of sites necessary to affect conservation using five different criteria based on various measures of diversity, including complementarity analysis as modified by Turpie (1995). Their findings differed dramatically based on five selection criteria. Notably, complementarity analysis showed that San Francisco Bay alone was sufficient to protect populations of 13 focal species, although the percentages of the flyway populations protected were quite low. By contrast, other selection

methods based on flyway population criteria (see the 1% rule of Ramsar and the Hemispheric designation under WHSRN) protected up to 38 estuaries and a much greater percentage of flyway populations. These findings indicate that algorithms such as complementarity analysis that are designed to maximize biodiversity are unlikely to be a successful approach to conservation of estuaries and wetlands necessary to maintain the migration systems of shorebirds. This is especially true considering that virtually all of the species studied by Page et al. (1999) rely on multiple wetlands along the Pacific Coast as necessary staging sites during migration. Protecting just a subset or one site (like San Francisco Bay) would almost certainly be detrimental to maintaining connectivity for migrants. This latter point was emphasized by Myers et al. (1987). These various attempts to prioritize shorebird habitat emphasize the need for a new model for conservation of migratory species based on reserve design for important habitats and connectivity of sites. In particular, conservation of migration habitat must recognize that the spatial arrangement and availability of resources (food) during migration provide unique opportunities and constraints in every flyway. Moreover, habitats in coastal regions are likely to differ greatly in the predictability of food resources compared with sites in the continental interior where the vagaries of weather create more ephemeral wetland habitats.

LANDSCAPE MANAGEMENT

On a regional scale, conservation planning may also benefit from clear articulation of fundamental and means objectives. In the San Francisco Bay, large areas of commercial salt ponds have been a part of the wetland habitat fringing the bay for well over a century. Salt ponds were recently sold to the U.S. government with the objective of restoring natural habitats. These salt ponds have value as overwintering and staging habitats for shorebirds; a few species also breed in salt pans or on levees and islands amid ponds. The area occupied by salt ponds, however, has the potential to enhance salt marsh habitats that have been lost and degraded elsewhere around the bay by development. Stralberg et al. (2009) used empirical data on waterbird abundance and habitat use in an integer programming exercise to understand the consequences of varying degrees of salt pond conversion to salt marsh for sensitive species and waterbird diversity. They determined that shallowly managed salt ponds provided the greatest benefit for the greatest diversity of bird species. However, maximizing abundance of sensitive salt marsh species (for example, the Salt Marsh Harvest Mouse, *Reithrodontomys raviventris*; and the California Clapper Rail, *Rallus longirostris obsoletus*) dictated conversion to greater than 50% salt marsh. The spatial context of various habitat types did not appear to influence the objective of maximizing diversity, but the optimal configurations were those with salt marsh and shallow, low-salinity ponds nearest to intertidal habitats with high-salinity ponds situated farther inland, which were of greatest benefit to waterbirds. Interestingly, Stralberg et al. (2009) suggested that the expert opinion, rather than extensive fieldwork, could suffice for empirical data in future attempts to use integer programming to solve conservation solutions.

WETLAND MANAGEMENT

A PRIMER IN WETLAND ECOLOGY

A general definition of a wetland is that it is a habitat where hydric soils and hydrophytic plants predominate under a hydrologic regime that promotes the delivery and persistence of water over varying periods (Mitsch and Gosselink 2007). As such, wetlands are transitional habitats between terrestrial and aquatic ecosystems, habitat in which water is at or near the surface such that it influences the soils, flora, and fauna during the growing season (Batzer and Sharitz 2007). Wetland soils are flooded or saturated with water for a minimum of 7 to 10 days annually, and often year round. As a

result, anaerobic conditions persist in the upper soil layers, and wetland species (such as sedges, cattails, and rushes) dominate the plant communities.

From the perspective of shorebirds, this simple definition does not acknowledge several significant features of wetlands that account for why shorebirds vary in their use of wetland habitats. First, wetlands offer rich food resources in the form of invertebrates that are readily available to shorebirds. This stems from the flow of energy from plants into herbivores and detritivores that consume autotrophs. As indicated in Chapter 5, shorebirds rely heavily on macroinvertebrates to sustain themselves during the winter, fuel their migrations, and provide energy and nutrients for breeding.

Classification systems (Stewart and Kantrud 1971; Cowardin et al. 1979) recognize a variety of wetland types, including marine, estuarine, lacustrine, and palustrine. Not surprisingly, shorebirds collectively use virtually all of these wetland types, although some are clearly more important than others to individual species. For example, ocean-fronting sandy beaches constitute an important component of the marine system used by wintering Sanderlings and migrant Red Knots. Worldwide, estuaries provide significant intertidal foraging habitats for large concentrations of shorebirds; these habitats are influenced to varying extents by tidal regimes and salinities of nearby ocean habitats. Throughout continental interiors, shorebirds frequent lacustrine wetlands, especially those with shallow water allowing birds access to invertebrates. Finally, palustrine habitats are used to a limited extent by some shorebirds, notably the Spotted Sandpiper, Wrybill, and Ibisbill.

Much of the management that manipulates wetland habitats for shorebirds and other waterbirds is directed at improving the quantity and availability of these invertebrates. To this end, ecomorphological relationships (the observation that interspecific variation in the depth of water in which shorebirds forage is constrained by hindlimb morphology) are the foundation of management activities that provide shallowly flooded habitats and gentle contours to the bathymetry of wetlands. For most shorebirds, wetland habitats that are most attractive consist of a mix of exposed substrates with sparse vegetation, either desiccating to varying degrees or flooded by shallow water such that individuals can wade and feed on benthic invertebrates, nekton, or plankton in the water column. A second consideration for the design and management of wetland habitats for shorebirds is to provide open, unobstructed views such that birds may effectively detect predators.

SHOREBIRD USE OF MANAGED WETLANDS

In North America, a long and rich tradition of wetland management has focused on creating and manipulating managed wetlands to benefit waterfowl. The original impetus behind wetland management was the improvement of habitats for wintering waterfowl, which was directly linked to a tradition of waterfowl hunting. This link between waterfowl populations and their wetland habitats was formalized in the mid-1980s, when biologists recognized that many populations of North American waterfowl were in decline. In an attempt to reverse these population trends, the Canadian and U.S. governments signed an agreement—the North American Waterfowl Management Plan (NAWMP)—to affect management that would recover populations of ducks, geese, and swans; Mexico entered the program in 1994. It was widely recognized that the decline of waterfowl populations was associated with a dramatic decrease in wetland habitats. To achieve the recovery of waterfowl populations (a fundamental objective of NAWMP), the U.S. Congress enacted the North American Wetland Management Act (NAWCA), which provides funding to create and enhance wetland habitats for waterfowl (a means objective). In most cases, however, the design and implementation of NAWCA projects is undertaken with broader goals of enhancing conservation for a broader array of wildlife, including shorebirds.

The underlying ecological premise of wetland management directed at shorebirds has its foundation in two bodies of empirical data. The first is the ecomorphological relationship between foraging habitat use and leg length. Specifically, a positive correlation exists between hindlimb length and foraging water depth across a wide range of species (e.g., Baker 1979; Ntiamoa-Baidu et al. 1998; Isola et al. 2000; Colwell and Taft 2000; Collazo et al. 2002). Consequently, food availability is constrained by water depth, especially when wetlands are flooded deeply. Even the largest shorebirds, however, are constrained to forage in relatively shallow water, compared with other waterbirds such as wading birds (herons and egrets) and waterfowl (Isola et al. 2000). So wetland design considerations necessitate providing sufficient habitat to accommodate shorebirds by providing moist or shallowly flooded habitats measuring only a few centimeters in depth (Fredrickson 1988; Helmers 1992; Taft et al. 2002). The second body of ecological data relevant to wetland management concerns the notion that food is in short supply during the nonbreeding season such that providing more of it by manipulating wetlands will result in increased survival and productivity of individuals. I reviewed evidence for this subject in the context of carrying capacity in Chapter 7.

There are two approaches to increasing the availability of foraging habitats for shorebirds on managed wetlands. The first involves the design of wetlands with varying bathymetry such that mudflats, shallow-water habitats, and deep-water habitats are available for foraging shorebirds and other waterbirds. A diverse bathymetry can provide wading or mudflat habitats across a wide range of water depths and flooding. The second approach involves the manipulation of water levels, including the depth of flooding, the seasonal timing of drawdowns, and the rate at which water is removed from a wetland. Fredrickson (1988) provides details on water level manipulation to achieve increased wetland use by various waterbird guilds under varying growing seasons and water regimes. Harrington (2003) summarizes approaches for shorebirds specifically.

MOIST-SOIL MANAGEMENT

Many state and federal refuges, as well as private lands such as waterfowl hunting clubs, have a system for delivering water that consists of canals and water control structures. This system allows managers to manipulate the timing of water application and dewatering as well as the depth of water covering a wetland. The management of these wetlands has been motivated primarily by interest in improving the foraging habitat for waterfowl during the nonbreeding season. To this end, moist-soil management practices are commonly employed to produce food plants that nourish overwintering waterfowl. Moist-soil management practices vary the timing of wetland drawdown to germinate specific waterfowl food plants such as swamp timothy (*Crypsis schoenoides*), watergrass (*Echinochloa crusgalli*), and smartweed (*Polygonum* spp.). Plants are irrigated by a series of water applications during the spring and summer, after which wetlands are flooded in late summer and autumn to provide these food plants to waterfowl. A secondary benefit of the fall flood-up is that decomposing material from emergent plants supports a diverse and abundant community of detritivores. These macroinvertebrates are a rich source of fat and protein to female waterbirds during late winter and early spring in anticipation of breeding. Shorebirds certainly benefit from these practices, but the availability of food may be constrained by the depth to which managers flood wetlands. Consequently, only minor changes to wetland management practices directed at waterfowl may benefit shorebirds. The following treatment of wetland management considers shorebird responses together with other waterbirds. The reason for this is that managers of wetlands rarely seek to manipulate habitats for only one species or guild (a group of species that feed on a common resource in a similar manner, such as shorebirds that wade in shallow water and probe for invertebrate prey). Rather, management objectives often are stated so as to maximize the use of wetlands by the greatest diversity of wildlife, which is synonymous with integrated wetland management (Laubhan and Fredrickson 1993).

Moist-soil management practices involve the seasonal flooding of wetland habitats and subsequent manipulations of water levels later in the nonbreeding season. In a typical moist-soil managed wetland, water is delivered via a system of canals and water control structures and applied to habitats sometime during the late summer or autumn. The timing of flooding often precedes waterfowl hunting seasons, so the benefits to southbound shorebird migrants are mistimed. For instance, the passage of many Arctic-breeding shorebirds through North America occurs from July through September, before the typical time when water managers flood moist-soil wetlands. However, shorebirds that overwinter in northern regions may benefit from these manipulations.

Early prescriptions for management of seasonal wetlands were quite general. Eldridge (1992) suggested that managers of wetlands in the midcontinent of North America could maximize the availability of invertebrate food to southbound migrants by a simple system combining water depths and wetland topography. Specifically, partially dewatering a wetland to provide a combination of open mudflat and shallow (2 to 5 cm) water in a wetland with gradually sloping sides and very little vegetation would make food most available to shorebirds. The production of invertebrates, however, relied on providing decomposing plant material for invertebrates, principally benthic chironomid larvae. Fredrickson (1988) offered general prescriptions for waterbirds across North America. Helmers (1992) provided a first treatment of wetland management specifically for shorebirds, using case studies from across the United States with an emphasis on water level manipulation in moist-soil managed wetlands. He provided regional details on the timing of wetland drawdowns (the dewatering of wetlands) to maximize the availability of mudflat and shallow water habitats to the greatest number of migrants.

Shorebirds are abundant in wetlands of the Central Valley of California for much of the year. The presence of moist-soil managed wetlands on state and federal refuges, as well as in the many private duck clubs, offers the opportunity to provide shorebirds with food and habitat that would otherwise be absent or of low quality. This is especially the case in the late summer and fall, when high temperatures have desiccated all natural wetlands. The availability of a reliable water source delivered via the Central Valley Project has enabled land managers to flood wetlands at times of year when the landscape normally would be dry. Water is applied before the opening of waterfowl hunting season, and it remains through the hunting season and into the spring.

Water is critical to the management of moist-soil wetlands in California, both in its influence on the plant communities and via depth on the availability of food to waterbirds. Beginning in the spring, water is removed from a wetland to stimulate the germination of moist-soil plants, such as smartweed, swamp timothy, and watergrass. Typically, a wetland is dewatered rapidly by removing all boards from a stop-log water control structure. This produces a rather uniform germination of desirable waterfowl food plants, depending on the bottom topography of a wetland. The timing of drawdowns determines the plant community and dominant species. In The Grasslands of the northern San Joaquin Valley, California, a series of four drawdown intervals spans late winter and early spring. Early drawdowns (January) stimulate germination of grasses and forbs and produce pasturelike habitats that are used to graze livestock. Successively later drawdowns yield smartweed (February to March), swamp timothy (March to April), and watergrass (April). During summer, sporadic irrigations are applied to wetlands to foster plant growth.

In late summer and autumn, moist-soil wetlands are flooded to provide habitat for waterfowl. During the winter and spring, the timing of flooding, depth of water, and speed and timing of subsequent drawdowns can have strong impacts on waterbird community composition. In The Grasslands, water depth was the strongest predictor of waterbird diversity and densities of most shorebirds, wading birds, and waterfowl (Colwell and Taft 2000). Highest species richness occurred in wetlands of 10 to 20 cm where underlying wetland topography provided additional variation to depth that could accommodate species of varying morphology and foraging behaviors. The only species that decreased in density with shallow water conditions were several diving ducks. This study was conducted during the winter when water flooded much of the landscape to varying depths but often deeply. Consequently, the response of various species to water depth may have been strong because shallow water conditions were rare on the landscape. Support for this was provided by Taft et al. (2002), who experimentally manipulated water depths in the same study area and detected different responses of waterbirds to water depth during winter (when mudflats were rare habitats) and spring (when deep wetlands were rare).

The results of correlative and experimental research examining the relationship between waterbird diversity and densities and water depth have yielded the following wetland management prescriptions (Taft et al. 2002). First, wetlands flooded with average water depths of 10 to 20 cm appear to support the greatest diversity of waterbirds. Shorebirds in particular responded positively to these shallow water conditions, and densities of most species of waterfowl (but not diving ducks, coots, and grebes) were unaffected by the shallow water conditions. Second, the greater the variation in water depth within a wetland, the greater the diversity of waterbirds. Habitat diversity depends on the underlying topography of a wetland (Fredrickson and Taylor 1982; Fredrickson 1988; Skagen and Knopf 1994; Colwell and Taft 2000). Conversely, greater topographic variation (varying bathymetry) of a wetland will increase the range of depths over which multiple habitat will be available to birds (Taft et al. 2002).

The success of these prescriptions will vary regionally in association with the availability of water and seasonal variation in waterbird assemblages that are present in an area during the nonbreeding season. For example, the Playa Lakes region of the southern Great Plains of North America hosts a diverse shorebird assemblage during migration. Few species, however, winter in this region. Accordingly, moist-soil management techniques are most likely to benefit shorebirds during the migration windows of spring, summer, and early autumn (Davis and Smith 1998).

Managed impoundments that go dry may be attractive to breeding shorebirds, especially those that nest in open, unvegetated habitats. Paton and Bachman (1997) discussed methods of enhancing nesting substrates for several shorebirds by supplementing gravel onto the barren salt flats of impoundments that lay adjacent to Great Salt Lake, Utah. Snowy Plovers used these enhanced habitats for courtship (as evidenced by scrapes) and a few nested on the gravel. Killdeer and American Avocets also nested on a few of the enhanced substrates.

AGRICULTURAL LANDS

Shorebirds use a diversity of agricultural lands, especially as foraging habitats during the nonbreeding season. However, there is rather limited use of these habitats during the breeding season, probably because the breeding distributions of so many shorebirds occur in remote northern latitudes. The attractiveness of agricultural lands to nonbreeding shorebirds is a common observation. Shorebirds prefer habitats with abundant food resources, especially macroinvertebrates, and openness that allows easy detection of predators. The simplified habitat features of many croplands suit these needs well. The following is a summary of the most important facets of agricultural land use practices as they relate to shorebirds. A detailed treatment of bird use of intensively farmed agricultural lands can be found in Taft and Elphick (2007).

RICE

Domesticated rice (*Oryza sativa*) is arguably the most important food for humans. Worldwide, rice provides the major source of calories for more than 50% of the world's population. It constitutes 30% of the world's grain supply as the second most consumed grain (behind *Maize*), and each year more that 145 million hectares are cultivated (U.S. Fish and Wildlife Service 2008). The world's principal rice-growing regions occur in areas formerly covered by extensive wetland. The conversion of these natural habitats to rice has promoted extensive use of these agricultural lands by large populations of waterbirds adhering to traditional flyways and wintering habitats (Blanco, et al. 2006; Taft and Elphick 2007). The most important regions for growing rice occur in temperate and tropical lowlands, where seasonal weather patterns (such as monsoonal rains) or pumped groundwater provide the abundant resource necessary to grow rice. Rice cultivation is especially prevalent in eastern Asia (especially China, Japan, and Korea), Southeast Asia, India, South America, Africa, and Australia. In Europe, rice is grown in areas bordering the Mediterranean Sea, such as northwestern Italy, the Rhone Delta (Camargue) of France, and the Ebro Delta of Spain. In other temperate regions, the availability of delivered water facilitates rice farming in drier climates. In the United States, for example, major areas of rice cultivation exist in the Mississippi Alluvial Valley, across coastal areas of the southern states bordering the Gulf of Mexico, and in the Sacramento Valley of California. In these temperate regions, rice agriculture is more energy-intensive than elsewhere around the world owing to mechanized cultivation and harvest.

Flooded rice fields offer attractive habitats for waterbirds over much of the year. During the growing season, rice provides foraging habitat for some shorebirds that nest in surrounding habitats. For example, in Europe, Pied Avocets, Black-tailed Godwits, and Black-winged Stilts occasionally nest in rice fields in northwestern Italy (Fasola and Ruiz 1997). In North America, American Avocets, Black-necked Stilts, and Killdeer are common shorebirds in rice fields during the breeding season (Shuford et al. 1998). In Japan, the Little Ringed Plover occurs at low densities in rice fields during the breeding season (Maeda 2001). In South America, a diverse community of Nearctic-breeding shorebirds uses rice fields during the austral summer (Blanco et al. 2006).

The importance of rice to shorebirds is much more pronounced during the nonbreeding season. In the northern hemisphere, large concentrations of shorebirds frequent rice fields during the late summer through early spring, especially when fields are flooded. In California's Sacramento Valley, most species of shorebird occur predominantly in flooded rice fields (Day and Colwell 1998; Shuford et al. 1998; Eadie et al. 2008). Along the Gulf Coast of the United States, nearly a quarter of a million shorebirds wintered in a 3,522 km² area of rice agriculture in densities of approximately 64 birds per km² (Remsen et al. 1991). High densities of foraging shorebirds result from abundant and available food coupled with open habitats, which reduces the risk of danger from predators (Elphick and Oring 1998).

In the Sacramento Valley, methods of harvest and the means by which farmers dispose of rice straw strongly influence patterns of use by a diverse assemblage of waterbirds (Day and Colwell 1998; Eadie et al. 2008). There are two methods of harvesting rice. Most farmers harvest rice using a method that cuts the rice plant low to the ground and produces significant waste grain. In an attempt to decrease the latter, some farmers have switched to a method that "strips" fields using a vibrating cylinder to remove the panicle from the rice plant; this process leaves taller, standing vegetation in the field. These harvest methods leave different amounts (350 versus 300 lbs per acre, respectively) of waste grain in the field and differ in the height of the remaining rice stalk. Stripped fields had

lower waterbird diversity and fewer shorebird species than the traditional method of harvesting rice (Day and Colwell 1998). Lower use by shorebirds may stem from the obstructive nature of the taller vegetation left in stripped fields; birds may avoid these habitats owing to the reduced ability to detect predators.

The postharvest treatment of rice straw has a more significant impact on shorebirds, and recent changes appear to have increased the attractiveness of rice fields to nonbreeding shorebirds. In California, legislation passed in 1991 dramatically reduced the once-common practice of burning to dispose of rice straw. This legislation was enacted to improve air quality, as most farmers burned straw, which amount to 140,000 to 180,000 hectares annually (Elphick 2004). As a result, burning caused substantial pollution during the autumn. Subsequently, farmers have used various methods to dispose of straw, including baling and removing it or incorporating it into the soil using various mechanical methods (disking and rolling). Importantly, in most cases, water is applied to fields to enhance straw decomposition, and this increases the availability of macroinvertebrates to shorebirds. The application of water, especially at shallow depths, clearly enhances the attractiveness of rice fields to waterbirds, including several species of shorebird (Day and Colwell 1998). Although waterbird diversity increased in fields categorized as dry, puddled, and flooded, the maximum diversity of shorebird use occurred in shallowly flooded fields. There was, however, no clear relationship between the various methods of rice straw treatment and waterbird diversity. Lastly, because rice paddies are uniform in topography and very gently sloped to facilitate efficient delivery and removal of water, the habitats created by flooding are quite uniform in depth within a field. Consequently, there is a predictable relationship between degrees of flooding and the use by species that comprise waterbird communities (Day and Colwell 1998).

Beyond flooding, it is difficult to say whether the varying methods of postharvest straw disposal have consequences for wintering water-birds (Day and Colwell 1998). Conventionally harvested fields are structurally simpler, with shorter vegetation and more waste grain compared with fields harvested using a vibrating cylinder to strip rice from the head of the plant. Consequently, bird communities in stripped fields are depauperate compared with those in conventionally harvested fields.

Several studies have compared the habitat quality of rice fields with seminatural wetlands to address the functional equivalency of these habitats. Rice fields are attractive habitats for waterbirds because they offer abundant food, which is available because of the shallow water. Flooded rice fields may also be attractive to waterbirds owing to their openness, which facilitates vigilance and early detection of predators. Elphick and Oring (1998) showed that predation threat was lower in rice fields compared with seminatural wetlands. Combined with food availability and behavior, these findings suggest that rice fields are functionally equivalent to seminatural wetlands (Elphick 2000).

In conclusion, rice agriculture has the potential to provide significant wetland habitats that may be functionally equivalent to natural wetlands. The value of rice to shorebirds is clearly linked to practices that flood fields shallowly to decompose straw. As a result, the availability of macroinvertebrates for nonbreeding shorebirds is enhanced. A secondary benefit to shorebirds of rice agriculture is that the openness of fields appears to reduce the danger posed by predators, making rice fields even more attractive as foraging habitats. An added benefit of waste grain and flooding is that it supports a strong winter economy based on waterfowl hunting. It is possible, however, that some rice farming practices pose significant threats to wildlife, including shorebirds. These include the abundant use of agrochemicals and pesticides to increase yield. It is noteworthy that postharvest methods of incorporating rice straw into the soil followed by flooding appear to decrease the need for some fertilizers (Bird et al. 2002).

ROW CROPS

Various row crops (corn, wheat, and potatoes, for example) are cultivated and harvested in a manner such that the fallow fields provide potential foraging habitat for shorebirds (Taft and Elphick 2007). Much like postharvest rice fields, these agricultural lands are attractive to foraging shorebirds because of their openness and availability of invertebrate prey. In many coastal areas of the world, estuarine habitats have been converted to farmland. This is especially the case in the deltas of major rivers of the world (such as the Tigris and Euphrates, the Rhine, and the Ganges) where seasonal flooding and river-borne sediments have produced fertile soils conducive to agriculture.

Important breeding habitats for shorebirds also exist in coastal lowlands of northern Europe. In Iceland, for instance, where significant percentages of the world's population of several species (for instance, 52% of Eurasian Golden-Plovers, 46% of Purple Sandpipers, and 40% of Whimbrels) breed, many shorebirds breed in coastal lowlands (Gunnarsson et al. 2006). It is difficult to know how the conversion of these habitats has impacted shorebirds because population monitoring has only been in place for a few decades. There must have been some negative impact, however, as loss of these estuarine habitats probably resulted in a decrease in intertidal foraging areas. Moreover, the conversion of natural habitats to agriculture undoubtedly affected the breeding habitats of native birds. Recent evidence strongly suggests that agricultural intensification in Europe over the past few decades has resulted in significant declines in bird populations (Donald et al. 2001). In some instances, however, the arable lands that have replaced coastal meadows and wetlands have provided significant feeding habitats for shorebirds during the nonbreeding season. It is possible that, for some species, these habitats offer better foraging areas such that local populations have increased where pastures and arable lands lie adjacent to estuaries. Bignall and McCracken (1996) offer a review of the importance of agricultural lands to wildlife with some examples of shorebirds.

Growing evidence indicates that agricultural lands provide important habitats for shorebirds during the nonbreeding season. Among the earliest to demonstrate this was Townshend's (1981) study of Eurasian Curlews. Smaller-bodied male curlews were more likely to be found in pastures and arable lands during the winter where they foraged on earthworms and other invertebrates. During winter, female curlews often defended feeding territories in intertidal habitats, which limited the access of males to food. Consequently, it is possible that cultivated lands enhanced the availability of food and increased the wintering population size of curlews. Similar arguments may apply to Long-billed Curlews wintering in coastal northern California (Leeman and Colwell 2005). In California, Mountain Plovers overwinter almost exclusively in agricultural fields. Elsewhere in Britain, Eurasian Golden-Plovers and Northern Lapwings are regularly observed in agricultural lands. Adjacent to The Wash in Norfolk, England, golden-plovers and lapwings selected arable lands where they foraged extensively (Gillings et al. 2007).

The Dunlin is an abundant wintering shorebird along the northern Pacific Coast of North America. From the Fraser River Delta in British Columbia, south through California, several studies have shown that agricultural lands are important to Dunlin populations. For example, large numbers of Dunlin move on a regular, seasonal basis between intertidal habitats and agricultural lands that may be more than 100 km from principal wintering habitats associated with coastal estuaries. For example, radio-marked Dunlin wintering on Bolinas Lagoon in northern California moved between the coast and interior agricultural fields coincident with seasonal rains (Warnock et al. 1995). Similar, albeit short-distance, movements have been reported for Dunlin wintering elsewhere along the Pacific Coast. For example, telemetry studies showed that Dunlin use of agricultural habitats increased during prolonged periods of winter rain (Shepherd and Lank 2004; Conklin and Colwell 2007; Conklin et al. 2008). Dunlin wintering in the Fraser River Delta moved regularly between their main foraging areas in intertidal habitats and nearby agricultural lands, which consisted of fallow row crops and pastures (Shepherd and Lank 2004). Shorebirds using fallow fields probably feed mostly on terrestrial invertebrates (such as lumbricidae and various arthropods such as beetles, ants, and spiders) associated with wet substrates (Taft and Elphick 2007). Isotopic analysis of blood showed that, on average, 38% of Dunlin diets were composed of food acquired in agricultural fields (Evans Odgen et al. 2006). Some individuals, however, obtained as much as 98% of their diet from nontidal habitats of the Fraser River Delta. Moreover, the use of agricultural habitats peaked during midwinter, when cold, rainy weather created the most energetically demanding conditions for overwintering birds. This suggests that some Dunlins supplement their diets during energetically stressful periods by foraging in agricultural lands. The Dunlin is probably the most abundant wintering shorebird along the northern Pacific Coast. Given that observers have documented extensive use of agricultural lands by individuals (Shepherd and Lank 2004; Conklin et al. 2008) and populations (Warnock et al. 1995; Long and Ralph 2001), especially during winter, it is reasonable to assume that agricultural lands are a critical element of the mosaic of habitats used and required by local populations of wintering Dunlin (Taft and Haig 2006).

PASTURES

Meadows and pastures offer important breeding habitats for shorebirds in some regions of the world. In Great Britain, Common Redshanks and Northern Lapwings are strongly associated with pastures and coastal meadows (Stillman and Brown 1998; Smart et al. 2006), and there is evidence that these habitats are significant at this time of the annual cycle. Eurasian Golden-Plovers have declined in population throughout their temperate breeding range. Pearce-Higgins and Yalden (2003) studied the use of pastures and arable lands by off-duty plovers that were incubating clutches. When not tending eggs, adults of both sexes spent appreciable time feeding in short-vegetation swards, often commuting several kilometers from nests to reach these habitats. Elsewhere in Europe, coastal meadows provide significant breeding areas for some waders (Milsom et al. 2002; Ottvall and Smith 2006), although the extent of this habitat is decreasing. Rönkä et al. (2006) suggested that a contributing factor in the precipitous decline of Temminck's Stint in the Baltic coastal region was degradation of coastal meadows owing to cessation of grazing and hay-making. In the interior of North America, pastures and rangeland may provide significant habitat for several species, including the Long-billed Curlew, Marbled Godwit, Killdeer, Wilson's Phalarope, Willet, and Upland Sandpiper.

The success of breeding shorebirds in agricultural landscapes may be influenced by the land-use practices. For instance, the harvest of hay in coastal meadows of Europe may contribute to the mortality for eggs and chicks. Recognition of these potentially negative impacts of agriculture on breeding birds has led to the formulation of "meadow bird agreements." In the Netherlands, agreements seek to ameliorate the negative effects of agriculture by prohibiting any agricultural activities on fields between April 1 and sometime in June or July. For instance, delaying the timing of haying may increase the productivity of meadow-breeding species such as the Black-tailed Godwit, and reducing grazing densities may decrease the number of nests trampled by livestock. Some level of grazing, however, seems to benefit shorebird use of coastal meadows; birds appear to breed in habitats with the shorter vegetation structures that result from grazing (Ottvall and Smith 2006), although the reduction in cover at nests may contribute to lowered nest survival from predation.

Other negative impacts of humans on breeding productivity arise from subtle effects of

structures positioned in the landscape such that predators are more likely to find shorebird nests. Wallander et al. (2006) examined the effects of man-made structures on the location and success of shorebird nests in coastal meadows in Sweden. They found no evidence that nest success differed in proximity to structures that provided perches for nest predators, but shorebirds avoided nesting near these structures. Consequently, structures effectively reduced the amount of suitable nesting habitat. Maintaining water in coastal meadows appears to be associated with higher densities and productivity of several species (Milsom et al. 2002; Smart et al. 2006). Early evidence suggests that the meadow bird agreements are not having the desired effect on populations (Kleijn and van Zuijlen 2004).

During the nonbreeding season, large numbers of shorebirds forage and roost in agricultural landscapes worldwide. The importance of pastures and arable lands to nonbreeding shorebirds can be gauged in a variety of ways. First, importance can be gauged by a species' abundance or the proportion of a population that use a habitat. Second, importance may be evaluated based on the behaviors of individuals. Lastly, time-activity budgets and intake rates may indicate how much of the daily energy budget is acquired from agricultural lands. Several examples illustrate the value of coastal pastures to shorebirds.

In some coastal regions, large numbers of shorebirds use agricultural lands, especially near the time of high tide when intertidal foraging areas are inundated, and during the winter when precipitation increases the availability of terrestrial invertebrates. For instance, around Humboldt Bay, use of agricultural lands was highest during midtide and high-tide intervals for the Dunlin, Western Sandpiper, Marbled Godwit, and Long-billed Curlew (Long and Ralph 2001). This suggests that arable lands are used as supplemental feeding areas by some species. Support for this comes from the observation that approximately 75% of observations of waterbirds (mostly shorebirds) using coastal pastures near Humboldt Bay were of foraging birds (Colwell and Dodd 1995). In the Fraser River Delta, Dunlins preferred intertidal habitats over agricultural lands. But 80% of individuals in a sample of radio-marked birds used agricultural lands, especially during high tide and at night (Shepherd and Lank 2004). Nocturnal use of pastures by wintering Dunlins occurs elsewhere along the Pacific Coast (Conklin and Colwell 2007).

In some cases, males and females of a species use agricultural habitats differently. Eurasian Curlews wintering in England forage in both intertidal habitats and nearby arable lands (Townshend 1981). In winter, females predominate in intertidal habitats, and males tend to make greater use of arable lands. These patterns correlate with differences in bill morphology. Longer billed females are able to garner more food by probing deeper in intertidal substrates than males; by contrast, males tend to use arable lands to a greater extent where they are more adept at feeding on surface invertebrates. Similar patterns probably occur in wintering populations of the Long-billed Curlew along the Pacific Coast (Leeman and Colwell 2005).

Shorebird use of pastures tends to increase during winter, when precipitation increases the availability of invertebrates, chiefly earthworms (Lumbricidae) brought to the surface when rain moistens the soil. This season of highest use also coincides with the time when the availability of intertidal invertebrates is lowest. Shorebird use of coastal pastures adjacent to Humboldt Bay was highest after the onset of winter rains (Colwell and Dodd 1995, 1997; Conklin and Colwell 2007). Similar to intertidal habitats, shorebirds prefer to feed in open areas with unobstructed views. In coastal pastures of northern California, five of six species that commonly used pastures (Killdeer, Dunlin, dowitchers, Wilson's Snipe, and Black-bellied Plover) were more likely to occur in pastures characterized by short vegetation (Colwell and Dodd 1995, 1997). The Marbled Godwit was the exception to this pattern, but they often

stood above the taller vegetation as they fed in pastures. Dunlin wintering in the Fraser River Delta tended to use habitats characterized by short vegetation (Shepherd and Lank 2004). In southern Portugal, Northern Lapwings and Eurasian Golden-Plovers avoided small fields with obstructed views and typically used expansive, open habitats distant from vegetation (Moreira et al. 2005).

SALT PONDS

In coastal areas worldwide, salt evaporation ponds, also known as salt pans or salinas, are an expansive and integral part of commercial salt production enterprises. Salt pans appear to be important for breeding and especially nonbreeding shorebirds, based on the concentrations of birds that occur there. In Australia, three of the most significant wetland complexes for nonbreeding shorebirds encompass salt ponds (Lane 1987). Along the Pacific Coast of North America, shorebirds use large expanses of salt evaporation ponds in Mexico (Page et al. 1997) and California (Page et al. 1999; Warnock et al. 2002). Salt pans in Puerto Rico provide important habitat for migrant shorebirds (Collazo et al. 1995). In Europe, large salinas occur adjacent to and supplement estuarine foraging habitat in Portugal (Rufino et al. 1984; Velasquez 1992). Similar findings indicate that salt pans are important habitats in South Africa (Velasquez 1993).

Salt ponds are ecologically stressful habitats in which variable but often very high salinities pose serious osmotic challenges to aquatic species. As a result, invertebrate communities in these hypersaline aquatic habitats are depauperate in diversity, and the few species present are both numerically dominant and offer an enormous biomass of potential food. Across the varying salinities, water boatmen (Corixicidae: *Trichocorixa* spp.) occur in ponds of low (<65 ppt) salinities (Tripp and Collazo 2003), whereas brine flies (*Ephydra* spp.) and brine shrimp (*Artemia* spp.) tolerate higher (>106 ppt) salinities. At natural hypersaline lakes (such as the Great Salt Lake and Mono Lake), these same taxa dominate invertebrate communities. In some regions, the seasonality of reproduction in these invertebrates occurs such that migrants capitalize on a superabundance of food and gorge themselves on energy-rich prey (Tripp and Collazo 2003; Caudell and Conover 2006).

SANDY, OCEAN-FRONTING BEACHES

High-energy, ocean-fronting beaches are dynamic ecosystems with the potential to be important foraging habitats for shorebirds (Hubbard and Dugan 2003). The physical impacts of wind and waves interact to alternately clean and replenish allochthonous detritus, which is distributed at varying tide lines in wrack. The detritus consists of plants, especially detached macrophytes (for example, brown algae such as *Fucus, Nereocystis, Egregia,* and *Postelsia*) and seagrasses (*Zostera* and *Heterozostera*) as well as decomposing animal matter that serves as the foundation of a comparatively simple food web (Kirkman and Kendrick 1997). On temperate beaches throughout the world, amphipods, isopods, and insects (such as Dipterans) capitalize on this detritus. Their abundances are directly related to the biomass of detritus available to them as food. Not surprisingly, shorebird abundances are higher in areas where wrack accumulates and provides the substrate for a rich community of detritivores. In southern California, for instance, Dugan et al. (2003) reported that the density of wrack-associated invertebrates varied over an order of magnitude, and this variation was positively correlated with the biomass of brown algae deposited on beaches. Moreover, abundances of two visually feeding shorebirds, the Snowy and Black-bellied Plovers, were positively correlated with invertebrate density in wrack. In some areas popular with sunbathers and swimmers, beaches are groomed regularly to remove the nuisance of flies that are associated with wrack. This grooming significantly depresses the food available to shorebirds, with a concomitant reduction in the density of shorebirds (Dugan et al. 2003). Elsewhere in the world, other algal macrophytes and seagrasses are harvested for commercial uses (Kirkman and Kendrick 1997), with similar reductions in the density and biomass of macrofaunal communities and their shorebird predators.

Human recreational use and commercial harvest of macroalgal detritus represent just a few of many stressors that impact sandy beach ecosystems and the shorebirds that occupy these habitats (Defeo et al. 2009).

In various regions of the world, high-energy beaches that are valuable feeding habitats for shorebirds are impacted by programs to replenish substrates eroded by winter storms and tides. These "nourishment" projects use dredged substrates to reinforce and fill sections of beach fronting highly valued coastal developments. An assumption of these projects is that recovery from these anthropogenic disturbances is rapid owing to the dynamic nature of the habitat under the influence of strong wave and tidal action (National Research Council 1995; Peterson et al. 2006). Peterson et al. (2006) took advantage of a scheduled replenishment project and available prior data to examine the impact of beach nourishment on physical and biological changes in beaches, culminating with shorebird predators and their prey. This study employed a before–after/control–impact (BACI) design. The dredged materials used to fill beaches resulted in a significant increase in the size of sediments, including considerable amounts of shell hash. Across filled beaches, substrate sizes were much more uniform compared with control sites. The abundances of several important shorebird prey species (such as the *Donax,* haustoriid amphipods, and *Emerita*) were dramatically lower (sometimes up to 100% density reduction) on filled beaches, and these depressed prey abundances persisted for months after completion of the project. The negative impact on shorebirds was equally dramatic, with a sevenfold reduction in the density of feeding shorebirds persisting for nearly a year after the end of the nourishment project. These findings stand in contrast to another study conducted at nearly the same time in the same region. Grippo et al. (2007) also used a BACI design to examine the effect on shorebirds of beach replenishment in North Carolina. Surprisingly, they reported a significant decrease in shorebird density for only one (Black-bellied Plover) of four species (including Sanderling, Willet, and Ruddy Turnstone). The survey methods, however, are poorly detailed, so it is difficult to evaluate findings. For instance, unlike Peterson et al. (2006) who surveyed during low tides, Grippo et al. (2007) appeared to survey during all daylight hours and did not indicate the timing of their surveys relative to tide level. Moreover, they vaguely describe a standardized effort to determine the density of shorebirds they observed along the four 1.6 km transects that were part of their study design. Regardless, replenishment projects appear to dramatically alter the foraging habitats of shorebirds on ocean-fronting beaches.

In summary, the coastal habitats (including estuaries, salt ponds, and ocean-fronting beaches) favored by shorebirds are increasingly impacted by humans owing to a continuing shift in population toward the coasts and development in these regions. As a result, stressors to ecological systems in coastal environs (Defeo et al. 2009) are many and are likely to increase over time. These stressors include development, recreation, pollution, sea level rise,

and global warming, to name a few. Accordingly, managing shorebirds and the ecological systems upon which they rely will become more challenging as the human population grows.

CONSERVATION IMPLICATIONS

Wetland habitats provide essential food resources for shorebirds throughout much of the annual cycle. The loss and degradation of wetland habitats has been associated with the decline of shorebird populations worldwide (Senner and Howe 1984), and direct evidence indicates that loss of wetland habitat can have consequences for individual mortality and local population sizes (Burton et al. 2006). Therefore, the management of wetlands to better provide food for shorebirds is justified. In many cases, land managers are able to manipulate water levels to increase the availability of food to nonbreeding shorebirds. Rice culture in many regions of the world offers additional wetland habitats to nonbreeding shorebirds. Pastures and agricultural lands may be important habitats for some species during winter because they offer the same suite of physical features that attract shorebirds to intertidal areas: open, unobstructed areas with abundant food. Meadow bird agreements are in place in European countries to ameliorate the negative effects on birds of agricultural intensification.

LITERATURE CITED

Baker, A. J., P. M. González, T. Piersma, L. J. Niles, I. de L. S. do Nascimento, P. W. Atkinson, N. A. Clark, C. D. T. Minton, M. K. Peck, and G. Aarts 2004. Rapid population decline in Red Knots: Fitness consequences of decreased refuelling rates and late arrival in Delaware Bay. *Proceedings of the Royal Society of London, Series B* 271: 875–882.

Baker, M. A. 1979. Morphological correlates of habitat selection in a community of shorebirds (Charadriiformes). *Oikos* 33: 121–126.

Batzer, D. P., and R. R. Sharitz, eds. 2007. *Ecology of freshwater and estuarine wetlands*. Berkeley: University of California Press.

Bignall, E. M., and D. I. McCracken. 1996. Low-intensity farming systems in the conservation of the countryside. *Journal of Applied Ecology* 33: 413–424.

Bird, J. A., A. J. Eagle, W. R. Horwath, M. W. Hair, E. E. Zilbert, and C. van Kessel. 2002. Long-term studies find benefits, challenges in alternative rice straw management. *California Agriculture* 56: 69–76.

Blanco, D. E., B. López-Lanús, R. A. Dias, A. Azpiroz, and F. Rilla. 2006. *Use of rice fields by migratory shorebirds in southern South America*. Buenos Aires: Wetlands International.

Breese, G., D. Smith, J. Nichols, J. Lyons, A. Hecht, N. Clark, S. Michels, et al. 2007. *Application of structured decision making to multiple species management of horseshoe crabs and shorebird populations in Delaware Bay.* U. S. Shepherdstown, WV: Fish and Wildlife Service National Training Center.

Burton, N. H. K., M. M. Rehfisch, N. A. Clark, and S. G. Dodd. 2006. Impacts of sudden winter habitat loss on the body condition and survival of redshanks *Tringa totanus. Journal of Applied Ecology* 43: 464–473.

Caudell, J. N., and M. R. Conover. 2006. Energy content and digestibility of brine shrimp (*Artemia franciscana*) and other prey items of Eared Grebes (*Podiceps nigricollis*) on the Great Salt Lake, Utah. *Biological Conservation* 130: 251–254.

Caughley, G., and A. Gunn. 1996. *Conservation biology in theory and practice.* Oxford: Blackwell Science.

Clemen, R. T. 1996. *Making hard decisions: An introduction to decision analysis.* 2nd ed. Belmont, CA: Duxbury Press.

Collazo, J. A., B. A. Harrington, J. S. Grear, and J. A. Colón. 1995. Abundance and distribution of shorebirds at the Cabo Rojo salt flats, Puerto Rico. *Journal of Field Ornithology* 66: 424–438.

Collazo, J. A., D. A. O'Hara, and C. A. Kelly. 2002. Accessible habitat for shorebirds: Factors influencing its availability and conservation implications. In *Managing wetlands for waterbirds: Integrated approaches,* ed. K. C. Parsons, S. C. Brown, R. M. Erwin, H. A. Czech, and J. C. Coulson. Special Publication 2, *Waterbirds* 25:13–24.

Colwell, M. A., and S. L. Dodd. 1995. Waterbird communities and habitat relationships in coastal pastures of northern California. *Conservation Biology* 9: 827–834.

———. 1997. Environmental and habitat correlates of pasture use by nonbreeding shorebirds. *The Condor* 99: 337–344.

Colwell, M. A., and O. W. Taft. 2000. Waterbird communities in managed wetlands of varying water depth. *Waterbirds* 23: 45–55.

Conklin, J. R., and M. A. Colwell. 2007. Diurnal and nocturnal roost site fidelity of Dunlin (*Calidris alpina pacifica*) at Humboldt Bay, California. *The Auk* 124: 677–689.

Conklin, J. R., M. A. Colwell, and N. W. Fox-Fernandez. 2008. High variation in roost use by Dunlin wintering in California: Implications for habitat limitation. *Bird Conservation International* 18: 275–291.

Cowardin, L. M., V. Carter, F. C. Golet, and E. T. LaRoe. 1979. *Classification of wetlands and deep-*water habitats of the United States. Washington, DC: U. S. Department of the Interior, U. S. Fish and Wildlife Service.

Davis, C. A., and L. M. Smith. 1998. Ecology and management of migrant shorebirds in the Playa Lakes region of Texas. *Wildlife Monograph* 140: 1–45.

Day, J. H., and M. A. Colwell. 1998. Waterbird communities in rice fields subjected to different post-harvest treatments. *Colonial Waterbirds* 21: 185–197.

Defeo, O., A. McLachlan, D. S. Schoeman, T. A. Schlacher, J. Dugan, A. Jones, M. Lastra, and F. Scapini. 2009. Threats to sandy beach ecosystems: A review. *Estuarine, Coastal and Shelf Science* 81: 1–12.

Donald, P. F., R. E. Green, and M. F. Heath. 2001. Agricultural intensification and the collapse of Europe's farmland bird populations. *Proceedings of the Royal Society of London, Series B* 268: 25–29.

Dugan, J. E., D. M. Hubbard, M. D. McCrary, and M. O. Pierson. 2003. The response of macrofauna communities and shorebirds to macrophyte wrack subsidies on exposed sandy beaches of southern California. *Estuarine, Coastal and Shelf Science* 58 (Suppl.): 25–40.

Eadie, J. M., C. S. Elphick, K. J. Reinecke, M. R. Miller. 2008. Wildlife values of North American ricelands. In *Conservation in ricelands of North America,* ed. M. Petrie, S. W. Manley, and B. Batts, 7–90. Memphis: Ducks Unlimited.

Eldridge, J. 1992. *Management of habitat for breeding and migrating shorebirds in the Midwest.* Leaflet 13.2.14. Washington, DC: U.S. Fish and Wildlife Service.

Elphick, C. S. 2000. Functional equivalency between rice fields and seminatural wetlands. *Conservation Biology* 14: 181–191.

———. 2004. Assessing conservation trade-offs: Identifying the effects of flooding rice fields for waterbirds on non-target birds species. *Biological Conservation* 117: 105–110.

Elphick, C. S., and L. W. Oring. 1998. Winter management of Californian rice fields for waterbirds. *Journal of Applied Ecology* 35: 95–108.

Erwin, R. M. 2002. Integrated management of waterbirds: Beyond the conventional. In *Managing wetlands for waterbirds: Integrated approaches,* ed. K. C. Parsons, S. C. Brown, R. M. Erwin, H. A. Czech, and J. C. Coulson. Special Publication 2, *Waterbirds* 25: 5–12.

Evans Ogden, L. J., K. A. Hobson, D. B. Lank, and S. Bittman. 2006. Stable isotope analysis reveals that agricultural habitat provides an important dietary component for nonbreeding Dunlin. *Avian Conservation and Ecology* 1: 3.

Fasola, M., and X. Ruiz. 1997. Rice farming and waterbirds: Integrated management in an artificial landscape. In *Farming and birds in Europe: The Common Agricultural Policy and its implications for bird conservation,* ed. D. J. Pair and M. W. Pienkowski, 210–235. London: Academic Press.

Fredrickson, L. H. 1988. Strategies for water level manipulations in moist-soil systems. In ed. D. H. Cross and P. Vohs, *Waterfowl management handbook,* Leaflet 13.4.6. Fort Collins, CO: U.S. Fish and Wildlife Service.

Fredrickson, L. H., and M. K. Laubhan. 1996. Managing wetlands for wildlife. In *Research and management techniques for wildlife and habitats,* 5th ed., ed. T. A. Bookout, 623–647. Bethesda, MD: The Wildlife Society.

Fredrickson, L. H., and T. S. Taylor. 1982. *Management of seasonally flooded impoundments for wildlife.* Resource publication 148. Washington, DC: U.S. Fish and Wildlife Service.

Gillings, S., R. J. Fuller, and W. J. Sutherland. 2007. Winter field use and habitat selection by Eurasian Golden Plovers *Pluvialis apricaria* and Northern Lapwings *Vanellus vanellus* on arable farmland. *The Ibis* 149: 509–520.

Grippo, M. A., S. Cooper, and A. G. Massey. 2007. Effect of beach replenishment projects on waterbird and shorebird communities. *Journal of Coastal Research* 23: 1088–1096.

Gunnarsson, T. G., J. A. Gill, G. F. Appleton, H Gíslason, A. Gardarsson, A. R. Watkinson, and W. J. Sutherland. 2006. Large-scale habitat associations of birds in lowland Iceland: Implications for conservation. *Biological Conservation* 128: 265–275.

Harrington, B. A. 2003. Shorebird management during the non-breeding season—an overview of needs, opportunities, and management concepts. *Wader Study Group Bulletin* 100: 59–66.

Helmers, D. L. 1992. *Shorebird management manual.* Manomet, MA: Western Hemisphere Shorebird Reserve Network.

Hubbard, D. M., and J. E. Dugan. 2003. Shorebird use of an exposed sandy beach in southern California. *Estuarine, Coastal and Shelf Science* 58 (Suppl.): 41–54.

Isola, C. R., M. A. Colwell, O. W. Taft, and R. J. Safran. 2000. Interspecific differences in habitat use of foraging shorebirds and waterfowl in managed wetlands of California's San Joaquin Valley. *Waterbirds* 23: 196–203.

Keeney, R. L. 1992. *Value-focused thinking.* Cambridge, MA: Harvard University Press.

Kirkman, H., and G. A. Kendrick. 1997. Ecological significance and commercial harvesting of drifting and beach-cast macro-algae and seagrasses in Australia: A review. *Journal of Applied Phycology* 9: 311–326.

Kleijn, D., and G. J. C. van Zuijlen. 2004. The conservation effect of meadow bird agreements on farmland in Zeeland, Netherlands, in the period 1989–1995. *Biological Conservation* 117: 443–451.

Lane, B. 1987. *Shorebirds in Australia.* Melbourne, Victoria, Australia: Nelson.

Laubhan, M. K., and L. H. Fredrickson. 1993. Integrated wetland management: Concepts and opportunities. *Transactions of the North American Wildlife and Natural Resources Conference* 58: 323–334.

Leeman, T. S., and M. A. Colwell. 2005. Coastal pasture use by Long-billed Curlews at the northern extent of their non-breeding range. *Journal of Field Ornithology* 76: 33–39.

Leopold, A. 1933. *Game management.* Madison: University of Wisconsin Press.

Long, L. L., and C. J. Ralph. 2001. Dynamics of habitat use by shorebirds in estuarine and agricultural habitats in northwestern California. *Wilson Bulletin* 113: 41–52.

Lyons, J. E., M. C. Runge, H. P. Laskowski, and W. L. Kendall. 2008. Monitoring in the context of structured decision-making and adaptive management. *Journal of Wildlife Management* 72: 1683–1692.

Maeda, T. 2001. Patterns of bird abundance and habitat use in rice fields of the Kanto plain, Central Japan. *Ecological Research* 16: 569–585.

Milsom, T. P., J. D. Hart, W. K. Parkin, and S. Peel. 2002. Management of coastal grazing marshes for breeding waders: The importance of surface topography and wetness. *Biological Conservation* 103: 199–207.

Mitsch, W. J., and J. G. Gosselink. 2007. *Wetlands.* 4th ed. New York: John Wiley.

Moreira, F., P. Beja, R. Morgado, L. Reino, L. Gordinho, A. Delgado, and R. Borralho. 2005. Effects of field management and landscape context on grassland wintering birds in southern Portugal. *Agriculture, Ecosystems and Environment* 109: 59–74.

Moser, M. E. 1988. Limits to the numbers of Grey Plovers (*Pluvialis squatarola*) wintering on British estuaries: An analysis of long-term population trends. *Journal of Applied Ecology* 25: 473–485.

Myers, J. P., R. I. G. Morrison, P. Z. Anatas, B. A. Harrington, T. E. Lovejoy, M. Sallaberry, S. E. Senner, and A. Tarak. 1987. Conservation strategy for migratory species. *American Scientist* 75: 19–26.

National Research Council. 1995. *Beach nourishment and protection*. Washington, DC: National Academy Press.

Niles, L. J., H. P. Sitters, A. D. Dey, P. W. Atkinson, A. J. Baker, K. A. Bennett, R. Carmona, et al. 2008. *Status of the Red Knot (Calidris canutus rufa) in the Western Hemisphere*, ed. C. D. Marti. Studies in Avian Biology No. 36. Camarillo, CA: Cooper Ornithological Society.

Ntiamoa-Baidu, Y., T. Piersma, P. Wiersman, M. Poot, P. Battley, and C. Gordon. 1998. Water depth selection, daily feeding routines and diets of waterbirds in coastal lagoons in Ghana. *The Ibis* 140: 89–103.

Ottvall, R., and H. G. Smith. 2006. Effects of an agri-environment scheme on wader populations of coastal meadows of southern Sweden. *Agriculture, Ecosystems and Environment* 113: 264–271.

Page, G. W., E. Palacios, L. Alfaro, S. Gonzalez, L. E. Stenzel, and M. Jungers. 1997. Numbers of wintering shorebirds in coastal wetlands of Baja California, Mexico. *Journal of Field Ornithology* 68: 562–574.

Page, G. W., L. E. Stenzel, and J. E. Kjelmyr. 1999. Overview of shorebird abundance and distribution in wetlands of the Pacific Coast of the contiguous United States. *The Condor* 101: 461–471.

Paton, P. W. C., and V. C. Bachman. 1997. Impoundment drawdown and artificial nest structures as management strategies for Snowy Plovers. *International Wader Studies* 9: 64–70.

Pearce-Higgins, J. W., and D. W. Yalden. 2003. Variation in the use of pastures by breeding European Golden Plovers *Pluvialis apricaria* in relation to prey availability. *The Ibis* 145: 365–381.

Peterson, C. H., M. J. Bishop, G. A. Johnson, L. M. D'Anna, and L. M. Manning. 2006. Exploiting beach filling as an unaffordable experiment: Benthic intertidal impacts propagating upward to shorebirds. *Journal of Experimental Marine Biology and Ecology* 338: 205–221.

Pressey, R. L., C. J. Humphreys, C. R. Margules, R. I. Wright, and P. H. Williams. 1993. Beyond opportunism: Key principles for systematic reserve selection. *Trends in Ecology and Evolution* 8: 124–128.

Remsen, J. V., M. M. Swan, S. W. Cardiff, and K. V. Rosenberg. 1991. The importance of the rice-growing region of south-central Louisiana to winter populations of shorebirds, raptors, waders, and other birds. *Journal of Louisiana Ornithology* 1: 35–47.

Rönkä, A., K. Koivula, M. Ojanen, V.-M. Paknen, M. Pohjoismäki, K. Rannikko, and P. Rauhala. 2006. Increased nest predation in a declining and threatened Temminck's Stint *Calidris temminckii* population. *The Ibis* 148: 55–65.

Rufino, R., A. Araujo, J. P. Pina, and P. Miranda. 1984. The use of salinas by waders in the Algarve, south Portugal. *Wader Study Group Bulletin* 42: 41–42.

Senner, S. E., and M. A. Howe. 1984. Conservation of Nearctic shorebirds. In *Shorebirds: Breeding behavior and populations,* ed. J. Burger and B. L. Olla, 379–421. New York: Plenum Press.

Shepherd, P. C. F., and D. B. Lank. 2004. Marine and agricultural habitat preferences of Dunlin wintering in British Columbia. *Journal of Wildlife Management* 68: 61–73.

Shuford, W. D., G. W. Page, and J. E. Kjelmyr. 1998. Patterns and dynamics of shorebird use of California's Central Valley. *The Condor* 100: 227–244.

Skagen, S. K., and F. L. Knopf. 1994. Migrating shorebirds and habitat dynamics at a prairie wetland complex. *Wilson Bulletin* 106: 91–105.

Smart, J., J. A. Gill, W. J. Sutherland, and A. R. Watkinson. 2006. Grassland-breeding waders: Identifying key habitat requirements for management. *Journal of Applied Ecology* 43: 454–463.

Stewart, R. E., and H. A. Kantrud. 1971. *Classification of natural ponds and lakes in the glaciated prairie region*. Bureau of Sport Fisheries and Wildlife, Resource Publication 92. Washington, DC/Jamestown, ND: U.S. Fish and Wildlife Service/Northern Prairie Wildlife Research Center Online. Available at: www.npwrc.usgs.gov/resource/wetlands/pondlake/index.htm (accessed May 10, 2010).

Stillman, R. A., and A. F. Brown. 1998. Pattern in the distribution of Britain's upland breeding birds. *Journal of Biogeography* 25: 73–82.

Stralberg, D., D. L. Applegate, S. J. Phillips, M. P. Herzog, N. Nur, and N. Warnock. 2009. Optimizing wetland restoration and management for avian communities using a mixed integer programming approach. *Biological Conservation* 142: 94–109.

Taft, O. W., M. A. Colwell, C. R. Isola, and R. J. Safran. 2002. Waterbird response to experimental drawdown: Implications for multispecies management of wetland mosaics. *Journal of Applied Ecology* 39: 987–1001.

Taft, O. W., and C. S. Elphick. 2007. *Waterbirds on working lands: Literature review and bibliography development*. Ivyland, PA: National Audubon Society/Monsanto Fund. Available at: http://

www.audubon.org/bird/waterbirds/downloads
.html (accessed May 10, 2010).

Taft, O. W., and S. M. Haig. 2006. Importance of wetland landscape structure to shorebirds in an agricultural valley. *Landscape Ecology* 21: 169–184.

Townshend, D. J. 1981. The importance of field feeding to the survival of wintering male and female Curlews *Numenius arquata* on the Tees estuary. In *Feeding and survival strategies of estuarine organisms*, ed. N. V. Jones and W. J. Wolff, 261–273. New York: Plenum Press.

Tripp, K. J., and J. A. Collazo. 2003. Density and distribution of water boatmen and brine shrimp at a major shorebird wintering area in Puerto Rico. *Wetland Ecology and Management* 11: 331–341.

Turpie, J. 1995. Prioritizing South African estuaries for conservation: A practical example using waterbirds. *Biological Conservation* 74: 175–185.

U.S. Fish and Wildlife Service. 2008. Rice, Water, and Birds. Available at: www.fws.gov/birds/waterbirds/rice/rice.html (accessed May 20, 2010).

Velasquez, C. 1992. Managing artificial saltpans as a waterbird habitat: Species' response to water level manipulation. *Colonial Waterbirds* 15: 43–55.

———. 1993. The ecology and management of waterbirds at commercial saltpans in South Africa. Ph.D. dissertation, University of Capetown, South Africa.

Wallander, J., D. Isaksson, and T. Lenberg. 2006. Wader nest distribution and predation in relation to man-made structures on coastal pastures. *Biological Conservation* 132: 343–350.

Warnock, N., G. W. Page, T. D. Ruhlen, N. Nur, J. Y. Takekawa, and J. T. Hanson. 2002. Management and conservation of San Francisco Bay salt ponds: Effects of pond salinity, area, tide and season on Pacific flyway waterbirds. In *Managing wetlands for waterbirds: Integrated approaches*, ed. K. C. Parsons, S. C. Brown, R. M. Erwin, H. A. Czech, and J. C. Coulson. Special Publication 2, *Waterbirds* 25: 79–92.

Warnock, N., G. W. Page, and L. E. Stenzel. 1995. Non-migratory movements of Dunlins on their California wintering grounds. *Wilson Bulletin* 107: 131–139.

12

Managing Predators

CONTENTS

PREDATION HAS INFLUENCED THE biology of shorebirds in myriad ways. Their social tendencies (Goss-Custard 1985), migratory habits (Lank et al. 2003), and especially breeding biology (Lack 1968) and parental behavior (Gochfeld 1984) have been shaped by a long evolutionary history of predation on adults, eggs, and young. For example, during the breeding season, the cryptic plumages of incubating sandpipers and the choice of nest site, clutch size, egg-laying intervals, and distraction displays of adults defending eggs and chicks may all be interpreted as a consequence of predation. In ecological time, predation also is known to have strong effects on population dynamics. A population's rate of change hinges on two principal demographic characteristics: annual survival of breeding adults and their productivity. It is difficult to directly enhance adult survival by reducing predation-based mortality, other than to provide the necessary habitat, food, and cover for individuals to survive from one breeding season to the next. By contrast, numerous techniques exist to improve productivity that has been compromised by predation.

The productivity of ground-nesting birds, including shorebirds, is strongly influenced by predation (Martin 1993; MacDonald and Bolton 2008b). A traditional approach to population management is to ameliorate the impact of predators as a factor limiting productivity (Leopold 1933). This approach is supported by the overwhelming effect that predators have on reproductive success, although there is substantial annual variation and differences among habitats in the magnitude of the effect that predators exert (MacDonald and Bolton 2008b). The negative impact of predators is especially problematic with rare shorebirds (e.g., Black Stilt: Dowding and Murphy 2001; Piping Plover: Murphy et al. 2003; Snowy Plover: Neuman et al. 2004; Temminck's Stint: Rönkä et al. 2006) such that predator management is an integral part of plans to recover populations. A common approach to increasing small populations has been to manage the predators on the breeding grounds. The methods of predator management vary from nonlethal techniques, such as the use of caged exclosures around nests, to lethal means that target "problem" individuals that have a disproportionately negative effect on a local breeding population or reduce the local abundance of predator populations. In this chapter, I summarize the literature addressing the impact of predators on breeding shorebirds (in terms of nesting success, chick survival, and breeding population size), and review studies that have used various lethal and nonlethal methods to control predators. I begin with a discussion of the ethical considerations and conflicts underlying the use of these methods.

ETHICAL CONSIDERATIONS AND DECISION MAKING

The management of predators may be one of the most challenging issues in wildlife conservation for several reasons. First, the scientific evidence linking population declines to predation is often lacking or debatable; this issue is addressed later in the chapter. A second, more common issue concerns acceptable methods of predator control, which depends on the diversity of backgrounds and ethical perspectives of stakeholders. Ultimately, the acceptability of various methods of predator control hinges on an individual's ethical stance, and recognition of this may go a long way toward resolving or avoiding conflicts (or at least agreeing to disagree). For example, local managers may have substantial information to support the contention that predators are the most important factor compromising a population's productivity. Although the public may agree with this fact, many private citizens would oppose lethal methods of control for ethical reasons.

Consider a scenario in which a native shorebird (such as the Black Stilt) has declined in abundance owing to negative effects imposed by an introduced vertebrate (such as the black rat or feral cat) on adult survival and breeding productivity. In this simple scenario, two important and divergent views often surface regarding predator control. Many biologists charged with managing wildlife espouse an ecocentric view of nature and argue that killing *individual* predators is necessary to recover a *population* threatened with extirpation. This perspective emphasizes the role that limiting factors play in the dynamics and downward trend of a population. In this case, predators pose a problem to the maintenance of a healthy population, and lethal control is justified. The observation that the predator is nonnative to the ecosystem further justifies the use of lethal control to eradicate the predator. Importantly, moral consideration is not given to the individual predator; rather, emphasis is on maintaining a viable prey (shorebird) population within a healthy, native ecosystem. A second perspective, often articulated by the public and advocates of animal rights and welfare, emphasizes a biocentric view of nature. In this case, emphasis is placed on the rights and humane treatment of the *individual* predator, rather than the shorebird *population*. Here, moral consideration weighs in favor of the individual (predator) based on intrinsic value arguments, commonly associated with sentience and intelligence. The complexity of this scenario increases as one considers an overabundant native predator or omnivore (such as corvids and foxes) and various nonlethal methods of control.

Putting aside challenging issues based on ethical perspectives, the decision to implement some form of predator management (nonlethal

NEW ZEALAND'S
SHOREBIRDS

Worldwide, anthropogenic habitat loss and destruction is the principal cause of avian extinctions and population declines. But for island birds, especially those that have evolved free of mammalian predators, the introduction of nonnative predators is equally or more important in avian extinctions. The shorebirds of New Zealand, like much of the archipelago's avifauna, have evolved many features that indicate a long evolutionary history free of mammalian (ground) predators: some have compromised flight ability, large size, and conservative reproductive traits such as delayed breeding and protracted fledging periods (Dowding and Murphy 2001). Consequently, the introduction of nonnative mammals to New Zealand, first by Polynesians (approximately 1,000 years ago) and more recently by Europeans (over the last 200 years), has resulted in serious problems for the persistence of the archipelagos shorebirds. The dogs and Polynesian rats arrived first, followed by cats, ferrets, stoats, and hedgehogs.

Dowding and Murphy (2001) reviewed the status of endemic New Zealand shorebirds and addressed factors contributing to their decline and extinction. The shorebirds consisted of 18 species or subspecies: three oystercatchers, one stilt, four plovers, and two snipes. One species (Chatham Island Snipe) and several subspecies are now extinct. Most other species have declined in range and/or abundance. As with other New Zealand species, predation of ground-nesting shorebirds appears to be the single most important factor contributing to population declines. Several taxa are confined to predator-free islands. In many instances, conservation plans for New Zealand's shorebirds include extensive lethal control of predators, both nonnative and native. In some cases, these efforts have resulted in significant increases in productivity and local population sizes. For example, intensive trapping in the braided river ecosystem of the Upper Witaki Basin resulted in significant increases in the nesting success of the Banded Dotterel and Wrybill. Still, other species have not recovered from low population sizes even with long-term, intensive lethal control of predators. For example, the Black Stilt survives today at critically low numbers (<20 birds in the wild and a similar number in captivity), despite over 20 years of lethal removal of predators (Keedwell et al. 2002).

versus lethal) is akin to any decision-making process with multiple objectives (Lyons et al. 2008). The fundamental objective of control is to increase the population size of a shorebird. The various means objectives may include measures that (1) decrease population size or alter distribution and behavior of predators so that they have less of a negative impact on shorebirds, (2) increase shorebird reproductive success by improving hatching and fledging success, and (3) are cost effective and morally justified in the eyes of stakeholders. Bolton et al. (2007) presented a decision tree for evaluating when and under what conditions to implement predator control. Although they did not state explicitly, I presume that control is synonymous with lethal methods. Their scheme considered control to be the last decision in a sequence that began by addressing the importance of the site to a population of shorebirds. If the site was the last known breeding location for a rare species, it justified control. After this decision, the quality of habitat was considered. If habitat was of high quality and the problem could be rectified by habitat management, then predator control was unnecessary. Additionally, Bolton et al. (2007) suggested that monitoring of nest survival and predator densities was an essential step before initiating control. Ultimately, this would include firm evidence for specific predators being the cause of population decline. In their decision-making framework,

Bolton et al. (2007) suggested that the decision to use predator control was a last resort that required evidence that alternative forms of management could not solve the problem.

Other factors may influence decisions to use different methods of predator control, including (1) whether the predator is native or introduced; (2) the likelihood that control will have significant positive effects on the prey population, including the geographical extent of the area where control is practiced; (3) the feasibility, costs, and logistics of implementing control over a set period of time; and (4) public perception of the particular method(s) used in control, especially lethal versus nonlethal means. For example, lethal control may be more justified when a nonnative predator is the main cause of a localized (island) population decline. A real-life example of this scenario occurred with the Black Stilt of New Zealand (Keedwell et al. 2002), which has suffered a dramatic population decline and range contraction owing to introduced cats (*Felis catus*), ferrets (*Mustela putorius*), and stoats (*M. erminea*). Lethal predator control is a critical element of the Black Stilt's conservation plan, which includes captive breeding and habitat management. It may be more difficult, however, to justify lethal control when the effect of predators on prey populations is less clear and not localized. Despite considerable effort to reduce nonnative populations of predators, Black Stilt productivity and population size and growth remain low.

Driscoll and Bateson (1988) articulated a "cube" model to facilitate decisions related to whether a behavioral biologist should proceed with research on animals. The three dimensions of their decision cube were (1) certainty of benefit, (2) quality of research, and (3) degree of animal suffering. A decision to proceed with research was justified when benefit and quality were high and suffering was low. It is possible to adapt this model to the decision to implement methods of predator control (Fig. 12.1). The dimensions of the decision cube are as follows. One axis represents the

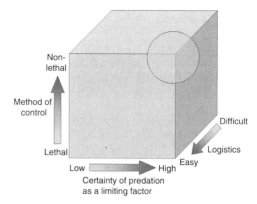

FIGURE 12.1. A decision cube to weigh the key factors influencing a predator control program to effect population recovery of a prey species. The decision to proceed with a program is easiest (transparent circle in upper foreground) when nonlethal methods are used, logistics are relatively easy, and the certainty of effecting the prey population recovery is high. After Driscoll and Bateson (1988).

certainty that predation is a critical factor influencing low population size: action taken would significantly reduce the threat and, therefore, increase the population size of the shorebird. A second axis represents the logistical challenges faced by implementing any given predator control program. In this element, the costs of a control program would need to consider the magnitude of the effect on the behavior or population of a predator. For instance, a local effort to reduce the effects of garbage or supplemental food sources in attracting predators to areas where birds nest would be easier than a regionwide program using lethal methods to reduce predator populations. A third axis represents the ethical dilemma faced in predator control associated with the use of lethal and nonlethal means. It is much easier to gain support for control programs that incorporate nonlethal methods, whereas lethal control is much more difficult. Moreover, lethal methods are easier to argue if the target is an individual animal known to eat eggs or chicks. In summary, a clear decision to proceed with a planned strategy of control is highest when the certainty of the relationship between predator and prey is high, logistics are easy, and the method of control is nonlethal. By contrast, it is more difficult to justify a control program incorporating lethal methods, which are logistically difficult to orchestrate; this decision is made increasingly difficult when the role of predators in limiting the population is unclear.

DO PREDATORS LIMIT SHOREBIRD POPULATIONS?

An important first step in evaluating the likely success of predator management is to understand the ecological role that predators play in limiting shorebird populations. Shorebird populations may be limited by predation during the breeding season if the strength and consistency of the negative effect on productivity is sufficient to depress abundance. The effect of predation may be strongest in populations that are less abundant, as may occur at the limit of a species' range. In essence, this is the argument made by Pienkowski (1984), who suggested that the southern range limit of the Common Ringed Plover was set by nest predation. The same argument was invoked to explain the declining population of Temminck's Stint at the southern limit of the species' range along the Baltic Coast of Sweden (e.g., Rönkä et al. 2006).

In Chapter 10, I reviewed evidence that shorebird populations are limited or regulated by predation. An important additional body of evidence to evaluate the extent to which predators limit shorebird populations comes from studies of the effects of introduction or removal of predators. In many cases, introductions involve nonnative predators, but increases in native vertebrates have also been implicated as a significant factor limiting the recovery of shorebird populations. Several examples with shorebirds are illuminating because they highlight the pronounced impact that introduced predators have on both rare and widespread species. These impacts generally involve predators severely reducing nesting success, productivity, and population densities.

PREDATOR INTRODUCTIONS

Perhaps the most dramatic effects of predators come from island ecosystems where introduced mammals have been responsible for many avian extinctions (Blackburn et al. 2004). Several species of shorebird have been eradicated from much of their range in Polynesia by introduced black rats (*Rattus rattus*) and feral cats. Much of the remnant population of the Tuamotu Sandpiper is currently confined to a few islands that remain free of predators (Pierce and Blanvillain 2004). In Hawaii, the Hawaiian Stilt population is approximately 1,600, and predation by rats, cats, and Indian mongooses (*Herpestes javanicus*) poses a major impediment to the species' continued existence. Persistence of Temminck's Stint at the southern extent of its European breeding range is threatened by the introduction of the raccoon dog (*Nyctereutes procyonoides*) from East Asia (Rönkä et al. 2006).

In the Outer Hebrides, Scotland, the western hedgehog (*Erinaceus europaeus*) was introduced in 1974. It has since spread northward, both naturally and further aided by deliberate introduction, into coastal meadows and adjacent heath, where significant populations of six species of shorebird breed. The hedgehog is well known as a predator of bird eggs. Jackson et al. (2004) compared shorebird populations before and after hedgehog expansion and showed that overall numbers had declined by 23%. This was the composite of a 9% increase in northern areas (which generally lacked hedgehogs) and a 39% decrease in southern regions (where hedgehogs were abundant). Furthermore, evidence indicates that hedgehogs have caused the reduction in shorebird populations by negatively impacting nesting success (Jackson and Green 2000; Jackson 2001).

PREDATOR REMOVAL

Nonlethal and lethal methods of predator control have been used to improve the productivity of rare shorebirds with mixed results. In New Zealand, the critically endangered Black Stilt was once widely distributed on the North and

South islands. However, by the late 1970s its population had declined to a low of 23 (about 10 breeding pairs). Presently, the population is threatened by introduced predators, habitat loss, and hybridization with the Pied Stilt. Intensive management, including lethal predator control of nonnative mammals such as the domestic cat, ferret, stoat, and rats was initiated in 1981. There was strong evidence that these introduced species were important predators of eggs and chicks. Moreover, trapping of predators was successful in boosting hatching and fledging success, but this effect was not evident across all years. Keedwell et al. (2002) summarized 20 years of data on fledging success and suggested that there was limited evidence that lethal control was effective in stilt conservation. The population of Black Stilts is marginally larger than it was after extensive lethal control of predators. Despite more than 20 years of intensive management and predator control, the Black Stilt population remains well below 100, with approximately 25 birds held in captivity. Keedwell et al. (2002) pointed out, however, that their conclusions were based on small sample sizes (owing to the species' rarity) and a lack of consistent effort to monitor control efforts. Moreover, two other shorebirds, the Banded Dotterel and Wrybill, in the same region at the same time, clearly benefited from predator control in other braided river systems of the South Island of New Zealand (Dowding and Murphy 2001). Although the effect on stilt population size has been marginal (a small increase in numbers), it may be that stilts would have gone extinct without the lethal control.

The Pacific Coast population of the Snowy Plover was listed in 1993 as threatened under the U.S. Endangered Species Act (U.S. Department of the Interior 1993). The U.S. Fish and Wildlife Service (2007) identified predation on eggs and chicks as one of several factors that are thought to limit the population. Accordingly, nonlethal and lethal predator control has been undertaken in a variety of areas to increase productivity (Neuman et al. 2004; Hardy and Colwell 2008). Neuman et al. (2004) reported on the results of

predator control efforts on productivity of Snowy Plovers around Monterey Bay, California. Their study compared productivity before (7 years) and after (9 years) predator control, which included the use of nest exclosures and lethal removal of nonnative mammals. In total, 118 red foxes (*Vulpes vulpes*) and 95 feral cats were removed from the study area over 7 years. In a before–after comparison, the average nesting success increased from 43% to 68%. However, the per capita fledging success was nearly identical before (0.86 ± 0.28) and after (0.81 ± 0.29) lethal predator removal began. Substantial annual variation in the per capita fledging success for the central California population (0.32 to 1.23) suggests that productivity is influenced by ecological factors additional to predation. For instance, high clutch survival where predator control is practiced may lead to lower fledging success owing to reduced food availability or inclement weather. Alternatively, the reason that lethal methods were not successful in increasing fledging success may stem from insufficient levels of predator removal. Overall, however, it seems that although lethal predator control may increase nesting success, it does not necessarily result in increased population size.

Perhaps the most thorough assessment of the success of lethal removal of native predators to boost shorebird productivity was conducted in Great Britain, where breeding shorebirds are increasingly concentrated on remnant grasslands and hence are vulnerable to predation. Bolton et al. (2007) presented an excellent example evaluating the objectives of an 8-year study of a lethal control program designed to mitigate the negative effects of red fox and Carrion Crow (*Corvus corone*) predation on Northern Lapwing breeding success. They conducted their study in lowland wet grasslands broadly distributed across Great Britain. The experiment included a crossover design in which study areas were assigned to one (lethal removal) or the other (no predator control) treatment for 4 years, after which the treatments were swapped for another 4 years. In general, lethal removal reduced the abundance of predators, especially in areas where fox and crow densities were highest. But this had mixed positive effects on lapwings. On the one hand, there was little difference between treatments in nest failure rate and chick survival, although lethal control appeared to increase nest survival in areas of high predator density. Moreover, the number of adults accompanied by young was high in years when predators were controlled. In the end, however, lethal control did not increase breeding population size of lapwings in respective study areas. This latter finding may be explained by higher emigration rates of lapwings in productive areas.

In summary, there is ample evidence that predators, especially nonnative, can have a significant impact on shorebird nesting success and productivity. In some cases, these negative effects cause a reduction in population size, either locally or regionally. Côté and Sutherland (1996) reviewed the literature on the effects of lethal predator removal on a variety of (mostly game) bird populations. Overall, there was a significant increase in nesting success in 75% of the 20 studies reviewed and an increase in postbreeding population size. However, breeding population size did not increase in many studies. Consequently, the need for predator control is often, but not always, justified.

METHODS OF CONTROL

The means of predator control are seemingly as diverse as the suite of predators they target. Management actions may include indirect measures taken to alter habitat to decrease the likelihood that predators will find eggs and chicks and to educate and change the behaviors of humans to minimize the attractive nuisance of garbage. Direct measures, either lethal or nonlethal, include altering the behavior of individual predators and effecting a change in the distribution and abundance of a predator population. In the following section, I provide specific examples of how conservationists have managed habitat, people, and predators to minimize the nega-

tive impacts of predators on shorebird breeding success.

HABITAT MANAGEMENT

As humans continue to encroach on wild spaces and degrade what remains of seminatural areas, shorebirds have become increasingly concentrated in remnant habitat patches. These remnants are likely to require greater management to offset the effects of predators, which may focus their foraging in patches with a resulting increase in the functional response (Bolton et al. 2007). Accordingly, various manipulations of habitat have been used to increase cover or enhance the crypsis of nests and chicks. On a large spatial scale, habitat restoration is often used to increase the overall amount of suitable nesting habitat. One underlying and untested assumption of these projects is that an increased amount of habitat and hence lower nesting densities (akin to the "needle in the haystack" argument) will yield higher nest survival, greater productivity, and increased population sizes. Little data exist to evaluate this assumption. What information does exist suggests that the opposite is true. Daily nest predation rates are sometimes correlated with high nesting densities, especially in species that collectively mob predators. For instance, Northern Lapwings breeding in loose colonies had higher nesting success than solitary breeders (Berg 1996; MacDonald and Bolton 2008a).

At a finer spatial scale, restoration may also increase the quality of breeding habitat by simply making it more difficult for a predator to find a nest or brood because they are better hidden. For sandpipers that rely on vegetation to conceal eggs and adults, this would involve increasing the vegetative cover at nests; for plovers and other taxa that nest in open, unvegetated habitats this would involve enhancing the crypsis afforded to nests and broods by increasing debris and substrate heterogeneity. Although some studies have shown that sandpipers select nest sites in taller or denser vegetation (Colwell 1992; Grant et al. 1999) or that plovers prefer open, unvegetated habitats (Muir and Colwell

2010), the assumption that nest survival increases with concealment or crypsis is generally not well supported by literature (e.g., Colwell 1992; Grant et al. 1999) nor has it commonly been evaluated by biologists studying shorebirds. Nevertheless, several management practices have been developed to increase nest concealment. For instance, discarded shells of bivalves (referred to as shell hash) are sometimes used to increase the clutter of nesting substrates for some temperate-breeding plovers.

Across Europe, lowland meadows provide important habitats for breeding shorebirds (Green and Robins 1993). A critical habitat element in these grasslands is shallowly flooded areas that provide food for adults and young. Shorebirds appear to nest preferentially near these wet areas. The management of water depth is via a system of canals that often are arranged in a linear manner on the landscape. There is concern that this linearity may cause predators to concentrate their foraging activity along canals and roads. Eglington et al. (2009) analyzed the movements of mammalian predators (principally foxes) of Northern Lapwing nests in association with linear features. They also quantified nesting success and causes of lapwing nest failures using thermistor probes inserted into nest cups. Foxes were determined to be the main nest predators based on a distinct nocturnal pattern of nest failure. Lapwing clutches survived poorly in areas with high predator activity, but predator movements did not appear to be influenced by wetland features or linear aspects of the landscape.

PEOPLE MANAGEMENT

Native and nonnative predators may be attracted to habitats occupied by breeding shorebirds owing to the presence of abundant food resources in the form of garbage. A first step in managing this problem is to educate people such that the attractants are minimized or eliminated. Accordingly, effective signage educating the public about the problem, coupled with garbage clean-up and refuse disposal, are integral parts of any predator management plan. It

is clear, however, that populations of some predators have been increasing dramatically in association with anthropogenic sources of food (at picnic and camping areas, eating establishments, and landfills) such that the problem of predator population size and growth should be addressed at a larger spatial scale than the individual site occupied by a breeding shorebird (Liebezeit and George 2002).

PREDATOR MANAGEMENT

A variety of methods have been developed to alter the behavior of predators so that they are less likely to depredate shorebirds eggs and chicks. Predators may be dissuaded from using areas frequented by breeding shorebirds by creating conditions that are unpleasant for them. Humans have hazed individual predators, especially corvids, using effigies or by otherwise scaring individuals using slingshots and electrified perches. Models or carcasses of predators, particularly corvids, have been used to scare or deter individual predators. Application of this form of predator control is localized and probably is best applied to circumstances where territorial corvids co-occur with high densities of breeding prey (Liebezeit and George 2002). For example, corvids avoided nesting areas of the Least Tern (*Sternula antillarum*) when crow carcasses and raven heads were placed near the breeding colony. Deterrence is most effective when models are lifelike and exhibit movement, but their effectiveness diminishes with time (Liebezeit and George 2002).

EXCLOSURES AND FENCING

A widespread technique used to increase nesting success is the use of caged exclosures surrounding a nest or fencing to limit access of predators to an area. The exclosure allows the incubating adult to come and go from the nest, which is located in the center of the cage and generally out of reach of predators. The design of exclosures varies greatly (Fig. 12.2). Often, large (3×3 m) cages anchored deeply in the substrate are built of hardware cloth and topped with netting designed to restrict entry of mam-

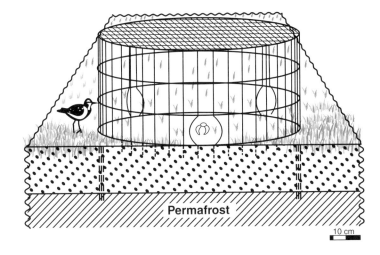

FIGURE 12.2. Small predator exclosure. Exclosures such as the one depicted here for use with Arctic-breeding shorebirds are sometimes used to increase hatching success. After Estelle et al. (1996).

Permafrost

10 cm

malian and avian predators in temperate areas. By contrast, smaller versions have been used for easy transport and deployment in Arctic habitats, where permafrost precludes anchoring the exclosure with deeply set posts.

Many studies have examined the effectiveness of exclosures by comparing hatching success between protected and unprotected nests (Table 12.1). In general, these studies suggest that exclosures are effective at increasing hatching success, but the benefits are often short term. The ineffectiveness of exclosures stems from a variety of issues. First, exclosures may not always exclude all predators. For instance, fence size does not keep small predators (such as snakes, beetles, and mice) from depredating eggs (Mabee and Estelle 2000). Second, exclosures may increase the likelihood of nest abandonment. Hardy and Colwell (2008) showed that Snowy Plover clutches were significantly more likely to be abandoned. Abandonment may be much more likely if exclosures are erected during the laying process because females will be reluctant to enter the cage to lay an egg. Exclosures are less likely to cause abandonment when they are erected during the incubation phase (Niehaus et al. 2004). A third, serious problem associated with exclosures is the risk that they increase the mortality of incubating adults. Adults detecting a predator may be delayed in leaving the exclosure by the cage itself.

This delay in evasive response may render individuals more prone to capture by raptors. Several studies have reported unusually high mortality of adults in exclosures (e.g., Neuman et al. 2004; Isaksson et al. 2007; Hardy and Colwell 2008).

A problem with interpreting the success of nest exclosures (see Table 12.1) is that most studies have lacked proper experimental design such as random allocation of nests to treatments (that is, receiving protection from an exclosure versus an unprotected control) (Mabee and Estelle 2000). In particular, there may be a strong bias in results demonstrating exclosure effectiveness if observers opportunistically place exclosures around all nests such that the sample of unexclosed nests consists of those that failed before exclosures could be erected. These problems were addressed by Isaksson et al. (2007). Working in Sweden, they randomly allocated nests of two species, the Northern Lapwing and Common Redshank, to treatments and compared the reproductive value of clutches using a combination of daily survival rates, hatchability, and partial clutch loss. In both species, exclosed nests had nearly double the reproductive value of unexclosed clutches. Pauliny et al. (2008) reported similar findings for Dunlin nests in Europe.

Although there are clear short-term advantages of using exclosures, the costs often

TABLE 12.1

Summary of the effectiveness of caged predator exclosures based on the percentage of protected and unprotected nests
that successfully hatched or daily survival rate (DSR)

SPECIES	PROTECTED NESTS		UNPROTECTED NESTS		LOCATION	SOURCE
	% SUCCESSFUL OR DSR	N	% SUCCESSFUL OR DSR	N		
Northern Lapwing	0.989		0.966		Sweden	a
Piping Plover	92	26	25	24	Massachusetts	b
	90	29	17	24	Massachusetts	c
	60	5	75	4	Colorado	d
Snowy Plover	69	13	57	14	Colorado	d
	57	14	54	13	Colorado	d
	0.985	137	0.876	133	California	e
Killdeer	33	12	29	17	Ontario	f
	14	7	22	9	Colorado	d
	38	52	13	53	California	g
Common Redshank	0.997		0.964		Sweden	a
Pectoral Sandpiper	77	13	3	39	Alaska	h
Dunlin	80	25	57	60	Sweden	i

Source: a. Isaksson et al. 2007; b. Rimmer and Deblinger 1990; c. Melvin et al. 1992; d. Mabee and Estelle 2000; e. Hardy and Colwell 2008; f. Nol and Brooks 1982; g. Johnson and Oring 2002; h. Estelle et al. 1996; i. Pauliny et al. 2008.

outweigh the benefits. These costs include compromised adult survival and influencing the settlement of birds in low-quality habitats where predation is a problem. Isaksson et al. (2007) showed that predators killed some redshanks incubating in exclosures but that lapwings did not suffer similar fates. This difference between species may stem from the tendency for sandpipers to sit tight on a nest and become easier prey during the approach of a predator, whereas plovers leave the nest at a great distance. Murphy et al. (2003) reported that Piping Plovers incubating in small exclosures experienced high levels of mortality compared with adults incubating unprotected nests. Unusually high mortality

of incubating adult Snowy Plovers was associated with use of exclosures in two studies (Neuman et al. 2004; Hardy and Colwell 2008).

An additional shortcoming of exclosures is that although their use may alleviate problems with clutch predation it may not improve productivity if the same predators that depredate eggs also prey on chicks. Hardy and Colwell (2008) showed that the positive effect of increasing hatching success was not accompanied by high fledging success in Snowy Plovers. Neuman et al. (2004) reached similar conclusions from a long-term study. Furthermore, if exclosures boost hatching (but not fledging) success, it may increase fidelity to poor-quality breeding sites. In other words, artificially increasing hatching success with exclosures may dupe individuals into returning to or remaining at sites of low quality in subsequent breeding attempts. Alternatively, individuals may be better off moving to other breeding locations where predation is less of a problem.

In summary, exclosures have the potential to temporarily boost productivity of a local population. However, because population variation is most sensitive to changes in adult survival (Sandercock 2003) and the use of exclosures has been shown to occasionally increase adult mortality, the benefits of exclosures should be weighed against substantial costs that may compromise a population's viability. Consequently, exclosures are viewed as a short-term management tool to increase reproductive success in habitats where predation is a problem and alternative, lethal methods are precluded by public opinion or safety.

AVERSIVE CONDITIONING

Poisons (such as copper oxalate, methiocarb, and cabachol) are occasionally used to condition predators by causing individuals that ingest tainted food to become ill; with sufficient exposure, individuals avoid these foods. In practice, a distasteful chemical applied to a shorebird egg eaten by a predator would elicit a negative response, and with sufficient exposure individual predators learn to avoid the prey (eggs).

Successful conditioning requires special circumstances. The individual predators responsible for depredating eggs or chicks must be few in number and localized such that the conditioning has its effect on the offending animals. For example, at Point Reyes National Seashore in coastal northern California, breeding Common Ravens defend territories from which they exclude conspecifics (Roth et al. 2004). Ravens occasionally prey on Snowy Plover eggs and chicks, although their main fare is a mix of seabird eggs and young, and carrion. In this case, successful use of poisoned eggs to condition resident, territorial ravens might be productive in arresting the predation problem. However,

raven populations in the western U.S. have grown dramatically in the past few decades, so large numbers of nonbreeding "floaters" exist in the population. It would be a challenging program of aversive conditioning to manage the predatory effect on the productivity of a shorebird, given the abundance, wide-ranging movements, and intelligence of corvids.

TRANSLOCATION AND HOLDING

Predators may be captured and either released at distances beyond which they are likely to return to the site where problems exist, or they may be held in captivity until the breeding season is over. These approaches avoid issues related to lethal control, but they may be costly and ineffective if individual predators simply return to a site and continue to have a negative effect on a population.

SUBSIDIZED FOOD

An unusual and seemingly contradictory practice of providing food for predators, especially corvids, has been used in some conservation programs, although not for shorebirds. The idea is that an unlimited and predictable food source will keep individual predators from preying on eggs and chicks. Moreover, territorial predators will keep conspecifics from the area. This practice may be effective with localized, small populations of Common Raven depredating nests (Liebezeit and George 2002), but the consequences for population growth of the predator may be problematic in that abundant food may increase productivity of the predator population. In the long run, this approach acts counter to the objective of reducing predator abundance and of ameliorating negative impacts on prey populations.

CONTRACEPTION/STERILANTS

Chemicals such as gametocides (triethylene-melamine) reduce the short-term and long-term reproductive success of individuals. They thus bring about population control by reducing the productivity of a predator population. They are applied by injection or ingested.

They have limited effectiveness in controlling predator populations, they are expensive, and the individual predators may develop aversions to the taste of chemicals. There is no reported use of contraceptives to benefit shorebird populations.

CONSERVATION IMPLICATIONS

Predator control is undeniably one of the more complex and challenging facets of wildlife management because it elicits strong responses from conservationists and the public alike. It should be undertaken only after evidence has been collected that predators limit a shorebird population (Bolton et al. 2007). Even then, it may represent one aspect of a multifaceted approach to increasing shorebird population size. In reality, predator control is one of several tools (including habitat restoration and education) that address the root causes of the vulnerability of shorebirds eggs and chicks to predators.

LITERATURE CITED

Berg, Å. 1996. Predation on artificial, solitary and aggregated wader nests on farmland. *Oecologia* 107: 343–346.

Blackburn, T. M., P. Cassey, R. P. Duncan, K. L. Evans, and K. J. Gaston. 2004. Avian extinction and mammalian introductions on oceanic islands. *Science* 305: 1955–1958.

Bolton, M., G. Tyler, K. Smith, and R. Bamford. 2007. The impact of predator control on Lapwing *Vanellus vanellus* breeding success on wet grassland nature reserves. *Journal of Applied Ecology* 44: 534–544.

Colwell, M. A. 1992. Wilson's phalarope nest success is not influenced by vegetation concealment. *The Condor* 94: 767–772.

Côté, I. M., and W. J. Sutherland. 1996. The effectiveness of removing predators to protect bird populations. *Conservation Biology* 11: 395–405.

Driscoll, J. W., and P. Bateson. 1988. Animals in behavioural research. *Animal Behavior* 36: 1569–1574.

Dowding, J. E., and E. C. Murphy. 2001. The impact of predation by introduced mammals on endemic shorebirds in New Zealand: A conservation perspective. *Biological Conservation* 99: 47–64.

Eglington, S. M., J. A. Gill, M. A. Smart, W. J. Sutherland, A. R. Watkinson, and M. Bolton. 2009. Habitat management and patterns of predation of Northern Lapwings on wet grasslands: The influence of linear habitat structures at different spatial scales. *Biological Conservation* 142: 314–324.

Estelle, V. B., T. J. Mabee, and A. H. Farmer. 1996. Effectiveness of predator exclosures for Pectoral Sandpiper nests in Alaska. *Journal of Field Ornithology* 67: 447–452.

Gochfeld, M. 1984. Antipredator behavior: Aggressive and distraction displays of shorebirds. In *Shorebirds: Breeding behavior and populations,* ed. J. Burger and B. L. Olla, 289–377. New York: Plenum Press.

Goss-Custard, J. D. 1985. Foraging behaviour of wading birds and the carrying capacity of estuaries. In *Behavioural ecology,* ed. R. M. Sibly and R. H. Smith, 169–188. Oxford: Blackwell.

Grant, M. C., C. Orsman, J. Easton, C. Lodge, M. Smith, G. Thompson, S. Rodwell, and N. Moore. 1999. Breeding success and causes of breeding failure of curlew *Numenius arquata* in Northern Ireland. *Journal of Applied Ecology* 36: 59–74.

Green, R. E., and M. Robins. 1993. The decline of the ornithological importance of the Somerset Levels and Moors, England, and changes in the management of water levels. *Biological Conservation* 66: 95–106.

Hardy, M. A., and M. A. Colwell. 2008. The impact of predator exclosures on Snowy Plover nesting success: A seven-year study. *Wader Study Group Bulletin* 115: 161–166.

Isaksson, D., J. Wallander, and M. Larsson. 2007. Managing predation on ground-nesting birds: The effectiveness of nest exclosures. *Biological Conservation* 136: 136–142.

Jackson, D. B. 2001. Experimental removal of introduced hedgehogs improves wader nest success in the Western Isles, Scotland. *Journal of Applied Ecology* 38: 802–812.

Jackson, D. B., R. J. Fuller, and S. T. Campbell. 2004. Long-term population changes among breeding shorebirds in the Outer Hebrides, Scotland, in relation to introduced hedgehogs (*Ericaeus europaeus*). *Biological Conservation* 117: 151–166.

Jackson, D. B., and R. E. Green. 2000. The importance of the introduced hedgehog (*Ericaeus europeus*) as a predator of eggs of waders (Charadrii) on the marchai in South Uist, Scotland. *Biological Conservation* 93: 333–348.

Johnson, M., and L. W. Oring. 2002. Are predator exclosures an effective tool in plover conservation? *Waterbirds* 25: 184–190.

Keedwell, R. J., R. F. Maloney, and D. P. Murray. 2002. Predator control for protecting kaki (*Himantopus novaezelandiae*): Lessons from 20 years of management. *Biological Conservation* 105: 369–374.

Lack, D. 1968. *Ecological adaptations for breeding in birds*. London: Methuen.

Lank, D. B., R. W. Butler, J. Ireland, and R. C. Ydenberg. 2003. Effects of predation danger on migration strategies of sandpipers. *Oikos* 103: 303–319.

Leopold, A. 1933. *Game management*. Madison: University of Wisconsin Press.

Liebezeit, J. R., and T. L. George. 2002. *A summary of predation by corvids on threatened and endangered species in California and management recommendations to reduce corvid predation*. Species Conservation and Recovery Program, Report 2002-02, Sacramento, CA: California Department of Fish and Game.

Lyons, J. E., M. C. Runge, H. P. Laskowski, and W. L. Kendall. 2008. Monitoring in the context of structured decision-making and adaptive management. *Journal of Wildlife Management* 72: 1683–1692.

Mabee, T. J., and V. B. Estelle. 2000. Assessing the effectiveness of predator exclosures for plovers. *Wilson Bulletin* 112: 14–20.

MacDonald, M. A., and M. Bolton. 2008a. Predation of Lapwing *Vanellus Vanellus* nests on lowland wet grassland in England and Wales: Effects of nest density, habitat and predator abundance. *Journal of Ornithology* 149: 555–563.

———. 2008b. Predation on wader nests in Europe. *The Ibis* 150 (Suppl.): 54–73.

Martin, T. E. 1993. Nest predation and nest sites: New perspectives on old patterns. *BioScience* 43: 523–532.

Melvin, S. M., L. H. MacIvor, and C. R. Griffin. 1992. Predator exclosures: A technique to reduce predation at Piping Plover nests. *Wildlife Society Bulletin* 20: 143–148.

Muir, J. J., and M. A. Colwell. 2010. Snowy Plovers select open habitats for courtship scrapes and nests. *The Condor*.

Murphy, R. K., I. M. G. Michaud, D. R. C. Prescott, J. S. Ivan, B. J. Anderson, M. L. French-Pombier.

2003. Predation on adult Piping Plovers at predator exclosure cages. *Waterbirds* 26: 150–155.

Neuman, K. K., G. W. Page, L. E. Stenzel, J. C. Warriner, and J. S. Warriner. 2004. Effect of mammalian predator management on Snowy Plover breeding success. *Waterbirds* 27: 257–263.

Niehaus, A. C., D. R. Ruthrauff, and B. J. McCaffery. 2004. Response of predators to Western Sandpiper predator exclosures. *Waterbirds* 27: 79–82.

Nol, E., and R. J. Brooks. 1982. Effects of predator exclosures on nesting success of Killdeer. *Journal of Field Ornithology* 53: 263–268.

Pauliny, A., M. Larsson, and D. Blomqvist. 2008. Nest predation management: Effects on reproductive success in endangered shorebirds. *Journal of Wildlife Management* 72: 1579–1583.

Pienkowski, M. W. 1984. Breeding biology and population dynamics of Ringed Plovers *Charadrius hiaticula* in Britain and Greenland: Nest-predation as a possible factor limiting distribution and timing of breeding. *Journal of Zoology, London* 202: 83–114.

Pierce, R. J., and C. Blanvillain. 2004. Current status of the endangered Tuamotu Sandpiper or Titi *Prosobonia cancellata* and recommended actions for its recovery. *Wader Study Group Bulletin* 105: 93–100.

Rimmer, D. W., and R. D. Deblinger. 1990. Use of predator exclosures to protect Piping Plover nests. *Journal of Field Ornithology* 61: 217–223.

Rönkä, A., K. Koivula, M. Ojanen, V.-M. Paknen, M. Pohjoismäki, K. Rannikko, and P. Rauhala. 2006. Increased nest predation in a declining and threatened Temminck's Stint *Calidris temminckii* population. *The Ibis* 148: 55–65.

Roth, J. E., J. P. Kelly, W. J. Sydeman, and M. A. Colwell. 2004. Sex differences in space use of breeding Common Ravens (*Corvus corax*) in western Marin County, California. *The Condor* 106: 529–539.

Sandercock, B. K. 2003. Estimation of survival rates for wader populations: A review of mark-recapture methods. *Wader Study Group Bulletin* 100: 163–174.

U.S. Department of the Interior. 1993. Threatened status for the Pacific coast population of the Western Snowy Plover. *Federal Register* 58: 12864–12874.

U.S. Fish and Wildlife Service. 2007. *Recovery plan for the Pacific Coast population of the Western Snowy Plover (*Charadrius alexandrinus nivosus*)*. 2 vols. Sacramento, CA: California/Nevada Operations Office, U.S. Fish and Wildlife Service.

Managing Human Disturbance

CONTENTS

SHOREBIRDS OCCUPY HABITATS that are highly valued by humans for commercial, recreational, and agricultural purposes. Consequently, human activity has the potential to negatively influence the behavior, local distribution and abundance, productivity, and survival as well as population dynamics of shorebirds in a variety of habitats. In meadows and prairies, the nests of shorebirds may be lost to trampling by livestock or run over by farm equipment used to harvest crops. In coastal regions, several species of shorebird breed on ocean beaches that are prized by humans for recreation, with negative consequences for incubation and reproductive success. Finally, in estuaries and interior wetlands, large flocks forage and roost amid habitats where the presence of humans may disturb feeding and roosting birds and thus degrade habitat quality. These human activities have the potential to negatively affect reproductive success, survival, and hence reduce population sizes. In fact, the U.S. Shorebird Conservation Plan (Brown et al. 2001) repeatedly lists disturbance as an important conservation issue confronting shorebirds. Surprisingly, however, remarkably little has been done to quantify the impact of human disturbance on shorebirds. In this chapter, I review evidence that humans directly and indirectly alter shorebird behaviors and distributions, and I assess evidence that these effects translate into population declines. I conclude with a review of management practices used to ameliorate the negative effects of human disturbance on shorebirds.

DEFINITIONS OF HUMAN DISTURBANCE

A rich and growing body of literature has addressed the negative impacts of human activity on wildlife. An important subset of the disturbance literature has focused on shorebirds throughout the annual cycle. Only recently, however, has a clear understanding of the measures and consequences of disturbance to birds been articulated (see papers in Drewitt 2007). It was Gill et al. (2001) who clearly made the case that effects measured in the changes in behavior and distribution are especially of conservation concern if they have negative consequences for populations. Shorebirds may be disturbed by humans and be flushed from a roost, for instance, but the effect of this disturbance on a local population may be minimal if birds have alternative habitats of comparable quality to which they can retreat. Under chronic disturbance, however, individuals may incur energy costs of evasive flight that translate into reduced survival, especially if food is short in supply and inclement weather taxes the birds' energy.

An important initial step in addressing the problem of negative impacts of human activity on wildlife requires a clear understanding of what constitutes "disturbance." Certainly, natural events commonly disturb wildlife by altering their behavior. A Peregrine Falcon hunting wintering sandpipers over an estuary imposes an energy cost on birds that encompasses increased vigilance, reduced intake rates, and costs of evasive flight. Taken to an extreme, a successful falcon causes some level of mortality. In the case of anthropogenic disturbance, however, we are concerned with the effect of human activity on shorebirds above and beyond this background level of natural disturbance. Senner and Howe (1984) defined disturbance as the "disruption of normal activity patterns (e.g., flushing a feeding bird), including the direct destruction of nests, eggs or young." This definition requires a clear distinction between what constitutes "normal," which is assumed to be related to natural or "background levels" of disturbance by predators

(perceived or real) and physical features of the environment (such as a crashing wave causing rocky intertidal shorebirds to take flight). This definition also mixes individual-based behavioral measures with fitness consequences. As such its utility is limited. For example, one could quantify incubation constancy in nesting birds under levels of increasing human activity and demonstrate a positive correlation with more incubation recesses of prolonged duration. But the consequences of altered reproductive behaviors for individual reproductive success and the resulting population effects are unknown in this case. In other words, simply demonstrating that behavior is increasingly affected by higher levels of human activity does not get at the fitness consequences. Do disturbed birds that experience more frequent interruptions to incubation incubate for longer and have higher hatching asynchrony, and does this increase clutch loss and lower reproductive success? And what are the relationships between disturbance and predation? For instance, plovers nesting in sandy substrates leave footprint impressions each time they come and go from a nest. A possible consequence of human disturbance is that incubating adults leave more tracks near a nest, which attracts the attention of scavengers such as corvids, which increases the likelihood of nest failure.

Others have offered slightly different definitions of disturbance. Burger (1981) distinguished between *direct* and *indirect* effects of human activity on shorebirds by drawing attention to whether human activity influenced the behavior, reproduction, and mortality of individuals versus altered a species' distribution, respectively. The usefulness of this distinction is unclear. In effect, it distinguishes between the immediate effects on individuals and the long-term population responses of shorebirds to chronic levels of human activity. For instance, Pfister et al. (1992) documented the indirect effects of human activity on migrant shorebird distribution on Cape Cod, Massachusetts. They used correlations to suggest that increased human activity had caused shorebirds to leave beach habitats and shift to back-dune wetlands where humans were less likely to occur. In this case, chronic human activity was argued to have affected the choice of habitat by shorebirds such that with increased disturbance birds had opted to use habitats where fewer humans were present. In Burger's (1981) scheme, this is an example of indirect effects of disturbance, yet presumably it stems from direct effects of chronic disturbance altering behavior.

Several works (e.g., Davidson and Rothwell 1993; Gill 2007) have clarified the conservation value of understanding disturbance by identifying four levels at which human activity affects wildlife (Fig. 13.1). First, the distribution of wildlife may be affected owing to long-term avoidance of areas of chronic, high levels of human activity or short-term, intensive use by humans. Second, individual behaviors (flight response, vigilance, incubation pattern, or foraging intake rates) may be altered owing to the presence of humans. Third, these behavioral changes may negatively impact a species' demography by reducing fecundity or survival. Last, there may be population consequences associated with these changes in demography.

Gill et al. (2001) illustrated the conservation implications stemming from the interrelationships between behavioral responses, habitat availability, and population consequences (Fig. 13.2). Consider a case in which human activity occurs in the vicinity of a shorebird roost. This activity may have some effect on shorebird behavior by increasing their vigilance and flight response. If disturbance persists and increases, birds may opt to move to another roost. The availability of alternative roosts to these birds is crucial to understanding the relative strength of impacts of this disturbance to the birds. If alternative roosts exist nearby (Fig. 13.2A and B), then the costs of disturbance to individuals are relatively minor, and the number of roosting birds at the site will decrease as individuals move to alternative roosts. In many cases, however, biologists would interpret this behavioral response of moving to another habitat as a negative impact

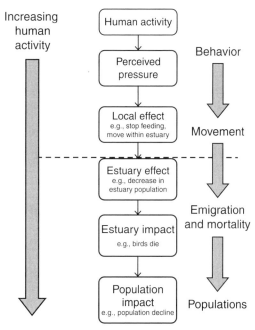

FIGURE 13.1. Impact of human activity on wildlife. The effect begins with altered behaviors owing to the perception of danger by animals. As the level of human activity increases, birds move more frequently within a local area (such as an estuary) to avoid areas of human activity. At sufficiently high or chronic levels, disturbance may cause the exodus or death of some individuals. A final consequence of human disturbance is the local decline of a population. Note that emigration may not result in a decline in a global population if emigrants find suitable habitat elsewhere of sufficient quality to allow them to survive and reproduce. This would represent a shift in distribution but not a decline in abundance. After Davidson and Rothwell (1993).

of disturbance. However, while it does alter behavior, the population effect is probably trivial, especially if it occurs infrequently. Now consider a scenario of similar levels of disturbance with the difference being that alternative roosts do not exist (Fig. 13.2C and D). Birds would behave similarly but their numbers would not change at the roost. In this case, the population consequences of disturbance may be substantial because individuals would repeatedly incur the energy costs of evasive flight but have no alternative location where they could roost. The number of roosting birds at the site will remain the same, but the fitness costs may be comparatively large.

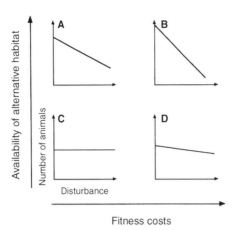

FIGURE 13.2 Effect of increasing fitness costs of disturbance (that is, increasing mortality or decreasing reproductive success) and availability of alternative habitat on the distribution of animals across sites that vary in disturbance levels. Four scenarios are presented that show how the relative number of animals across sites varies with the degree of disturbance in those sites. If the fitness costs of disturbance are high (B and D), individuals move away from disturbed sites; hence, numbers are lower in sites with greater disturbance. Similarly, a high availability of alternative habitat elsewhere (A and B), allowing individuals to move readily, will result in a strong decrease in numbers in disturbed sites, even when the fitness costs of disturbance are low (A). If there is little excess habitat to move to (C and D), there will be little change in numbers with increasing disturbance, even if there is a large fitness cost in terms of reduced survival or fecundity (D). From Gill et al. (2001).

Consider the same reasoning applied to foraging habitat where human activity disturbs foraging birds. If alternative foraging habitats are available nearby, then disturbance may be inconsequential compared with when alternative foraging habitat is unavailable or at a great enough distance that transit costs are high. In the latter case, shorebirds must incur the cost of disturbance while remaining to feed in the only available location. Application of the conceptual model of Gill et al. (2001) to breeding circumstances is rather straightforward, especially because the availability of alternative habitats is considered low and the fitness costs of disturbance are high. Consider an incubating shorebird with a nest situated on a beach with high recreational activity. The availability of alternative habitat is always low because a bird cannot simply move its nest to another site. Consequently, any level of human activity in close proximity to the nest will elicit a behavioral response by an individual, which may include absences from the nest for varying durations. This behavioral response of an adult may have subtle or not so subtle effects on embryo development and hatching synchrony, and may increase the risk of egg predation. The population consequences of these behavioral changes in incubation behavior may be more difficult to demonstrate.

CHARACTERIZING DISTURBANCE

One of the most challenging facets of researching the impact of human activity on wildlife involves quantifying disturbance itself. The response variables of altered distributions, behavior (vigilance and incubation), and demographic variables are relatively straightforward. However, it is more difficult to characterize the various types of disturbance that cause changes in the behavior of individuals. Disturbance varies in type, distance to a shorebird, and the loudness of sound produced. As a contrast, consider the differences that exist in disturbances, ranging from the report of a hunter's gun some distance away to the chronic noise and motion from passing vehicles on a road. It is easier to quantify the response of the individual shorebird than to characterize the disturbance itself.

Humans undertake a variety of activities in habitats frequented by breeding and nonbreeding shorebirds. On coastal beaches worldwide, recreational activities such as fishing, walking and jogging, horseback riding, driving vehicles, and dog-walking on and off leash occur to varying extents. In the coastal plain of Alaska's North Slope, noise and vehicle traffic associated with oil and gas exploration occur across prime shorebird breeding habitats. In temperate grasslands habitats, farming practices may disrupt nesting shorebirds. At wetlands and coastal estuaries, activities associated with commerce and recreation may compromise the quality of foraging habitat by altering intake rates of shorebirds during winter or at staging sites. It is clear, however, from a review of the literature that the various types of human activity are perceived differently by breeding and nonbreeding shorebirds.

Disturbance may vary in type (such as the load report of a shotgun or a dog running off leash), intensity (decibels or speed of movement), and duration. Shorebirds appear to be more sensitive to loud, fast-moving stimuli (such as aircraft, dogs, and vehicles) in close proximity compared with distant, slow-moving objects (such as pedestrians). A variety of studies suggest that dogs, both on and off leash, have a greater tendency to alter the normal behavior of shorebirds. Nesting shorebirds are especially susceptible to disturbance because the stationary nests constrain their ability to avoid human activity by moving to another habitat. Lord et al. (2001) examined responses of incubating New Zealand Dotterels to three types of experimental approach by humans: walkers, joggers, and a person walking a leashed dog. Dotterels reacted most strongly to dogs—they ceased incubation and left the nest at approximately 100 m, and they remained off the nest for nearly 5 minutes. Wintering shorebirds also react strongly to dogs. Lafferty (2001) examined the Snowy Plover response to varying types of disturbance at wintering locations in southern California. At all distances up to 80 m, Snowy Plovers reacted (became vigilant or took flight) more strongly to dogs than people.

RESPONSES TO DISTURBANCE

The responses of wildlife to disturbance can be monitored with a variety of methods, some focused on the physiology, condition, and behaviors of individual animals and others quantifying changes in distribution and abundance over time. Relatively few studies have effectively quantified the effect of disturbance on demography and population size (Gill 2007). Table 13.1 summarizes some studies that have examined the various impacts of human activity on the behavior, distribution, and demography of shorebirds.

PHYSIOLOGICAL AND CONDITION MEASURES

Researchers have examined plasma and fecal levels of corticosterone, the stress hormone produced by the adrenal glands, and have shown higher levels associated with greater disturbance (e.g., Wasser et al. 1997). In response to disturbance and elevated corticosterone, plasma levels of triglycerides decreased 25% to 45% in captive European Starlings (Remage-Healey and Romero 2001). Finally, heart-rate monitors permit an immediate assessment of the stress imposed by human activity, which may last for several hours (e.g., Weimerskirch et al. 2002). After prolonged stress, individuals may have varied indices of condition (such as decreased mass). No papers have examined the physiological responses of shorebirds to human activity.

BEHAVIOR

A straightforward means of examining disturbance is to measure the behavioral responses of individuals in association with varying levels of human activity. For example, vigilance may increase and foraging rate decrease with greater human activity, and birds subjected to disturbance in close proximity may take flight. The difficulty in using behaviors as an index of disturbance is in distinguishing between baseline levels of vigilance and flight in association with the presence of natural predators (Sutherland 2007).

Shorebirds have high metabolic rates compared with other birds (Kersten and Piersma 1987). During migration, staging shorebirds require relatively undisturbed foraging conditions to fuel their long-distance migrations. Consequently, the presence of humans in close proximity may adversely affect their ability to forage and may reduce their daily energy intake. Burger et al. (2007) examined behavior and avoidance of disturbance in several species of shorebird feeding on eggs of horseshoe crabs

TABLE 13.1

Summary of selected studies evaluating direct and indirect responses of shorebirds to disturbance during the breeding and nonbreeding periods

	MEASURED RESPONSE TO HUMAN ACTIVITY				
	BEHAVIOR	DISTRIBUTION	DEMOGRAPHY	COMMENTS	SOURCE
BREEDING					
Common Ringed Plover		✓		Breeding mostly confined to protected areas in U.K.	Prater (1989)
Piping Plover	✓	✓		Foraging time decreased with greater numbers of humans, and habitat use negatively correlated with human numbers	Burger (1994)
	✓		✓	Adult and chick behavior altered in presence of humans; lower fledging success in areas of higher disturbance	Flemming et al. (1988)
Snowy Plover			✓	Higher chick mortality on weekends versus weekdays	Ruhlen et al. (2003)
	✓	✓	✓	Fenced area resulted in decreased disturbance, increased abundance, and higher reproductive success compared with before fence established	Lafferty et al. (2006)
			✓	Lower chick survival on beaches favored by humans for recreation; higher fledging success within fenced area	Colwell et al. (2007)

Species			Effect	Reference
White-fronted Plover	✓	✓	Decreased nest attentiveness with increased disturbance; habituated birds allowed closer approach and returned to incubate sooner at sites with higher disturbance; higher chick mortality at site with higher human activity	Baudains and Lloyd (2007)
Eurasian Oystercatcher		✓	Disturbance in foraging areas decreased incubation and increased time feeding; food delivered to chicks decreased with disturbance	Verhulst et al. (2001)
NONBREEDING				
Eurasian Oystercatcher		✓	Birds allowed closer approach and returned to feed sooner during energetically demanding times of winter	Stillman and Goss-Custard (2002)
Snowy Plover		✓	Birds disturbed less often in protected areas and on weekends (versus weekdays)	Lafferty (2001)
Eurasian Oystercatcher, E. Curlew, Dunlin, Common Redshank		✓	Reduced feeding and lower densities associated with construction adjacent to intertidal areas	Burton et al. (2002)
Eurasian Curlew, Northern Lapwing, Black-tailed Godwit		✓	Lower curlew (but not Lapwing or Godwit) densities associated with higher road densities in agricultural grasslands	Milsom et al. (1998)
Dunlin		✓	Roosting birds more "nervous" in larger flocks, but no clear relationship to variation in human activity among six locations	Conklin et al. (2008)

TABLE 13.1. (*continued*)

| | MEASURED RESPONSE TO HUMAN ACTIVITY | | | | |
	BEHAVIOR	DISTRIBUTION	DEMOGRAPHY	COMMENTS	SOURCE
			NONBREEDING		
Red Knot, Sanderling, Semipalmated Sandpiper, Ruddy Turnstone	✓	✓		Foraging birds left beaches in association with disturbance, especially by dogs, and did not return to predisturbance abundance after as much as 10 min; education and restricted access to beaches resulted in lowered disturbance	Burger et al. (2007); Burger et al. (2004)
Semipalmated Plover, Least Sandpiper	✓			Foraging rates decreased with increased human activity	Yasué (2005, 2006)
Sanderling	✓	✓		Foraging time decreased with greater numbers of humans within 100 m, and birds altered foraged more at night	Burger and Gochfeld (1991)

during the birds' spring stopover in Delaware Bay. Various human activities, including vehicles, pedestrians, and dogs, caused shorebirds to abandon stretches of beach where they foraged. In a before–after comparison of bird use of beaches, the densities of shorebirds remained low for up to 10 minutes after a disturbance, especially those involving dogs. It was unclear, however, whether shorebirds had simply moved to an alternative foraging site where disturbance was less, which is critical to evaluating population-level consequences (Gill et al. 2001).

An insightful way of gauging impacts of disturbance on foraging shorebirds would be to examine changes in time-activity budgets and intake rates associated with varying levels of human activity. Yasué (2005) examined the effect of human activity on the intake rates of Semipalmated Plovers and Least Sandpipers at a migration stopover site in coastal British Columbia. He controlled for the possible effects of flock size (that is, interference) and prey density and found that plovers had lower intake rates in the presence of humans; the effect of humans on sandpipers feeding on amphipods was less clear. Given that plovers use visual cues to locate prey, it is plausible that the decrease in intake rates resulted from an increase in vigilance by plovers feeding in close proximity to humans. Tactile-feeding sandpipers, by contrast, may have been able to continue feeding simultaneously with maintaining high vigilance near humans.

The behavioral impacts of disturbance may be of greatest concern when associated with breeding behavior, especially incubation and brooding. The incubation behavior of shorebirds, especially open-nesting species such as plovers, can be negatively affected by the presence of humans. Flemming et al. (1988) showed that humans disturbed foraging adult Piping Plovers at a greater distance than the natural disturbances (predators). Moreover, when humans approached to within 160 m of young plovers, they exhibited decreased feeding rates and increased vigilance. Incubating female Snowy

Plovers were more likely to cease incubation for longer intervals under conditions of high human activity within 100 m of the nest (Hoffmann 2005). Finally, Verhulst et al. (2001) conducted experiments in which they disturbed breeding Eurasian Oystercatchers tending chicks. Adults spent less time feeding themselves and delivered less food to their young when disturbed by humans. In each of these examples, disturbance translates into lowered reproductive success, but the mechanism by which this occurs varies from decreased incubation constancy in plovers to reduced feeding and lowered mass in oystercatchers.

Although shorebird young are precocial and nidifugous, adults tending young may be constrained in their abilities to avoid human disturbance because chicks, especially young ones, lack the mobility necessary to evade disturbance. In eastern Australia, the Hooded Plover breeds in beach habitats, which are recreationally important to humans. Weston and Elgar (2005) showed that disturbance reduced feeding and disrupted brooding of chicks, but there was no clear relationship to reproductive success.

DISTRIBUTIONAL PATTERNS

Given chronic levels of (even low-intensity) disturbance, shorebirds may opt to move to alternative locations that afford greater foraging opportunities, safer resting habitats, or more productive breeding conditions. Pfister et al. (1992) showed that over 17 years and under conditions of increasing human activity, several species of southbound migrants staging on Cape Cod had shifted their principal roosting locations from ocean-fronting beaches to back beaches. Burton et al. (2002) showed that activity associated with construction of a tidal barrage in Cardiff Bay, Wales, resulted in reduced densities and decreased foraging in several species of shorebird over 11 years. Other studies have suggested that human recreational activity occasionally has minimal adverse effects on shorebirds. For instance, shorebird abundance, diversity, and foraging behavior (the percentage of individuals feeding) did not differ between locations where trails ran adjacent to San Francisco Bay tidal flats and where sites lacked trails (Trulio and Sokale 2008). In coastal northern California, variation in human activity on 40 ocean-fronting beaches was not correlated with shorebird densities (Colwell and Sundeen 2000), but the level of human activity on these beaches was undoubtedly far less than that recorded on Cape Cod by Pfister et al. (1992).

DEMOGRAPHIC PARAMETERS

Human disturbance degrades habitat quality by affecting the ability of individuals to meet their daily energy budgets, reproduce, and survive. Consequently, the reproductive success and survivorship of individuals should differ among habitats with varying levels of human activity. Flemming et al. (1988) suggested that fledging success of Piping Plovers was lower on beaches with higher levels of disturbance. Ruhlen et al. (2003) studied Snowy Plover chick mortality on ocean beaches of Point Reyes National Seashore, California, in association with weekends and weekdays. They did not directly measure human activity, however. Plover chicks were more likely to die on weekends than weekdays. Melvin et al. (1994) reported on 19 mortalities of Piping Plovers caused by off-road vehicles running over adults (2) and chicks (17) on Atlantic Coast beaches. On rocky shores of southeast Alaska, Morse et al. (2006) showed nest success was compromised by high tides and that survival of Black Oystercatcher chicks was generally low. Human activity in the national park study area was relatively low and was concentrated at beach locations that offered easy access to boaters. Human activity was not clearly linked to the reduced productivity of oystercatchers, but disturbance was generally low.

The seemingly conflicting results of various studies seeking to quantify the negative impacts of human activity on shorebirds may be related to a variety of factors, some methodological and others specific to the shorebird species or the habitats they occupy at different times of year. First, species (large versus small) have been shown to differ in their response to human activity (Smit and Visser 1993; Davidson and Rothwell 1993). Second, the responses of individuals almost certainly vary with time of year and the principal activities engaged in by breeding and nonbreeding shorebirds. Third, studies have been conducted in habitats that vary greatly in type, intensity, and proximity and in level of human activity. The failure to find a correlation between varying levels of disturbance in one geographic region (such as northern California) may stem from the generally low levels of human activity there compared to other locations (Cape Cod). Moreover, the subtleties of habitat context (narrow versus wide beaches and the ability of shorebirds to move away from human activity) are often not considered. Finally, the availability of alternative habitats that allow individuals to evade disturbance has rarely been considered in many studies demonstrating behavioral responses (Gill et al. 2001).

INTERSPECIFIC DIFFERENCES

The behavioral responses of birds to human activity vary among species, and this variation is correlated with species' size (mass) (Smit and Visser 1993; Davidson and Rothwell 1993). This difference may be explained by greater wing-loading of large than small shorebirds and the fact that individuals of greater mass require longer to take flight in response to danger. Remarkably, this result has even been shown for individual Western and Least Sandpipers that differed in mass (Burns and Ydenberg 2002). Consequently, smaller species allow humans to approach more closely than larger shorebirds. This has obvious implications for the effects of disturbance on different species. In particular, larger shorebirds may be less tolerant of the presence of even a few humans. Hence, the amount of tidal flat "degraded" by the presence of humans (who may be clamming, for instance) is greater for large than small taxa (Smit and Visser 1993; Davidson and Rothwell 1993). Moreover, Davidson and Rothwell (1993) showed that the adverse impact exerted by varying numbers of people (evenly dispersed) on tidal flats differed among species. Negative impacts were greatest for the Eurasian Curlew, followed by the Common Redshank and Dunlin.

BEHAVIORAL CONTEXT

Clearly, the response of individuals to disturbance varies with activity patterns that vary seasonally. During the nonbreeding season, Stillman and Goss-Custard (2002) showed that Eurasian Oystercatchers were more tolerant of disturbance during the middle of winter when food shortages made foraging and maintaining energy stores more difficult. During the breeding season, single disturbances may have greater and more immediate fitness consequences (a failed nest attempt) than during the nonbreeding season (lowered intake rate). While incubating, species also differ in response to disturbance based on their life-history characteristics. For example, shorebirds that nest in open habitats (such as plovers, stilts and avocets, oystercatchers, and thick-knees) are more prone to disturbance while incubating than species that conceal themselves in vegetation (such as sandpipers). Several studies have shown that plovers of a variety of species stop incubation when humans approach to within 50 to 100 m (Lord et al. 2001; Lafferty 2006). By contrast, sandpipers may tolerate much closer approach before they flush from a nest.

HABITUATION

Under some circumstances, the impact of human activity on behavior and distribution of shorebirds may be decreased over time as individuals become habituated to humans, especially when activities pose minimal perceived risk. Habituation results when individuals, exposed to chronic disturbance (stimuli), alter their behavior to become increasingly tolerant of human activity. The behaviors of vigilance and flight response decrease in foraging or roosting birds; incubation and brooding behavior persists upon closer approach by humans. Any assessment of disturbance using behavioral measures must consider that individuals may become habituated to disturbance. This is especially true when individuals may not be able to avoid disturbance, as is the case with nesting shorebirds that are forced to incubate a clutch and cannot simply move away in response to increasing human activity nearby.

Demonstrating that individuals have become habituated to humans is a challenging task. Ideally, it involves detailed prior knowledge of individually marked birds with known behavioral and reproductive histories. In this way, a before–after/control–impact study is possible with individuals of known age and experience serving as controls in a situation where they may breed multiple times in areas varying in disturbance. Unfortunately, no study has satisfactorily addressed these difficult measures to demonstrate habituation. Nevertheless, several studies offer results suggesting that shorebirds may become habituated to human activity. Lord et al. (2001) noted that New Zealand Dotterels seemed to be habituated to humans because the average distance of response and the duration off the nest were shorter at sites with more human activity. Still, there is some limit to which individuals will tolerate human activity, and this response threshold probably varies among species and depends on the context with which disturbance occurs, as well as the type, frequency, and intensity of human activity.

MANAGING DISTURBANCE

A first step in minimizing the adverse effects of human disturbance on wildlife is to educate the public with effective outreach programs that communicate the negative impact that humans have on shorebirds. In many instances, humans are unaware of the presence of breeding shorebirds when they engage in recreational activities. For instance, kayakers in Alaskan coastal waters are largely unaware of the presence of

FIGURE 13.3. Restricting human activity near shorebirds. In coastal areas where human recreational use of beaches occurs in close proximity to breeding shorebirds, roped fencing and signs are used to educate the public and reduce disturbance to breeding shorebirds. In this image, a Snowy Plover nest in the background has been protected by a caged predator exclosure, and fencing limits the approach of humans. Photo credit: Sean McAllister.

nests and young of Black Oystercatchers breeding in rocky intertidal habitats. Similar observations apply to human recreational use in the vicinity of breeding Piping and Snowy Plovers, both of which nest on ocean-fronting beaches that are favored by humans for a variety of recreational activities. In many instances, beachgoers are ignorant of their impact on shorebirds, and simple signage meant to educate the public appears inadequate to ameliorate this observation. Docents, especially those interacting with the public during periods of high human recreational use, may be a more effective means of educating the public.

Aside from education, the various ways of managing the negative impact of disturbance on shorebirds all involve some level of restriction on the activity of humans within shorebird habitats. As human populations increase, these restrictive measures may become increasingly important in managing disturbance. For breeding shorebirds, areas may be temporarily closed

to human access to minimize the extent to which incubating and brooding adults are kept from tending eggs or young. For example, the increase in the Atlantic Coast population of the Piping Plover stems, in part, from the closure of beaches to humans and increased productivity of plovers in these closed areas. Beach closure is an extreme measure that is often met with considerable resistance from the public. A less restrictive approach to minimizing disturbance is the use of "symbolic fencing" to limit access of humans to particular habitats that are important for breeding shorebirds (Fig. 13.3). Along the Pacific Coast, the local increase in the Snowy Plover at Coal Oil Point, California, was directly related to efforts to educate the public and restrict access to important wintering habitat along beaches. Once wintering habitat was protected, a rapid increase in breeding plovers ensued (Lafferty et al. 2006).

CONSERVATION IMPLICATIONS

Disturbance represents a form of habitat degradation in which the behavior of individuals is altered (increased incubation recesses, compromised parental care of chicks, and lowered intake rate) owing to the presence of humans. These changes in behavior have the potential to reduce productivity and survival and, consequently, to decrease the size of shorebird populations. Various means of reducing disturbance entail restrictive measures to reduce human activity in close proximity to shorebirds. The behavioral effects of disturbance may not, however, translate directly into reduced survival or productivity, especially when alternative habitats are available. Even protected areas set aside specifically for shorebirds may be increasingly impacted by human activities just outside reserve boundaries (Burton 2007). This requires that remaining habitats be managed to minimize the effects of disturbance.

LITERATURE CITED

Baudains, T. P., and P. Lloyd. 2007. Habituation and habitat changes can moderate the impacts of human disturbance on shorebird breeding performance. *Animal Conservation* 10: 400–407.

Brown, S., C. Hickey, B. Harrington, and R. Gill. 2001. *United States Shorebird Conservation Plan*. 2nd ed. Manomet, MA: Manomet Center for Conservation Sciences.

Burger, J. 1981. The effect of human activity on birds at a coastal bay. *Biological Conservation* 21: 231–241.

———. 1994. The effect of human disturbance on foraging behavior and habitat use in Piping Plover (*Charadrius melodus*). *Estuaries* 17: 695–701.

Burger, J., S. A. Carlucci, C. W. Jeitner, and L. Niles. 2007. Habitat choice, disturbance, and management of foraging shorebirds and gulls at a migratory stopover. *Journal of Coastal Research* 23: 1159–1166.

Burger, J., and M. Gochfeld. 1991. Human activity influence and diurnal and nocturnal foraging of Sanderlings (*Calidris alba*). *The Condor* 93: 259–265.

Burger, J., C. Jeitner, K. Clark, and L. J. Niles. 2004. The effect of human activities on migrant shorebirds: Successful adaptive management. *Environmental Conservation* 31: 283–288.

Burns, J. G., and R. C. Ydenberg. 2002. The effects of wing-loading and gender on escape flights of Least (*Calidris minutilla*) and Western (*Calidris mauri*) Sandpipers. *Behavioral Ecology and Sociobiology* 52: 128–136.

Burton, N. H. K. 2007. Landscape approaches to studying the effects of disturbance on waterbirds. *The Ibis* 149 (Suppl. 1): 95–101.

Burton, N. H. K., M. M. Rehfisch, and N. A. Clark. 2002. Impacts of disturbance from construction work on the densities and feeding behavior of waterbirds using the intertidal mudflats of Cardiff Bay, U.K. *Environmental Management* 30: 865–871.

Colwell, M. A., S. J. Hurley, J. N. Hall, and S. J. Dinsmore. 2007. Age-related survival and behavior of Snowy Plover chicks. *The Condor* 109: 638–647.

Colwell, M. A., and K. R. Sundeen. 2000. Shorebird distributions on ocean beaches of northern California. *Journal of Field Ornithology* 71: 1–15.

Conklin, J. R., M. A. Colwell, and N. W. Fox-Fernandez. 2008. High variation in roost use by Dunlin wintering in California: Implications for habitat limitation. *Bird Conservation International* 18: 275–291.

Davidson, N. C., and P. I. Rothwell. 1993. Disturbance to waterfowl on estuaries: Conservation and management implications of current knowledge. *Wader Study Group Bulletin* 68: 97–105.

Drewitt, A., ed. 2007. "Birds and recreational disturbance." Special issue, *The Ibis* 149 (Suppl. 1).

Flemming, S. P., R. D. Chiasson, P. C. Smith, P. J. Austin-Smith, and R. P. Bancroft. 1988. Piping Plover status in Nova Scotia related to its reproductive and behavioral responses to human disturbance. *Journal of Field Ornithology* 59: 321–330.

Gill, J. A. 2007. Approaches to measuring the effects of human disturbance on birds. *The Ibis* 149 (Suppl. 1): 9–14.

Gill, J. A., K. Norris, and W. J. Sutherland. 2001. Why behavioural responses may not reflect the population consequences of human disturbance. *Biological Conservation* 97: 265–268.

Hoffmann, A. 2005. Incubation behavior of female Snowy Plovers (*Charadrius alexandrinus nivosus*) on sandy beaches. M.Sc. thesis, Humboldt State University.

Kersten, M., and T. Piersma. 1987. High levels of energy expenditure in shorebirds: Metabolic adaptations to an energetically expensive way of life. *Ardea* 75: 175–187.

Lafferty, K. D. 2001. Disturbance to wintering Western Snowy Plovers. *Biological Conservation* 101: 315–325.

Lafferty, K. D., D. Goodman, and C. P. Sandoval. 2006. Restoration of breeding by snowy plovers following protection from disturbance. *Biodiversity and Conservation* 15: 2217–2230.

Lord, A., J. R. Waas, J. Innes, and M. J. Whittingham. 2001. Effects of human approaches to nests of northern New Zealand Dotterels. *Biological Conservation* 98: 233–240.

Melvin, S. M., A. Hecht, and C. R. Griffin. 1994. Piping Plover mortalities caused by off-road vehicles on Atlantic coast beaches. *Wildlife Society Bulletin* 22: 409–414.

Milsom, T. P., D. C. Ennis, D. J. Haskell, S. D. Langton, and H. V. McKay. 1998. Design of grassland feeding areas for waders during winter: The relative importance of sward, landscape factors and human disturbance. *Biological Conservation* 84: 119–129.

Morse, J. A., A. N. Powell, and M. D. Tetreau. 2006. Productivity of Black Oystercatchers: Effects of recreational disturbance in a national park. *The Condor* 108: 623–633.

Pfister, C., B. A. Harrington, and M. Levine. 1992. The impact of human disturbance on shorebirds at a migration staging area. *Biological Conservation* 60: 115–126.

Prater, A. J. 1989. Ringed Plover *Charadrius hiaticula* breeding population of the United Kingdom in 1984. *Bird Study* 36: 154–159.

Remage-Healey, L., and M. Romero. 2001. Corticosterone and insulin interact to regulate glucose and triglyceride levels during stress in a bird. *American Journal of Physiology—Regulatory, Integrative, and Comparative Physiology* 281: 994–1003.

Ruhlen, T. D., S. Abbott, L. E. Stenzel, and G. W. Page. 2003. Evidence that human disturbance reduces Snowy Plover chick survival. *Journal of Field Ornithology* 74: 300–304.

Senner, S. E., and M. A. Howe. 1984. Conservation of Nearctic shorebirds. In *Shorebirds: Breeding behavior and populations,* ed. J. Burger and B. L. Olla, 379–421. New York: Plenum Press.

Smit, C. J., and G. J. M. Visser. 1993. Effects of disturbance on shorebirds: A summary of existing knowledge from the Dutch Wadden Sea and Delta area. *Wader Study Group Bulletin* 68 (Suppl.): 6–19.

Stillman, R. A., and J. D. Goss-Custard. 2002. Seasonal changes in the response of oystercatchers *Haematopus ostralegus* to human disturbance. *Journal of Avian Biology* 33: 358–365.

Sutherland, W. J. 2007. Future directions in disturbance research. *The Ibis* 149 (Suppl. 1): 120–124.

Trulio, L. A., and J. Sokale. 2008. Foraging shorebird response to trail use around San Francisco Bay. *Journal of Wildlife Management* 72: 1775–1780.

Verhulst, S., K. Osteerbeek, and B. J. Ens. 2001. Experimental evidence for effects of human disturbance on foraging and parental care in oystercatchers. *Biological Conservation* 101: 375–380.

Wasser, S. K., K. Bevis, G. King, and E. Hanson. 1997. Noninvasive physiological measures of disturbance in the Northern Spotted Owl. *Conservation Biology* 11: 1019–1022.

Weimerskirch, H., S. A. Schaffer, G. Mabiklle, J. Martin, O. Boutard, and J. L. Rouanet. 2002. Heart rate and energy expenditure of incubating wandering albatrosses: Basal levels, natural variation, and the effects of human disturbance. *Journal of Experimental Biology* 205: 475–483.

Weston, M. A., and M. A. Elgar. 2005. Disturbance to brood-rearing Hooded Plover *Thinornis rubricollis*: Responses and consequences. *Bird Conservation International* 15: 193–209.

Yasué, M. 2005. The effects of human presence, flock size and prey density on shorebird foraging rates. *Journal of Ethology* 23: 199–204.

———. 2006. Environmental factors and spatial scale influence shorebirds' response to human disturbance. *Biological Conservation* 128: 47–54.

Education and Outreach

CONTENTS

THE SUCCESS OF CONSERVATION lies in ameliorating the negative impact of humans and preserving and managing the habitats necessary to sustain wildlife populations for generations to come. All the science and management in the world will go for naught if the general populace does not understand the consequences of lost biodiversity, whether it is in the form of species' extinctions or loss of ecological services provided by intact communities of native species assembled within healthy ecosystems. The goal of environmental education, outreach, and interpretation (see Definitions) is to communicate these ideas and relationships in a way that alters the behaviors of individuals so that conservation is enhanced. In essence, people become increasingly literate about the environmental consequences of their actions, with the outcome that they will change their lives and reduce their impact on the natural world.

Shorebirds exist as a subset of this natural world. Still, they can be effective "tools" to accomplish environmental education. To this end, there are many notable programs around the globe that utilize the technologies of the Internet to advance communication between biologists studying shorebirds and learners of a variety of ages. Significantly, the U.S. Shorebird Conservation Plan, which represents a form of outreach, includes a section on education and outreach (Johnson-Shultz et al. 2000). This document is rich with information, including online, video, and other sources of educational materials (such as curricula tailored to

DEFINITIONS

Much like any discipline, environmental education is replete with definitions, and clarifying what is meant in a discussion is essential to understanding the successes (and failures) of an outreach program. First, EDUCATION is the process of developing an individual's knowledge, values, and skills. At its heart, environmental education seeks to foster an informed populace with an understanding of how their actions influence the shorebirds and their habitats. By doing so, education will enhance the likelihood that conservation will be effective. OUTREACH is the communication of a group's mission and goals to a wide variety of audiences. For example, the Sister Schools Shorebird Program (SSSP) employs online media and a classroom curriculum to enable educators to teach children about the ecology of shorebirds, with an emphasis on conservation. Another example of outreach comes from various birding festivals strung out along the Pacific Coast of North America that highlight the magnificence of shorebird migration. Increasingly, this same message is communicated to the public via Web sites that track the daily movements of large-bodied shorebirds fitted with satellite transmitters. The process of INTERPRETATION often is viewed as mission based; it attempts to forge an emotional and intellectual connection between the interests of the audience and the meanings inherent in the resource. For example, the SSSP curriculum strongly endorses field trips during which a timely encounter between students and a massive flock of Western Sandpipers can produce a life-changing experience as well as a "teachable moment."

An effective program of environmental education identifies its target audience and establishes a set of tools (media, curricula, exhibits, programs, and demonstrations) to achieve an objective or outcome. In the case of shorebird education, the objective depends on the audience. For instance, one target audience of a scientifically based international shorebird conservation plan (such as the U.S. Shorebird Conservation Plan) is the policy-makers who may be responsible for funding various elements of the plan. To this end, coordinators of the USSCP continue to work hard to develop population estimates for all Nearctic species, with the idea that these estimates would provide benchmarks to evaluate the success of conservation actions aimed at reversing population declines. This approach is similar to the North American Waterfowl Management Plan, which has target population sizes as benchmarks for success in various management and conservation actions directed at populations of swans, geese, and ducks. Similarly, Shorebirds 2020, the education and outreach program of Birds Australia, uses various online and other media to educate volunteers and scientists involved in the fieldwork that serves as the backbone of shorebird conservation. A recently produced compact disc provides useful information on the identification of shorebirds commonly observed in Australia, and it has a well-constructed exercise in estimating the size of shorebird flocks in a variety of settings such as at roosts, while foraging, and in flight.

specific grade levels) as well as case studies exemplifying educational programs. In other regions of the world, sophisticated and rich resources are also available to enhance the understanding and appreciation of the populace to the conservation challenges faced by shorebirds. Birds Australia has a classy Web site, Shorebirds 2020 (www.shorebirds.org.au), which engages the visitor in a variety of learning exercises aimed at conservation.

In this chapter, I review various means by which information on shorebirds is shared among scientists, educators, and the public. I emphasize environmental education programs that have been developed to educate young and old on the wonder of shorebirds. These programs portray the amazing migratory feats of shorebirds as well as the vulnerability and ecological importance of the habitats on which they rely throughout the year. I also draw attention to several birding festivals that have shorebirds as the figurehead for annual celebrations. Finally, I offer an overview of some of the best online resources available to enhance an enthusiast's knowledge and understanding of shorebirds.

PROFESSIONAL GROUPS

The communication of information about shorebirds to affect their conservation is accomplished via a variety of media, each targeting a different audience. Audiences range from the community of scientists and conservationists researching shorebird ecology to policy makers in government positions that may fund conservation measures and environmental educators of children in public schools. For each of these groups, education and outreach are accomplished differently.

NONPROFIT GROUPS

Many professionals working on shorebirds are members of the Wader Study Group (WSG), which claims approximately 450 members in 50 countries. The WSG is an organization of "enthusiastic professionals and amateurs researching waders, also called shorebirds." The WSG had its origins in Europe in the 1970s when scientists and members of local ringing groups came together to share their information and experiences at annual conferences. The WSG is active in conservation, issuing documents such as those resulting from the 2003 conference in Cadiz, Spain. The conclusions of the Cadiz conference addressed the decline of shorebird populations worldwide within the context of two international summits (in 2001, the European Union Heads of State in Göteborg, Sweden; and in 2002, the World Summit on Sustainable Development in Johannesburg, South Africa) that addressed the losses of biodiversity worldwide. The Cadiz conclusions spelled out the declining status of many shorebird populations, the challenges of reversing these trends, and the actions needed to accomplish the goals.

The principal way in which scientists communicate their findings is by publishing in peer-reviewed scientific journals. The WSG publishes the *Wader Study Group Bulletin* three times a year. It also publishes occasional issues under the name *International Wader Studies* (IWS). To date, 19 special issues have been published on topics including effects of disturbance on waterbirds, regional summaries of shorebird population sizes, and accounts of migration in specific estuaries. The WSG also maintains an informative Web site (Table 14.1) with downloadable copies of many issues of WSGB and IWS. Finally, the WSG annually hosts a conference in Europe, the first of which was held in Liverpool, England, in 1975. The New World counterpart to the WSG is the Western Hemisphere Shorebird Group. This group formed in 2006 when researchers convened for a meeting (Shorebird Science in the Western Hemisphere) in Boulder, Colorado. The WHSG does not publish a journal or maintain a Web site.

Several nonprofit conservation groups actively maintain informative Web sites with sections focused on shorebird conservation. Notable among these are the Manomet Center for Conservation Sciences and Point Reyes Bird Observatory (PRBO) Conservation Science (see Table 14.1).

RAMSAR AND WHSRN

The Web sites associated with the 1971 Ramsar treaty and the Western Hemisphere Shorebird Reserve Network (see Table 14.1) are effective tools that communicate information to scientists, conservationists, and the public.

ENVIRONMENTAL EDUCATION

If humans are to reverse the current path of environmental degradation and concomitant loss of biodiversity, it will be accomplished over the next several generations by educating people, especially children, about the intrinsic and instrumental value of earth's biota. A principal objective of environmental education is to cre-

TABLE 14.1

Summary of online resources for shorebird identification, education, and pleasure

SITE	URL	COMMENTS
IDENTIFICATION		
Surfbirds	www.surfbirds.com	A great resource for improving birding skills
American Bird Guide	www.americanbirdguide.com/shorebirds.shtml	North American identification
Ocean Wanderers Guide to Shorebirds	www.oceanwanderers.com/Shorebirds.html	Annotated list of world's shorebirds
The Peep's Puddle	www.gpnc.org/peeps.htm	Shorebirds of the U.S. Great Plains
Shorebirds 2002 (Birds Australia)	www.shorebirds.org.au	Focus on Australian species
EDUCATION AND CONSERVATION		
Ramsar Convention on Wetland Conservation	www.ramsar.org	
International Wader Study Group	www.waderstudygroup.org	
Manomet Center for Conservation Sciences	www.manomet.org	
PRBO Conservation Science	www.prbo.org	
Canadian Shorebird Conservation Plan	www.ec.gc.ca/Publications/default.asp?lang=En&xml=4A90A2A1-1260-41CC-B4F2-4E736D6F6E0E	
Winging Northward: A Shorebird's Journey	shorebirds.pwnet.org/migration/resource_center_sister.htm	Migratory ecology: Pacific flyway shorebirds
Migration Science and Mystery: A Distance Learning Adventure	migration.pwnet.org	

Name	URL	Description
Greenshank: The Migration Story	www.greenshank.info	
Western Hemisphere Shorebird Reserve Network	www.whsrn.org	
Waterbirds around the World (2007)	www.jncc.gov.uk/Default.aspx?page=3891	
U.S. Fish and Wildlife Service, Migratory Birds Program	www.fws.gov/migratorybirds/	General information on shorebirds
Shorebirds 2020 Shorebird Conservation (Birds Australia)	www.shorebirds.org.au	Conservation in Australia and beyond
Shorebird Sister Schools	www.fws.gov/sssp/index.html	
Delaware Bay Shorebirds	www.state.nj.us/dep/fgw/ensp/shorebird_info.htm	Delaware Bay (New Jersey) focus
Hawaii Nature Center: Kolea Watch	hawaiinaturecenter.org/koleawatch.html	Education and research on the Pacific Golden-Plover

BIRDING FESTIVALS

Name	URL	Description
Godwit Days	www.godwitdays.com	Arcata, California, spring birding festival
Grays Harbor Shorebird Festival	www.shorebirdfestival.com	Hoquiam, Washington, spring shorebird festival
Kachemak Bay Shorebird Festival	www.homeralaska.org/events/kachemakBayShorebirdFestival/	Homer, Alaska, spring birding festival
Copper River Delta Shorebird Festival	cordovachamber.com/index.php?option=com_content&task=view&id=57&Itemid=44	Cordova, Alaska, spring shorebird festival

TRACKING MIGRATION

Name	URL	Description
Pacific Shorebird Migration Program	alaska.usgs.gov/science/biology/shorebirds/pacific_migration.html	Real-time movements of Nearctic breeders
Shorebirds 2020: Counting Shorebirds	www.shorebirds.org.au/flash/counting/index.html	Lesson in estimating shorebird numbers

FIGURE 14.1. Outdoor experiences can serve as an important educational tool. Here, university students learn how to use plumage characteristics to age Least Sandpipers. Photo credit: Luke Eberhart-Phillips.

ate an emotional response in an individual such that they recognize the value of something. In doing so, they are more likely to take action that will result in meaningful changes to their life and to alter the world around them. In the case of shorebirds, outdoor experiences (Fig. 14.1), combined with a formal curriculum, can instill wonder and appreciation for the amazing feats accomplished during migrations and the vulnerability of populations to anthropogenic effects on ecosystems. To this end, several programs have been developed with resources to provide educational opportunities for children and curricula for teachers.

SHOREBIRD SISTER SCHOOLS PROGRAM

One of the most often cited shorebird education projects is the Shorebird Sister Schools Program (SSSP). An overview of the SSSP is provided by Chapman and colleagues (2005, 2006). The SSSP began in 1994 as an offshoot of the migratory shorebird festival held each spring in Homer, Alaska. In its infancy, the program was conceived by a grade school teacher to provide an opportunity for children to understand the ecological linkages between the shorebirds they

observed on Kachemak Bay and the wintering and breeding areas the birds occupied at other times of year. The initial emphasis was in providing a curriculum that could be enhanced by birdwatching visits to coastal habitats during the peak of spring migration. As such, the SSSP emphasized the Pacific Americas Flyway. Since its inception, however, the program has grown substantially to include at least five other flyways, spanning North, Central, and South America, Asia, and Europe. In 2000, the success of the program resulted in its transfer to the U.S. Fish and Wildlife Service National Conservation Training Center in Shepherdstown, West Virginia (Chapman and Johnson 2005). The SSSP has been guided by a national and eight regional coordinators located in each of the U.S. Fish and Wildlife Service administrative regions.

The SSSP is an Internet-based, interactive system of activities that links students with one another, educators, and researchers along shorebird flyways to study the progress made by birds as they wend their way north to Arctic breeding areas. The program currently consists of an Internet site and a curriculum that can be applied to grades K-12. Through the years, various projects have been implemented as part of the SSSP, such as an electronic mailing list, teacher workshops, shorebird tracking projects, and a pen-pal program. An engaging and thorough treatment of the ecology of migrant shorebirds was established in 2001 with an online approach to convey the messages of the SSSP. A Web site (Winging Northward: A Shorebird's Journey) was activated in 2002 to coordinate students, educators, and shorebird enthusiasts along the Pacific Americas Flyway. The educational message was embodied in Maya, a female Western Sandpiper, whose annual wanderings were tracked from her wintering area in Bahia Santa Maria, Sinaloa, to her breeding ground in western Alaska. Various Web site chapters convey the ecological needs of Maya (and shorebirds in general) throughout the year. For instance, each spring, Maya hopscotches northward, moving among estuaries and laying down the energy reserves necessary to complete her journey. There is even a section on predation by raptors!

The success of an environmental education program is evaluated by the quality of the information delivered to participants, the changes in behavior prompted by the curriculum, and the satisfaction of users with the information available to them. These metrics are balanced by the costs of the program. The Web site for the SSSP states its mission is to "encourage public participation in the conservation of shorebirds and their habitats by connecting people along flyways and increasing their awareness and knowledge of local natural resources to inspire community conservation." Ultimately, evaluating the success of the SSSP in accomplishing this mission would be achieved by understanding whether the attitudes of learners, especially schoolchildren, changed with regard to their perspectives on shorebirds, wetlands, and natural resources. Unfortunately, no evaluation of this facet of the SSSP has been conducted. The one attempt to evaluate the program (Horr 2007) focused on educators and one administrator. Horr (2007) used qualitative and quanti-

tative methods applied to a sample of users, all of whom were educators. Unfortunately, her sample size was small (23 respondents out of 117 Internet users) and was potentially biased by only the enthusiastic educators responding to her survey. There was no attempt to evaluate the most important user group, the schoolchildren, which severely limits the conclusions of the study regarding its effectiveness: there is no information to suggest that children's attitudes toward shorebirds changed. Nevertheless, educators rated the SSSP highly with regard to content. The exception to this was the Internet site, which could have had enhanced technologies to deliver its message. At this time, the various Internet sites associated with the SSSP have not added new information for over 5 years. Horr (2007) concluded that the SSSP is currently underfunded by the U.S. Fish and Wildlife Service and has had little opportunity for growth. As of 2009, the program was not effectively acting to provide educational materials online, and the reasons for this are unclear.

SHOREBIRD SISTER CITIES PROGRAM

The Shorebird Sister Cities Program (SSCP) recently was developed under the auspices of the SSSP (Johnson-Shultz et al. 2000). Development of linkages between cities, often from different countries, is not a unique concept, as sister cities are found throughout the world. The concept of using common shorebird species to link cities along a flyway is the basis of a successful partnership between San Blas, Nayarit, Great Salt Lake, Utah, and Chapin Lake, Saskatchewan. Educational programs in local schools, community bird festivals, biologist exchanges, and monitoring and research projects on shared shorebird species have been successful in raising awareness of the conservation needs of shorebirds in all three geographic regions. Unfortunately, the SSCP does not appear to have been widely implemented, although the intent is to expand the program geographically to include the populace of cities in the East Asia/Australasian Flyway.

SHOREBIRD EDUCATION AUSTRALIA

The Convention on Wetlands (Ramsar, Iran, 1971) established programs to enhance awareness of wetland values and functions. To achieve this objective, in 1999 the Australian government initiated an educational program, Shorebird Education Australia (SEA), emphasizing shorebirds and the wetland habitats they occupy in the East Asia/Australasian Flyway. The first activities of this program began in 2001. This program has similarities to the SSSP and acknowledges this on its Web site, with a curriculum emphasizing the reliance of migratory shorebirds on the nexus of wetland habitats that link the breeding and nonbreeding areas of the world. The objectives of the program are to increase communication among three principal groups: site managers, educators, and students at schools. Site managers are local experts situated near key estuaries; they provide up-to-date information on the movements of shorebirds as well as detailed knowledge of the ecology of shorebirds. Educators occur principally at schools throughout the flyway and rely on site managers for the details that they communicate among themselves regarding shorebird movements. The targeted student audience consists of middle-school children.

The SEA program uses several tools and exercises to instill appreciation and wonder in schoolchildren for shorebirds. The objective is to elicit a positive emotional response in students such that they will be more interested in conserving nature. To accomplish this, SEA uses a system of postcards inscribed with artwork by students and sent to a Web site where postcards are subsequently posted. Additional materials to improve shorebird identification skills exist online.

OTHER LOCAL EFFORTS

The Bay of Fundy has long been recognized as an important area for migrant shorebirds. Several recent programs have worked to purchase lands, enhance habitats, and provide stewards to benefit staging shorebirds in the Minas Basin.

These programs were organized by the Nature Conservancy of Canada, working with the governments of New Brunswick and Nova Scotia as well as with federal assistance from the Canadian Wildlife Service. In 2004, the Bay of Fundy Shorebird Project (BOFSP) began in an attempt to minimize human disturbance of shorebirds during their southward migration. The BOFSP is a community-based effort to expand stewardship, increase communication, and establish a community-based foundation for continued stewardship. The effort specifically sought to enlist members of the tourism industry in communicating to visitors the value of coastal habitats to shorebirds. To accomplish this, the project built interpretive centers, promoted and delivered educational materials and programs, and conducted surveys of shorebirds in the area. The timing of program activities was specifically targeted to ameliorate the expected impact of thousands of tourists arriving during peak shorebird migration during the third World Acadian Congress held during August 2004. To accomplish this, the program conducted workshops for members of the tourism/hospitality industry to enhance their ability to share knowledge with tourists; this was done well in advance of the festival. Additionally, project staff led field trips during the August period of peak tourism and shorebird abundance.

In England, local bird banders (Farlington ringing group) created an online project to educate the public and communicate with local schoolchildren about the migrations of shorebirds. The group had been banding birds in the local marshes since 1967; in 1992 they expanded their efforts and focused on color-marking Common Greenshanks. The project, *Greenshank: The Migration Story*, capitalizes on the individual histories of banded Greenshanks to convey appreciation for shorebirds and nature in general. It includes information on the history of the project, details on the activities of researchers capturing and tracking the color-marked sandpipers, as well as a curriculum for use in schools. The curriculum includes elements of biology (diversity of life, reproduction, and growth), geography (tracking migratory greenshanks with reports from wintering areas on marked individuals), and ethical considerations such as the ecological benefits of maintaining healthy ecosystems.

In April 2005, Boundary Bay and the Fraser River Delta in British Columbia were recognized as a Hemispheric Site under the Western Hemisphere Shorebird Reserve Network. Subsequent to this designation, the online program *Birds on the Bay* began, with an emphasis on shorebirds. The Web site includes links to scientists, conservationists, and artists, as well as a link to a Web site (Migration Science and Mystery) with educational materials. This site has similarities to the SSSP site that tracks Maya, the Western Sandpiper, during her movements between wintering and breeding grounds.

In Hawaii, the U.S. Fish and Wildlife Service worked with Hawaii Nature Center and the Hawai'i Audubon Society to collaborate on a project to engage the public in monitoring the nonbreeding ecology of the Pacific Golden-Plover, known to native Hawaiians as the Kōlea. The project, Kōlea Watch, began in 2003 with online records posted by individuals who were responsible for tracking marked plovers. These records included the spring departure dates of individuals for their Alaskan breeding grounds. In some cases, radio-marked plovers were relocated on breeding territories, and this information was also posted to the Web site. For instance, one plover ("Makoa," marked in Hawaii) departed for Alaska on April 27, 2005, based on regular observations and the bird's absence from its winter territory at the Hawaii State Veterans Cemetery in Kaneohe. Approximately 6 days later, Makoa was observed at its breeding territory near Egegik, Alaska, where it remained through the breeding season.

The success of the various educational programs, whether local or spanning an entire flyway, lies in dedicated funding, community support and involvement, and participation by individuals who have a vested interest in the outcomes and accomplishments of the program. It is noteworthy that several programs have out-of-date Web sites, and the cause of this deficiency is unknown. However, without dedicated individuals to maintain the information, Web sites that rely on real-time conveyance of information quickly become useless. The same can be said for the maintenance and vigor of the overarching programs—without leadership and funding, they quickly become stagnant and fail to achieve their educational objectives.

ECOTOURISM AND BIRDING FESTIVALS

The boom of birding has spawned a plethora of regional festivals tapping into ecotourism. By 2006, there were more than 200 birding festivals in the United States alone (Masurkewich 2006), which represents substantial growth from the 79 celebrations reported in 1997 (Scott et al. 1999) and the handful of events that existed in the early 1990s (Lawton 2009). The growth of birding and bird festivals has proven to be a major stimulus to some local economies (Leonard 2008; Pullis La Rouche 2001).

In many instances, birding festivals are marketed under the banner of ecotourism. Three core criteria characterize ecotourism (Weaver 2005; Lawton 2009): (1) attractions involve the natural environment; (2) the focus is on education and learning; and (3) environmental and sociocultural sustainability is emphasized. Birding festivals have the potential to accomplish significant results in the area of environmental education. However, they vary greatly in the extent to which they satisfy these three criteria and achieve educational outcomes. At one end of the continuum are festivals that achieve little of significance in education. These minimalist approaches involve shallow learning (Weaver 2005). By contrast, comprehensive learning is sometimes achieved by deeper experiences that enhance the understanding of sustainability. Lawton (2009) evaluated 108 birding festivals in the United States by surveying event organizers (not participants) and found that many reported that their festivals provided high values for immersion into the natural environment.

This suggests the potential for significant educational results. It is important to note, however, that the profile of the typical participant in birding festivals is a highly educated, older tourist with higher income (Lawton 2001). Consequently, the educational benefit of birding festivals in achieving broad environmental goals is probably diminished by the observation that these festivals "preach to the choir." Finally, Lawton's (2009) survey sampled event organizers and not participants; consequently, results may be biased and do not represent the experiences of the people attending the festivals.

Some birding festivals have taken shorebirds as their namesake, and several have had relatively long runs, dating to the early or mid-1990s. Interestingly, most of these festivals occur along the Pacific Coast of North America, and they emphasize the spring passage of migrant shorebirds. As mentioned earlier, the origin of the SSSP followed shortly after the inaugural Kachemak Bay Shorebird Festival in Homer, Alaska. It may be that the strong representation along the Pacific Americas Flyway stems from the coordinated efforts of individuals in organizing communities and schools to participate in the SSSP. From south to north, these festivals celebrating migratory shorebirds include: Godwit Days in Arcata, California (first celebration in 1995); Grays Harbor Shorebird Festival in Hoquiam, Washington (1995); Tofino Shorebird Festival in Tofino, British Columbia (1997); Copper River Delta Shorebird Festival in Cordova, Alaska (1990); and Kachemak Bay Shorebird Festival in Homer, Alaska (1992). On the Atlantic Coast, the Horseshoe Crab and Shorebird Festival is held in Milton, Delaware (2003). Other festivals occur during fall migration: the Oregon Shorebird Festival in Charleston, Oregon (1986); and the Jamaica Bay Shorebird Festival in New York City (2005). In 2009 in Mexico, San Blas, Nayarit, initiated a migratory bird festival in partnership with other communities in North America, and their festival emphasizes shorebirds.

BOOKS AND ONLINE RESOURCES

An abundance of resources are available online to enhance an enthusiast's understanding of shorebirds (see Table 14.1). These resources include several recent books covering behavioral ecology (van der Kam et al. 2004) and field identification (Hayman et al. 1986; Paulson 2005; Message and Taylor 2006; O'Brien et al. 2006; Chandler 2009). There are also Web sites managed by individuals with photo quizzes testing identification skills, educational tools, and links to international programs sponsored by both nongovernmental groups and governmental agencies. The Web sites vary greatly in quality, and many appear to be updated infrequently. However, some sources are truly exceptional. Probably the single best online resource is *Shorebirds 2020*, a program meant to reinvigorate shorebird conservation in Australia. This Web site is sponsored by Birds Australia, World Wildlife Fund–Australia, the Australasian Wader Study Group, and the Australian government's Natural Heritage Trust. The visually pleasing, engaging, and well-constructed site includes unique features such as training modules on shorebird identification and estimating flock sizes. It includes details on various international conventions dedicated to conservation of migratory birds in general and links to take the viewer to other Web sites. There is even a section on a decision-making framework to provide evaluative features of the success of the program!

CONSERVATION IMPLICATIONS

Conservation will be enhanced by the populace at all levels if they possess a deep understanding of the consequences of human actions in the form of lost biodiversity. Efforts such as the Sister Schools Shorebird Program and Shorebird Education Australia appear to be filling an empty niche by communicating the wonder of shorebirds and the challenges posed by anthropogenic degradation and loss of habitats. The success of these and other programs has not been evaluated critically. Consequently, it is difficult to know whether the various educational efforts and programs undertaken in the name of shorebirds have achieved their missions. Critical evaluation of learning outcomes is necessary to determine whether a curriculum or outreach project is effective in meeting conservation objectives. If not, then programs should be terminated, restructured, and revamped to achieve their desired objectives.

LITERATURE CITED

Chandler, R. 2009. *Shorebirds of North America, Europe, and Asia: A photographic guide.* Princeton, NJ: Princeton University Press.

Chapman, H., B. A. Andres, and S. Fellows. 2006. Shorebird Sister Schools Program—Shorebird education in North America and beyond. In *Waterbirds around the world,* ed. G. C. Boere, C. A. Galbraith, and D. A. Stroud, 832. Edinburgh: The Stationery Office.

Chapman, H., and H. Johnson. 2005. Linking shorebird conservation and education along flyways: An overview of the Shorebird Sister Schools Program. *U.S. Department of Agriculture, General Technical Report* PSW-GTR-191: 443–445.

Horr, E. E. T. 2007. Determining a more complete program valuation: Integrating tools from program theory and economics to better inform program decisions. Ph.D. dissertation, The Ohio State University.

Johnson-Shultz, H., J. Burton, N. Cirillo, and S. Brown, eds. 2000. *National shorebird education and outreach plan.* Manomet, MA: Manomet Center for Conservation Science.

Lawton, L. J. 2001. A profile of older adult ecotourists in Australia. *Journal of Hospitality and Leisure Marketing* 9: 113–132.

———. 2009. Birding festivals, sustainability, and ecotourism. *Journal of Travel Research* 48: 135–145.

Hayman, P., J. Marchant, and T. Prater. 1986. *Shorebirds: An identification guide to the waders of the world.* Boston: Houghton Mifflin.

Leonard, J. 2008. *Wildlife watching in the U.S.: The economic impacts on national and state economies in 2006.* Report 2006-1. Arlington, VA: Wildlife and Sport Fish Restoration Programs, U.S. Fish and Wildlife Service.

Masurkewich, K. 2006. Wild west is now for the birds: The soaring business of bird watching transforming tiny Arizona towns. *Wall Street Journal,* April 1, 2006, online.wsj.com/article/SB114384419787613937.html?mod=todays_us_pursuits (accessed May 9, 2010).

Message, S., and D. Taylor. 2006. *Shorebirds of North America, Europe, and Asia: A guide to field identification.* Princeton, NJ: Princeton University Press.

O'Brien, M., R. Crossley, and K. Karlson. 2006. *The shorebird guide.* Boston: Houghton Mifflin.

Paulson, D. 2005. *Shorebirds of North America: The photographic guide.* Princeton, NJ: Princeton University Press.

Pullis La Rouche, G. 2001. *Birding in the United States: An economic and demographic analysis.* Report 2001-1. Washington, DC: Division of Federal Aid, U.S. Fish and Wildlife Service.

Scott, D., S. M. Baker, and C. Kim. 1999. Motivations and commitments among participants in the Great Texas Birding Classic. *Human Dimensions of Wildlife* 4: 50–67.

van de Kam, J., B. Ens, T. Piersma, and L. Zwarts. 2004. *Shorebirds: An illustrated behavioural ecology.* Utrecht, the Netherlands: KNNV Publishers.

Weaver, D. B. 2005. Comprehensive and minimalist dimensions of ecotourism. *Annals of Tourism Research* 32: 439–455.

APPENDIX

LIST OF THE WORLD'S SHOREBIRDS, THEIR BREEDING RANGES, AND POPULATION SIZE ESTIMATES

Taxonomy follows Clements, J. F. 2007. *The Clements checklist of birds of the world*. Ithaca, NY: Cornell University Press, and Gill, F., and M. Wright. 2006. *Birds of the world*. Princeton, NJ: Princeton University Press.

Sources for population estimates: Brown, S., C. Hickey, B. Harrington, and R. Gill, eds. 2001. *The U.S. shorebird conservation plan*. 2nd ed. Manomet, MA: Manomet Center for Conservation Sciences; Delaney, S., D. Scott, T. Dadman, and D. Stroud, eds. 2009. *An atlas of wader populations in Africa and Western Eurasia*. Wageningen, The Netherlands: Wetlands International; BirdLife International (http://www.birdlife.org)

FAMILY/COMMON NAME	LATIN NAME	BREEDING RANGE	POPULATION SIZE
Jacanidae			
Lesser Jacana	*Microparra capensis*	Africa south of the Sahara	25,000–100,000
African Jacana	*Actophilornis africanus*	Africa south of the Sahara	1,000,000
Madagascar Jacana	*Actophilornis albinucha*	Madagascar	1,000–10,000
Comb-crested Jacana	*Irediparra gallinacea gallinacea*	Borneo, Sulawesi, Philippines	25,000–1,000,000
	Irediparra gallinacea novaeguinae	New Guinea	
	Irediparra gallinacea novaehollandiae	New Guinea, Australia	
Pheasant-tailed Jacana	*Hydrophasianus chirurgus*	India, SE Asia, Philippines	100,000
Bronze-winged Jacana	*Metopidius indicus*	India, SE Asia, Java, Sumatra	50,000–100,000
Northern Jacana	*Jacana spinosa gymnostoma*	Mexico	500,000–5,000,000
	Jacana spinosa spinosa	Central America	
	Jacana spinosa violacea	Cuba, Jamaica, Hispaniola	
Wattled Jacana	*Jacana jacana hypomelaena*	Panama, N. Colombia	
	Jacana jacana melanopygia	Colombia, Venezuela	
	Jacana jacana jacana	Colombia, Brazil, Uruguay, Argentina	
	Jacana jacana intermedia	Venezuela	
	Jacana jacana scapularis	Ecuador, Peru	
	Jacana jacana peruviana	Peru, Brazil	
Rostratulidae			
Greater Painted-snipe	*Rostratula benghalensis*	Africa, Madagascar, Asia	36,000–1,000,000
Australian Painted-snipe	*Rostratula australis*	Australia	
American Painted-snipe	*Rostratula semicollaris*	Brazil, Paraguay, Argentina, Chile	Unknown
Dromadidae			
Crab Plover	*Dromas ardeola*	Indian Ocean	60,000–80,000
Haematopodidae			
Magellanic Oystercatcher	*Haematopus leucopodus*	Chile, Argentina	46,000–140,000
Blackish Oystercatcher	*Haematopus ater*	Peru, Chile	22,000–120,000

Common Name	Scientific Name	Range	Population
Black Oystercatcher	*Haematopus bachmani*	Western U.S., Canada	10,000
American Oystercatcher	*Haematopus palliatus palliatus*	North, Central, South America	37,000–110,000
	Haematopus palliatus galapagensis	Galapagos Islands	
African Oystercatcher	*Haematopus moquini*	Namibia, South Africa	6,000
Eurasian Oystercatcher	*Haematopus ostralegus ostralegus*	Iceland, Scandinavia, Europe	1,020,000
	Haematopus ostralegus longipes	Russia, Siberia, Caspian and Aral seas	100,000–200,000
	Haematopus ostralegus osculans	Kamchatka, North Korea, China	
Pied Oystercatcher	*Haematopus longirostris*	Australia, Tasmania, New Zealand	11,000
South Island Oystercatcher	*Haematopus finschi*	New Zealand	110,000
Chatham Oystercatcher	*Haematopus chathamensis*	Chatham Islands, New Zealand	50–250
Variable Oystercatcher	*Haematopus unicolor*	New Zealand	4,000–4,300
Sooty Oystercatcher	*Haematopus fuliginosus ophthalmicus*	N. Australia	12,000
	Haematopus fuliginosus fuliginosus	S. Australia, Tasmania	
Ibidorhynchidae			
Ibisbill	*Ibidorhyncha struthersii*	Central Asia	
Recurvirostridae			
Black-winged Stilt	*Himantopus himantopus*	Mediterranean, sub-Saharan Africa, Asia	450,000–780,000
Pied Stilt	*Himantopus leucocephalus*	Indonesia, Australia, New Zealand	
Black Stilt	*Himantopus novaezelandiae*	New Zealand	40
Black-necked Stilt	*Himantopus mexicanus mexicanus*	North, Central and South America	175,000 (U.S. only)
	Himantopus mexicanus knudseni	Hawaiian Islands	1,400
White-backed Stilt	*Himantopus melanurus*	Chile, Peru, Brazil, Argentina	
Banded Stilt	*Cladorhynchus leucocephalus*	Australia	210,000
Pied Avocet	*Recurvirostra avosetta*	N. Africa, Eurasia	210,000–460,000
American Avocet	*Recurvirostra americana*	North America	450,000 (U.S./Canada)
Red-necked Avocet	*Recurvirostra novaehollandiae*	Australia	110,000
Andean Avocet	*Recurvirostra andina*	Peru, Argentina, Chile	

Appendix *(continued)*

FAMILY/COMMON NAME	LATIN NAME	BREEDING RANGE	POPULATION SIZE
Burhinidae			
Water Thick-knee	*Burhinus vermiculatus buettikoferi*	W. Africa	25,000
	Burhinus vermiculatus vermiculatus	Zaire, Somalia to South Africa	100,000
Eurasian Thick-knee	*Burhinus oedicnemus distinctus*	W. Canary Islands	900–1,200
	Burhinus oedicnemus insularum	E. Canary Islands	700–4,700
	Burhinus oedicnemus saharae	Mediterranean, Greece, Turkey, Iran, Iraq	10,000–100,000
	Burhinus oedicnemus oedicnemus	Britain, W. Europe, Balkans, Caucasus	122,000–206,000
	Burhinus oedicnemus harterti	W. Russia, Turkestan, Pakistan, NW India	
	Burhinus oedicnemus indicus	India, Sri Lanka, SE Asia	
Senegal Thick-knee	*Burhinus senegalensis*	Sub-Saharan Africa	25,000–100,000
Spotted Thick-knee	*Burhinus capensis maculosus*	Senegal, Somalia, Uganda, Kenya	46,000–160,000
	Burhinus capensis dodsoni	Somalia, Saudi Arabia	10,000–25,000
	Burhinus capensis capensis	Kenya, South Africa, Zambia, Angola	40,000–80,000
	Burhinus capensis damarensis	Namibia, Botswana, South Africa	5,000–10,000
Double-striped Thick-knee	*Burhinus bistriatus bistriatus*	Mexico, Central America	500,000–5,000,000
	Burhinus bistriatus dominicensis	Hispaniola	
	Burhinus bistriatus pediacus	Colombia	
	Burhinus bistriatus vocifer	Venezuela, Guyana, Brazil	
Peruvian Thick-knee	*Burhinus superciliaris*	Ecuador, Peru	
Bush Thick-knee	*Burhinus grallarius*	Australia, New Guinea	10,000–20,000
Great Thick-knee	*Burhinus recurvirostris*	Iran, India, SE Asia	1,000–25,000
Beach Thick-knee	*Burhinus magnirostris*	Malay Peninsula, Philippines, Australasia	6,000
Glareolidae			
Egyptian Plover	*Pluvianus aegyptius*	Sub-Saharan Africa, Zaire, Angola	20,000–50,000
Cream-colored Courser	*Cursorius cursor bogolubovi*	Turkey, Iran, Afghanistan, Pakistan, India	50,000–200,000

Common name	Scientific name	Distribution	Population
	Cursorius cursor cursor	Canary Islands, N. Africa, Arabian Peninsula	
	Cursorius cursor exsul	Cape Verde Islands	150–350
	Cursorius cursor somalensis	Eritrea, Ethiopia, Somalia	
	Cursorius cursor littoralis	Sudan, Kenya, Somalia	
Burchell's Courser	*Cursorius rufus*	Angola, Namibia, Botswana, South Africa	
Temminck's Courser	*Cursorius temminckii*	Sub-Saharan Africa	61,000–2,100,000
Indian Courser	*Cursorius coromandelicus*	Pakistan, India, Sri Lanka	10,000–25,000
Double-banded Courser	*Smutsornis africanus raffertyi*	Eritrea, Ethiopia, Djibouti	
	Smutsornis africanus hartingi	Ethiopia, Somalia	
	Smutsornis africanus gracilis	Kenya, Tanzania	
	Smutsornis africanus bisignatus	Angola	
	Smutsornis africanus traylori	Namibia, Botswana	
	Smutsornis africanus sharpei	Namibia	
	Smutsornis africanus africanus	Namibia, South Africa	
	Smutsornis africanus granti	South Africa	
Three-banded Courser	*Rhinoptilus cinctus cinctus*	Sudan, Ethiopia, Somalia, Kenya	25,000–1,000,000
	Rhinoptilus cinctus emini	Kenya, Tanzania, Zambia	10,000–25,000
	Rhinoptilus cinctus seebohmi	Angola, Namibia, Zimbabwe, South Africa	5,000–10,000
Bronze-winged Courser	*Rhinoptilus chalcopterus*	Sub-Saharan Africa	50,000–1,100,000
Jerdon's Courser	*Rhinoptilus bitorquatus*	SE India	Rediscovered 1986
Australian Pratincole	*Stiltia isabella*	Australia	25,000–1,000,000
Collared Pratincole	*Glareola pratincola pratincola*	Europe, Turkey, Iran, Pakistan	170,000–600,000
	Glareola pratincola erlangeri	Somalia, Kenya	<10,000–25,000
	Glareola pratincola fuelleborni	Senegal, Kenya, Zaire, Namibia, South Africa	100,000–350,000
Oriental Pratincole	*Glareola maldivarum*	E. Asia	2,900,000–3,000,000
Black-winged Pratincole	*Glareola nordmanni*	Romania, Russia, Kazakstan	100,000–300,000
Madagascar Pratincole	*Glareola ocularis*	Madagascar	5,000–10,000

FAMILY/COMMON NAME	LATIN NAME	BREEDING RANGE	POPULATION SIZE
Glareolidae (cont.)			
Rock Pratincole	*Glareola nuchalis liberiae*	Sierra Leone, Cameroon	100,000–300,000
	Glareola nuchalis nuchalis	Chad, Ethiopia, Zambia, Namibia, Mozambique	25,000–100,000
Gray Pratincole	*Glareola cinerea cineria*	W. Africa	10,000–25,000
	Glareola cinerea colorata		<10,000
Small Pratincole	*Glareola lactea*	Afghanistan, Pakistan, India, SE Asia	50,000–100,000
Charadriidae			
Northern Lapwing	*Vanellus vanellus*	Palearctic	5,500,0000–9,500,000
Long-toed Lapwing	*Vanellus crassirostris crassirostris*	Sudan, Zaire, Malawi, Angola	30,000–100,000
	Vanellus crassirostris leucopterus	Tanzania, Zaire, Angola, Botswana, South Africa	25,000–60,000
Blacksmith Plover	*Vanellus armatus*	E. and S. Africa	100,000–1,000,000
Spur-winged Plover	*Vanellus spinosus*	Sub-Saharan Africa, Middle East, Mediterranean	130,000–800,000
River Lapwing	*Vanellus duvaucelii*	India, SE Asia	1,000–25,000
Yellow-wattled Lapwing	*Vanellus malabaricus*	Pakistan, India, Bangladesh, Sri Lanka	1,000–25,000
Black-headed Lapwing	*Vanellus tectus tectus*	Sub-Saharan Africa	25,000–200,000
	Vanellus tectus latifrons	Somalia, Kenya	<10,000–25,000
White-headed Lapwing	*Vanellus albiceps*	Sub-Saharan Africa	56,000–130,000
Senegal Lapwing	*Vanellus lugubris*	Sub-Saharan Africa	25,000–70,000
Black-winged Lapwing	*Vanellus melanopterus melanopterus*	Sudan, Ethiopia, Kenya, Tanzania	10,000–50,000
	Vanellus melanopterus minor	South Africa, Mozambique	<10,000
Crowned Lapwing	*Vanellus coronatus coronatus*	S. and E. Africa	400,000–900,000
	Vanellus coronatus demissus	Somalia	10,000–100,000

Common Name	Scientific Name	Distribution	Population
Wattled Lapwing	Vanellus senegallus senegallus	Sub-Saharan Africa	25,000–60,000
	Vanellus senegallus major	Eritrea, Ethiopia	5,000–15,000
	Vanellus senegallus lateralis	Congo, Angola, Mozambique, South Africa	10,000–100,000
Spot-breasted Lapwing	Vanellus melanocephalus	Ethiopia	<10,000
Brown-chested Lapwing	Vanellus superciliosus	Ghana, Cameroon, Zaire	<1,000–25,000
Gray-headed Lapwing	Vanellus cinereus	China, Japan	25,000–100,000
Red-wattled Lapwing	Vanellus indicus aigneri	Turkey, Iraq, Iran, Afghanistan, Pakistan	51,000–60,000
	Vanellus indicus indicus	Pakistan, India, Bangladesh	
	Vanellus indicus lankae	Sri Lanka	
	Vanellus indicus atronuchalis	India, SE Asia, Malaysia	
Sunda Lapwing	Vanellus macropterus	Sumatra, Java	Probably extinct
Banded Lapwing	Vanellus tricolor	Australia, Tasmania	25,000–1,000,000
Masked Lapwing	Vanellus miles miles	New Guinea, Australia	200,000–2,000,000
	Vanellus miles novaehollandiae	Australia, Tasmania, New Zealand	
Sociable Lapwing	Vanellus gregarius	Russia, Kazakstan	3,200–11,000
White-tailed Lapwing	Vanellus leucurus	Turkey, Iran, Iraq, Afghanistan	10,000–100,000
Pied Lapwing	Vanellus cayanus	Venezuela, Ecuador, Brazil, Peru, Argentina	
Southern Lapwing	Vanellus chilensis cayennensis	N. South America	2,000,000
	Vanellus chilensis lampronotus	S. Brazil, N. Chile, Argentina	
	Vanellus chilensis chilensis	Argentina, Chile	
	Vanellus chilensis fretensis	S. Argentina, S. Chile	
Andean Lapwing	Vanellus resplendens	Colombia, Ecuador, Peru, Chile, Argentina	1,000–10,000
Red-kneed Dotterel	Erythrogonys cinctus	Australia, New Guinea	25,000–1,000,000
Pacific Golden-Plover	Pluvialis fulva	Siberia, Alaska (Palaearctic)	42,500 (50,000–100,000)
American Golden-Plover	Pluvialis dominica	North America	200,000
Eurasian Golden-Plover	Pluvialis apricaria albifrons	Greenland, Iceland, N. Eurasia, C. Siberia	640,000–1,200,000
	Pluvialis apricaria apricaria	Britain, Baltic Peninsula	140,000–210,000

FAMILY/COMMON NAME	LATIN NAME	BREEDING RANGE	POPULATION SIZE
Charadriidae (cont.)			
Black-bellied Plover	*Pluvialis squatarola squatarola*	Palearctic, Alaska	390,000
	Pluvialis squatarola cynosurae	Nearctic	150,000
Red-breasted Dotterel	*Charadrius obscurus aquilonius*	North Island, New Zealand	2,000
	Charadrius obscurus obscurus	Stewart Island, New Zealand	
Common Ringed Plover	*Charadrius hiaticula hiaticula*	NE Canada, Greenland, Scandinavia	10,000
	Charadrius hiaticula tundrae	Russia, Siberia	100,000–1,000,000
Semipalmated Plover	*Charadrius semipalmatus*	Nearctic	150,000
Long-billed Plover	*Charadrius placidus*	E. Asia (Russia, China, Japan)	1,000–25,000
Little Ringed Plover	*Charadrius dubius curonicus*	Palearctic	280,000–530,000
	Charadrius dubius jerdoni	India, SE Asia	
	Charadrius dubius dubius	Philippines, New Guinea, Bismarck Archipelago	
Wilson's Plover	*Charadrius wilsonia wilsonia*	U.S., Mexico, Belize, Greater Antilles	6,000
	Charadrius wilsonia beldingi	Pacific Coast W. Mexico, Central America, South America	
	Charadrius wilsonia cinnamominus	N. South America, Lesser Antilles	
Killdeer	*Charadrius vociferus vociferus*	Canada, U.S., Mexico	1,000,000
	Charadrius vociferus ternominatus	Greater Antilles	
	Charadrius vociferus peruvianus	Peru, Chile	
Piping Plover	*Charadrius melodus melodus*	Atlantic coast of North America	3,000
	Charadrius melodus circumcinctus	Great Lakes, North American prairie	3,000
Madagascar Plover	*Charadrius thoracicus*	Madagascar	3,100
Kittlitz's Plover	*Charadrius pecuarius*	Sub-Saharan Africa, NE Egypt, Madagascar	130,000–480,000
St. Helena Plover	*Charadrius sanctaehelenae*	St. Helena Island	200–220

Common Name	Scientific Name	Range	Population
Three-banded Plover	*Charadrius tricollaris tricollaris*	S. and E. Africa	81,000–170,000
	Charadrius tricollaris bifrontatus	Madagascar	10,000–30,000
Forbes' Plover	*Charadrius forbesi*	W. and C. Africa	10,000–50,000
White-fronted Plover	*Charadrius marginatus mechowi*	Sub-Saharan Africa, Botswana, Mozambique	25,000–35,000
	Charadrius marginatus marginatus	Angola, South Africa	10,000
	Charadrius marginatus arenaceus	Mozambique, South Africa	18,000–22,000
	Charadrius marginatus tenellus	Madagascar, E. Africa	20,000–40,000
Chestnut-banded Plover	*Charadrius pallidus venustus*	Kenya, Tanzania	6,500
	Charadrius pallidus pallidus	Southern Africa	11,000–16,000
Snowy Plover	*Charadrius alexandrinus alexandrinus*	W. Palearctic, Asia, China	100,000–200,000
	Charadrius alexandrinus dealbatus	Japan, E. China	
	Charadrius alexandrinus seebohmi	SE India, Sri Lanka	
	Charadrius alexandrinus nivosus	Nearctic	17,800
	Charadrius alexandrinus tenuirostris	Caribbean	2,500
	Charadrius alexandrinus occidentalis	Peru, Chile	
Javan Plover	*Charadrius javanicus*	Java, Bali, Kangean Islands	
Red-capped Plover	*Charadrius ruficapillus*	Australia, Tasmania	95,000
Malaysian Plover	*Charadrius peronii*	SE Asia, Philippines	10,000–25,000
Collared Plover	*Charadrius collaris*	Mexico, Central and South America	1,000–10,000
Puna Plover	*Charadrius alticola*	Peru, Argentina, Chile	25,000–1,000,000
Two-banded Plover	*Charadrius falklandicus*	Falkland Islands, Chile, Argentina	46,000–140,000
Double-banded Plover	*Charadrius bicinctus bicinctus*	New Zealand, Chatham Islands	51,000
	Charadrius bicinctus exilis	Auckland Islands	
Lesser Sandplover	*Charadrius mongolus pamirensis*	S. Asia, W. China	310,000–390,000
	Charadrius mongolus atrifrons	Himalayas, Tibet	
	Charadrius mongolus schaeferi	Tibet, Mongolia	
	Charadrius mongolus mongolus	E. Siberia and Russia	
	Charadrius mongolus stegmanni	Kamchatka Chukotsk Peninsula	

Appendix *(continued)*

FAMILY/COMMON NAME	LATIN NAME	BREEDING RANGE	POPULATION SIZE
Charadriidae (cont.)			
Greater Sandplover	*Charadrius leschenaultii columbinus*	Turkey, Jordan, Iran, Afghanistan	<10,000
	Charadrius leschenaultii crassirostris	Caspian Sea, Kazakstan	25,000–100,000
	Charadrius leschenaultii leschenaultii	W. China, Mongolia, S. Russia	25,000–50,000
Caspian Plover	*Charadrius asiaticus*	Caspian Sea, S. Asia, W. China	40,000–55,000
Oriental Plover	*Charadrius veredus*	Siberia, Mongolia, Manchuria	70,000
Eurasian Dotterel	*Charadrius morinellus*	W. Alaska, N. Palearctic	50,000–220,000
Rufous-chested Dotterel	*Charadrius modestus*	S. Argentina, S. Chile, Falkland Islands	130,000–1,100,000
Mountain Plover	*Charadrius montanus*	W. Nearctic	12,500
Hooded Plover	*Thinornis cucullatus*	S. Australia, Tasmania, New Zealand	7,000
Shore Plover	*Thinornis novaeseelandiae*	Rangitara (Chatham Islands, New Zealand)	160–166
Black-fronted Dotterel	*Elseyornis melanops*	Australia, Tasmania, New Zealand	17,000
Inland Dotterel	*Peltohyas australis*	Australia	14,000
Wrybill	*Anarhynchus frontalis*	South Island, New Zealand	4,500–5,000
Diademed Sandpiper-Plover	*Phegornis mitchellii*	Peru, Argentina, Chile	2,500–10,000
Tawny-throated Dotterel	*Oreopholus ruficollis pallidus*	Ecuador, Peru, Bolivia, Chile	1,000–10,000
	Oreopholus ruficollis ruficollis	Peru, Chile	
Pluvianellidae			
Magellanic Plover	*Pluvianellus socialis*	S. Argentina, S. Chile	
Scolopacidae			
Eurasian Woodcock	*Scolopax rusticola*	N. Palearctic	10,000,000–25,000,000
Amami Woodcock	*Scolopax mira*	Ryukyu Islands, Japan	2,500–10,000
Bukidnon Woodcock	*Scolopax bukidnonensis*	Philippines	
Dusky Woodcock	*Scolopax saturata saturata*	Sumatra, Java	
	Scolopax saturata rosenbergii	New Guinea	
Sulawesi Woodcock	*Scolopax celebensis*	Sulawesi	

314

Common name	Scientific name	Distribution	Population
Moluccan Woodcock	*Scolopax rochussenii*	Obi and Bacan Islands, Moluccas	2,500–10,000
American Woodcock	*Scolopax minor*	E. Nearctic	
Chatham Islands Snipe	*Coenocorypha pusilla*	Chatham Islands, New Zealand	
Subantarctic Snipe	*Coenocorypha aucklandica iredalei*	Islands off Stewart, New Zealand	29,000
	Coenocorypha aucklandica huegeli	Snares Islands, New Zealand	
	Coenocorypha aucklandica aucklandica	Auckland Islands, New Zealand	
	Coenocorypha aucklandica meinertzhagenae	Antipodes Islands, New Zealand	
	Coenocorypha aucklandica C. a. ssp	Campbell Islands, New Zealand	
Jack Snipe	*Lymnocryptes minimus*	N. Palearctic	>1,000,000
Solitary Snipe	*Gallinago solitaria solitaria*	C. Asia	11,000–110,000
	Gallinago solitaria japonica	Sakhalin Peninsula, Russia, NE China	
Latham's Snipe	*Gallinago hardwickii*	Sakhalin Peninsula, Japan	25,000–100,000
Wood Snipe	*Gallinago nemoricola*	Himalaya	2,500–10,000
Pintail Snipe	*Gallinago stenura*	Siberia	50,000–2,000,000
Swinhoe's Snipe	*Gallinago megala*	Siberia	25,000–100,000
African Snipe	*Gallinago nigripennis aequatorialis*	Ethiopia, Zaire, Tanzania, Malawi, Mozambique	10,000–100,000
	Gallinago nigripennis angolensis	Angola, Namibia, Zambia, Zimbabwe	10,000–100,000
	Gallinago nigripennis nigripennis	Mozambique, South Africa	10,000–25,000
Madagascar Snipe	*Gallinago macrodactyla*	Madagascar	<10,000
Great Snipe	*Gallinago media*	N. Palearctic	100,000–1,000,000
Common Snipe	*Gallinago gallinago faeroeensis*	Iceland, Faeroe, Orkney, and Shetland Islands	570,000
	Gallinago gallinago gallinago	N. Palearctic, Aleutian Islands	>2,500,000
Wilson's Snipe	*Gallinago delicata*	N. Nearctic	
South American Snipe	*Gallinago paraguaiae paraguaiae*	Trinidad, Colombia, Guyana, Brazil, Argentina	40,000–1,000,000
	Gallinago paraguaiae magellanica	Chile, Argentina, Falkland Islands	
Puna Snipe	*Gallinago andina*	Peru, Bolivia, Argentina, Chile	
Noble Snipe	*Gallinago nobilis*	Columbia, Venezuela, Ecuador, Peru	
Giant Snipe	*Gallinago undulata undulata*	Colombia, Venezuela, Guianas, Brazil	
	Gallinago undulata gigantea	Bolivia, Paraguay, Brazil, Argentina	

Appendix (continued)

FAMILY/COMMON NAME	LATIN NAME	BREEDING RANGE	POPULATION SIZE
Scolopacidae (cont.)			
Fuegian Snipe	Gallinago stricklandii	S. Argentina, S. Chile	2,500–10,000
Andean Snipe	Gallinago jamesoni	Venezuela, Colombia, Ecuador, Peru, Bolivia	
Imperial Snipe	Gallinago imperialis	Colombia, Ecuador, Peru	10,000
Short-billed Dowitcher	Limnodromus griseus caurinus	Alaska, Yukon, British Columbia	75,000
	Limnodromus griseus hendersoni	C. Canada	78,000
	Limnodromus griseus griseus	E. Canada	
Long-billed Dowitcher	Limnodromus scolopaceus	NE Siberia, W. and N. Alaska, McKenzie River delta	500,000
Asian Dowitcher	Limnodromus semipalmatus	Siberia, Manchuria	23,000
Black-tailed Godwit	Limosa limosa islandica	Iceland, Faeroe and Shetland Islands	47,000
	Limosa limosa limosa	W. Palearctic	250,000–500,000
	Limosa limosa melanuroides	E. Palearctic	
Hudsonian Godwit	Limosa haemastica	N. Canada	70,000
Bar-tailed Godwit	Limosa lapponica lapponica	Lapland, NE Russia	120,000
	Limosa lapponica menzbieri	N. Siberia	
	Limosa lapponica baueri	NE Siberia, NW Alaska	150,000
Marbled Godwit	Limosa fedoa beringiae	Alaskan Peninsula	2,000
	Limosa fedoa fedoa	C. Canada, U.S., James Bay, Canada	168,000
Eskimo Curlew	Numenius borealis	N. Nearctic	Extinct
Little Curlew	Numenius minutus	Siberia	180,000
Whimbrel	Numenius phaeopus phaeopus	NW Palearctic	250,000–1,000,000
	Numenius phaeopus alboaxillaris	C. Asia	<10,000
	Numenius phaeopus variegatus	Siberia	
	Numenius phaeopus hudsonicus	N. Nearctic	18,000

Common name	Scientific name	Range	Population
Bristle-thighed Curlew	*Numenius tahitiensis*	W. Alaska	10,000
Slender-billed Curlew	*Numenius tenuirostris*	SW Siberia, N. Kazakstan	<50
Eurasian Curlew	*Numenius arquata arquata*	British Isles, NW Palearctic	700,000–1,000,000
	Numenius arquata orientalis	E. Russia, Manchuria	25,000–100,000
Long-billed Curlew	*Numenius americanus parvus*	SW Canada, W. United States	40,000
	Numenius americanus americanus	W. and C. United States	
Far Eastern Curlew	*Numenius madagascariensis*	NE Asia	38,000
Upland Sandpiper	*Bartramia longicauda*	NW and W. Nearctic	350,000
Terek Sandpiper	*Xenus cinereus*	N. Palearctic	160,000–1,000,000
Common Sandpiper	*Actitis hypoleucos*	Palearctic	>1,500,000
Spotted Sandpiper	*Actitis macularius*	Nearctic	150,000
Green Sandpiper	*Tringa ochropus*	N. Palearctic	>1,000,000
Solitary Sandpiper	*Tringa solitaria cinnamomea*	Alaska, W. Canada	50,000
	Tringa solitaria solitaria	W., C., E. Canada	100,000
Gray-tailed Tattler	*Tringa brevipes*	Siberia	40,000
Wandering Tattler	*Tringa incana*	Alaska, NW Canada	17,500
Spotted Redshank	*Tringa erythropus*	N. Palearctic	70,000–150,000
Greater Yellowlegs	*Tringa melanoleuca*	N. Nearctic	100,000
Common Greenshank	*Tringa nebularia*	Palearctic	300,000–1,250,000
Nordmann's Greenshank	*Tringa guttifer*	E. Siberia	500–1,000
Willet	*Tringa semipalmata inornata*	C. and W. Nearctic	160,000
	Tringa semipalmata semipalmata	E. and S. Nearctic, West Indies	90,000
Lesser Yellowlegs	*Tringa flavipes*	N. Nearctic	400,000
Marsh Sandpiper	*Tringa stagnatilis*	Palearctic	64,000–140,000
Wood Sandpiper	*Tringa glareola*	N. Palearctic	>1,000,000
Common Redshank	*Tringa totanus robusta*	Iceland, Faeroes, Scotland	250,000–535,000
	Tringa totanus totanus	W. Europe	750,000–1,700,000
	Tringa totanus ussuriensis	C. and E. Palearctic	100,000–1,000,000

FAMILY/COMMON NAME	LATIN NAME	BREEDING RANGE	POPULATION SIZE
Scolopacidae (cont.)			
	Tringa totanus terrignotae	S. Manchuria	
	Tringa totanus craggi	NW China	
	Tringa totanus eurhinus	C. Asia, N. India, Himalayas	
Tuamotu Sandpiper	*Prosobonia cancellata*	Tuamotu Islands, French Polynesia	1,300
Ruddy Turnstone	*Arenaria interpres interpres*	NE Canada, Palearctic, N. Alaska	55,000
	Arenaria interpres morinella	Canada, NE Alaska	180,000
Black Turnstone	*Arenaria melanocephala*	W. Alaska	95,000
Surfbird	*Aphriza virgata*	C. Alaska, NW Canada	70,000
Great Knot	*Calidris tenuirostris*	NE Siberia	380,000–390,000
Red Knot	*Calidris canutus canutus*	Siberia	400,000
	Calidris canutus piersmai	New Siberia Islands, Russia	
	Calidris canutus rogersi	Chukotsk Peninsula, Russia	42,500
	Calidris canutus roselaari	Wrangell Island, Russia	
	Calidris canutus rufa	N. Canada	22,500
	Calidris canutus islandica	NE Canada, Greenland (Palearctic)	80,000–450,000
Sanderling	*Calidris alba*	N. Holarctic (Palearctic)	620,000–700,000)
Semipalmated Sandpiper	*Calidris pusilla*	N. Nearctic	3,500,000
Western Sandpiper	*Calidris mauri*	W. Alaska, NE Siberia	3,500,000
Red-necked Stint	*Calidris ruficollis*	NE Siberia, N. Alaska	320,000
Little Stint	*Calidris minuta*	N. Palearctic	>1,000,000
Temminck's Stint	*Calidris temminckii*	N. Palearctic	>1,000,000
Long-toed Stint	*Calidris subminuta*	NE Palearctic	25,000
Least Sandpiper	*Calidris minutilla*	N. Nearctic	700,000

Common Name	Scientific Name	Range	Population
White-rumped Sandpiper	*Calidris fuscicollis*	N. Nearctic	1,120,000
Baird's Sandpiper	*Calidris bairdii*	N. Nearctic	300,000
Pectoral Sandpiper	*Calidris melanotos*	NE Siberia, N. Nearctic	400,000
Sharp-tailed Sandpiper	*Calidris acuminata*	NE Siberia	16,000
Curlew Sandpiper	*Calidris ferruginea*	N. Siberia	>1,000,000
Dunlin	*Calidris alpina arctica*	NE Greenland	21,000–45,000
	Calidris alpina schinzii	Greenland, Iceland, British Isles, S. Scandinavia	1,000,000–1,200,000
	Calidris alpina alpina	Scandinavia, E. Russia	1,330,000
	Calidris alpina sakhalina	Russia, Chukotsk Peninsula	
	Calidris alpina actites	Sakhalin Peninsula, Russia	
	Calidris alpina kistchinskii	Kuril Islands, Russia	
	Calidris alpina articola	N. Alaska, NW Canada	750,000
	Calidris alpina pacifica	W. Alaska	550,000
	Calidris alpina hudsonia	N. Canada	225,000
Purple Sandpiper	*Calidris maritima*	NE Canada, Greenland, N. Palearctic	70,000–130,000
Rock Sandpiper	*Calidris ptilocnemis quarta*	Kuril Islands, Kamchatka Peninsula	
	Calidris ptilocnemis tschuktschorum	Chukotsk Peninsula, W. Alaska	50,000
	Calidris ptilocnemis ptilocnemis	Pribilof, St. Matthew and Hall Islands, Alaska	25,000
	Calidris ptilocnemis couesi	Aleutian Islands, Alaska Peninsula	75,000
Stilt Sandpiper	*Calidris himantopus*	N. Nearctic	820,000
Spoon-billed Sandpiper	*Eurynorhynchus pygmeus*	NE Siberia	450–1,000
Broad-billed Sandpiper	*Limicola falcinellus falcinellus*	Scandinavia, NW Russia	61,000–64,000
	Limicola falcinellus sibirica	N. Siberia	
Buff-breasted Sandpiper	*Tryngites subruficollis*	N. Nearctic	20,000
Ruff	*Philomachus pugnax*	N. Palearctic	1,000,000–1,500,000
Wilson's Phalarope	*Phalaropus tricolor*	C. and W. Nearctic	1,500,000
Red-necked Phalarope	*Phalaropus lobatus*	N. Holarctic	>2,500,000
Red Phalarope	*Phalaropus fulicarius*	N. Holarctic	>1,250,000

Appendix *(continued)*

FAMILY/COMMON NAME	LATIN NAME	BREEDING RANGE	POPULATION SIZE
Pedionomidae			
Plains-wanderer	*Pedionomus torquatus*	Australia	2,400–8,000
Thinocoridae			
Rufous-bellied Seedsnipe	*Attagis gayi latreillii*	Ecuador	
	Attagis gayi simonsi	Peru, Bolivia, Chile, Argentina	
	Attagis gayi gayi	Chile, Argentina	
White-bellied Seedsnipe	*Attagis malouinus*	S. Argentina, S. Chile	
Gray-breasted Seedsnipe	*Thinocorus orbignyianus ingae*	Peru, Bolivia, Chile, Argentina	
	Thinocorus orbignyianus orbignyianus	Chile, Argentina	
Least Seedsnipe	*Thinocorus rumicivorus pallidus*	Ecuador, Peru	100,000–1,000,000
	Thinocorus rumicivorus cuneicauda	Peru	
	Thinocorus rumicivorus bolivianus	Peru, Bolivia, Chile, Argentina	
	Thinocorus rumicivorus rumicivorus	Argentina, Chile	
Chionididae			
Snowy Sheathbill	*Chionis albus*	Argentina, Chile, Antarctic Peninsula, Falkland Islands	20,000
Black-faced Sheathbill	*Chionis minor marionensis*	Marion and Prince Edward Islands, South Africa	13,000–20,000
	Chionis minor crozettensis	Crozet Islands, France	
	Chionis minor minor	Kerguelen Islands, France	
	Chionis minor nasicornis	Heard and McDonald Islands, Australia	

INDEX

An 'f' after a page number indicates a figure; a 't' indicates a table.

size dimorphism, 6, 56, 60, 73
size ratios, 173–175
skeletal system, 27–29
skull musculature, 28
Slender-billed Curlew, 7, 62, 229
Snowy Plover, 7, 23, 30, 47t, 49, 53t–54, 56–57t, 61, 69t, 71t, 75, 77–80, 84, 88–89, 92, 94, 107, 163, 209–210, 214, 220, 227, 245t, 258, 266, 270, 273–275, 283–285t, 288, 290
Solitary Sandpiper, 72, 75, 78
Solway Firth, Scotland
Southern Lapwing, 53t
Spartina (cordgrass), 167, 228
spatial distributions, 181–201
specialization in diet, 132–133, 153
sperm morphology and storage, 39
Spoon-billed Sandpiper, 5, 7, 62
Spotted Redshank, 171
Spotted Sandpiper, 46–50, 53t, 56–59, 68–69t, 72, 82, 84, 87–88, 107, 212–213
Spotted Thick-knee, 93
spring arrival schedules, 49–50, 71
Spur-winged Plover, 83
staging sites, 111–114
sterilants for predators, 275–276
stomach mass, 36f–37
stopover duration, 114–118
stopover sites, 111–114
Subantarctic Snipe, 69t, 71t
subsidized food for predators, 275
substrate size and water-holding capacity, 147–148
supplemental feeding hypothesis, 142–143
supraorbital gland, 38–39
survival estimates, 206–210, 224, 229
symbolic fencing, 290f
synchrony of egg hatching, 90–91
systematics, 11–17

tactile (touch) feeding, 34, 287
Tagus estuary, Portugal, 149
Taimyr Peninsula, Russia, 93, 223f–224
Tees estuary, England, 171
Temminck's Stint, 47, 56–57t, 86, 89, 206, 210, 212, 266, 269
territoriality in winter, 186–187
testes size, 39
thermoregulation, 37–38
Thinocoridae, 4t, 15f, 21t, 85t, 137t
tidal flat(s), 148–149f
tidal influence on feeding, 143–144
time-activity budgets, 139–141
timing of incubation, 87–88

toe arrangement, 29
Tofino, British Columbia, 302
Tokyo Bay, Japan, 142
Tomales Bay, California, 243
tongue structure, 34–35
trace elements in feathers, 33
traditional vs. ephemeral use of roosts, 197–199f
triglycerides, 113, 283
Tringines, 4t, 21t, 55, 132
   taxonomy, 16
Tuamotu Sandpiper, 84, 269
Turnicidae, 15f
turnover rate during migration, 114
Two-banded Plover, 124

uniparental care, 54–58
Upland Sandpiper, 76, 79, 219, 226, 229, 256
Upper Witaki Basin, New Zealand, 267
uropygial gland, 29, 35
U.S. Endangered Species Act, 7, 23, 71, 94, 214–215, 220, 232, 270
U.S. National Wildlife Refuges, 242
U.S. Shorebird Conservation Plan, 215, 218, 279, 293–294

Vanellus, 21t, 73
variance in reproductive success, 58–59, 62
vigilance in flocks, 189, 191–193, 283
vision, 33–36, 89
visual feeding, 287
vital (birth, death) rates, 62, 213–214
vocalizations, 72

Wadden Sea, 111, 117t, 124, 176, 195–196
Wader Study Group, 295
Wader Study Group Bulletin, 295
The Wash, England, 124–125, 139, 144, 188, 197, 243, 255
water depth, 253
Wattled Jacana, 53t, 57t
weather-related mortality, 228–229
West Asia/East Africa Flyway, 121f, 124
Western Hemisphere Shorebird Group, 295
Western Hemisphere Shorebird Reserve Network (WHSRN), 7, 115, 218, 232, 243–247, 295–297t
Western Sandpiper, 33–34f, 53t, 69t, 75, 93, 109, 111, 113–116, 120, 135, 170, 172, 187, 191, 198, 208t, 211–213, 256, 288, 294, 301

Compositor: Westchester Book Group
Text: 9.5/13 Scala
Display: Scala Sans and Scala Sans Caps
Printer and binder: Thomson-Shore